# Research Advances in
# Alcohol and Drug Problems

## Volume 9

# RESEARCH ADVANCES IN ALCOHOL AND DRUG PROBLEMS

Series Editors:

Howard D. Cappell
Frederick B. Glaser
Yedy Israel
Harold Kalant
Wolfgang Schmidt
Edward M. Sellers
Reginald G. Smart

# Research Advances in
# Alcohol and Drug Problems

## Volume 9

Edited by

Howard D. Cappell, Frederick B. Glaser,

Yedy Israel, Harold Kalant, Wolfgang Schmidt,

Edward M. Sellers and Reginald C. Smart

Addiction Research Foundation and
University of Toronto
Toronto, Ontario, Canada

PLENUM PRESS · NEW YORK AND LONDON

The Library of Congress cataloged the first volume of this title as follows:

Research advances in alcohol & drug problems. v. 2–

New York [etc.] J. Wiley, 1974–
   v. 24 cm. annual.
   "A Wiley biomedical health publication."
   ISSN 0093-9714

1. Alcoholism — Periodicals. 2. Narcotic habit — Periodicals.
RC565.R37                          616.8′6′005                          73-18088

ISBN 0-306-42426-6

## ADVISORY PANEL

# Contributors

JOSEPH J. BARBORIAK, Department of Pharmacology and Toxicology, The Medical College of Wisconsin, and Biochemistry Section, Research Service, Veterans Administration Medical Center, Milwaukee, Wisconsin

NEAL L. BENOWITZ, Medical Service, San Francisco General Hospital Medical Center, and Department of Medicine and Langley Porter Psychiatric Institute, University of California, San Francisco, California

USOA BUSTO, Addiction Research Foundation, Toronto, Ontario, Canada

HOWARD CAPPELL, Addiction Research Foundation, Toronto, Ontario, Canada

THOMAS V. DUNWIDDIE, Medical Research Service, Denver Veterans Administration Hospital, Denver, Colorado

SHAHIN HASHTROUDI, Department of Psychology, George Washington University, Washington, D.C.

BARRY J. HOFFER, Department of Pharmacology, University of Colorado Health Sciences Center, Denver, Colorado

JAY G. HULL, Department of Psychology, Dartmouth College, Hanover, New Hampshire

LAWRENCE A. MENAHAN, School of Pharmacy, University of Wisconsin, Madison, Wisconsin

CHARLES P. O'BRIEN, Psychiatry Service, Philadelphia Veterans Administration Medical Center and University of Pennsylvania, Philadelphia, Pennsylvania

MICHAEL R. PALMER, Department of Pharmacology, University of Colorado Health Sciences Center, Denver, Colorado

ELIZABETH S. PARKER, Alcohol Research Center, Department of Psychiatry and Behavioral Sciences, UCLA School of Medicine, Los Angeles, California

EDWARD M. SELLERS, Addiction Research Foundation, Toronto, Ontario, Canada

SHEPARD SIEGEL, Department of Psychology, McMaster University, Hamilton, Ontario, Canada

RONALD R. VAN TREUREN, Department of Psychology, Dartmouth College, Hanover, New Hampshire
GEORGE E. WOODY, Psychiatry Service, Philadelphia Veterans Administration Medical Center and University of Pennsylvania, Philadelphia, Pennsylvania

# Preface

This, the ninth volume in the series, appears some 13 years after the first. Like most of its predecessors, Volume 9 is deliberately eclectic, covering a range of topics that the editors think worthy of inclusion. Some of the chapters, such as the review of the literature on benzodiazepines, represent areas that have received relatively little attention in previous volumes—largely because the literature has not previously been "ripe" for review. Others represent literatures that have been reviewed in the past but which continue to advance in sufficient measure that their ripening never ceases. Shepard Siegel's contribution represents a relative rarity in previous volumes: a chapter not laden with a consideration of current empirical work, but a reflective essay designed to stir thought with some provocative ideas.

The editors trust that readers will continue to find *Research Advances* to be an important repository of knowledge in the alcohol and drug fields.

The Editors

*Toronto*

# Contents

9.  ALCOHOL AND OPIATE DEPENDENCE: RE-EVALUATION OF THE
    VICTORIAN PERSPECTIVE     279

Shepard Siegel

# The Human Pharmacology of Nicotine

NEAL L. BENOWITZ

## 1. INTRODUCTION

Nicotine has been consumed in various forms of tobacco and from other plants for many hundreds of years. In the past century, it has been one of the three most widely used drugs, rivaling caffeine and ethanol. More recently, nicotine itself in the form of nicotine-containing chewing gum has become available as a pharmaceutical agent for therapy of tobacco addiction. Nicotine is important in human biology for two reasons. It appears to be the primary reason why people consume tobacco products, and it may contribute to causation of some tobacco-related diseases.

Because of its usefulness as a probe for studying cholinergic receptors, much research on the basic pharmacology of nicotine has been conducted over the past 50 years. In contrast, studies of the human pharmacology of nicotine have been conducted for the most part in the last decade. It has been only during this time that analytical methods adequate to measure nicotine concentrations in blood, which are necessary for human pharmacokinetic and pharmacodynamic studies, have been available.

I will review recent research on the human pharmacology of nicotine, concentrating on studies in which the effects, disposition, and metabolism of nicotine per se (as opposed to effects of cigarette smoking) have been studied. I will discuss the importance of nicotine in determining effects of cigarette smoking and its implications with respect to causation of human disease.

---

NEAL L. BENOWITZ ● Medical Service, San Francisco General Hospital Medical Center, and Department of Medicine and Langley Porter Psychiatric Institute, University of California, San Francisco, California 94110.

## 2.  NICOTINE IN TOBACCO PRODUCTS

In considering the clinical pharmacology of any drug, one of the first questions is the dose to the consumer. The case of nicotine is unique because of the complexity of the smoking process and the ability of the smoker to adjust intake on a puff-by-puff basis. Thus, intake of nicotine from a given product is dependent upon puff volume, depth of inhalation, extent of dilution with room air, puffing rate, and intensity of puffing (Herning et al., 1983 *b*). In addition, in the case of certain cigarettes, intake depends on whether or not ventilation holes within the filter are occluded by the smoker (Hoffmann et al., 1983; Kozlowski et al., 1982*a*).

The nicotine content of a tobacco product is obviously the limiting step in self-dosing. The nicotine content of cigarettes is not specified by manufacturers or by government testing agencies. Because tobacco is a plant product, there are differences in nicotine content among and within different strains of tobacco. In order to get an idea of the content of typical cigarettes, we measured the amount of nicotine in the tobacco of 15 American cigarette brands, of differing machine-determined yields (Benowitz et al., 1983*a*). On average, the tobacco contained 1.5% nicotine by weight, and lower-yield cigarettes tended to have higher concentrations of nicotine than higher-yield cigarettes (Fig. 1). However, lower-yield cigarettes contain less tobacco per cigarette, so the total amount of nicotine per cigarette averages 8.4 mg and is similar for different brands. Thus, low-yield cigarettes are not low yield because they contain less nicotine per cigarette, but because of filtration and ventilation characteristics of the cigarette which remove tar and nicotine by filtration and/or dilute smoke with air. The smoker, in seeking a particular dose of nicotine, is limited not by the quantity of nicotine but by his ability to manipulate cigarette-smoking conditions.

Nicotine concentrations in other tobacco products are similar to those of cigarettes (Gritz et al., 1981; Kozlowski et al., 1982*b*). A more recently available product is nicotine gum (Nicorette®, Merrell-Dow Pharmaceuticals, Inc.), which is nicotine bound to an ion exchange resin in a gum base. Nicotine is released as the gum is chewed. Nicotine gum is available in 2 mg and 4 mg preparations in Canada and the United Kingdom and 2 mg in the United States.

The dose of nicotine given during intravenous infusions, as described in the medical experimental literature, seems obvious. However, a source of confusion in some of those reports is whether the dose is expressed as weight of nicotine base or as the weight of its bitartrate salt. In some reports neither is specified. The molecular weights of nicotine base and nicotine bitartrate dihydrate are 162.2 and 498.4, respectively. The dose of nicotine base is 32.5% that of the salt. The salt rapidly dissociates in the body and the dose of the base is the appropriate one.

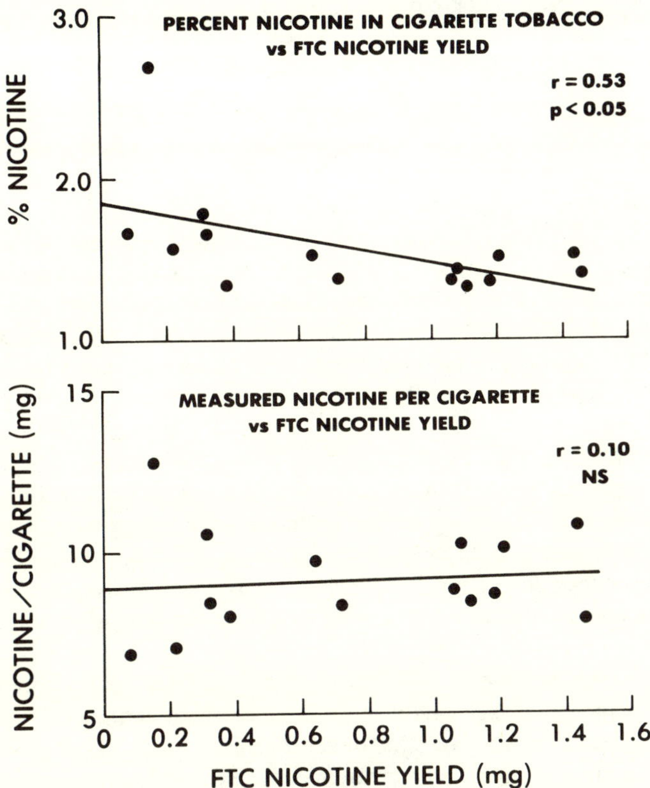


Figure 1.   Nicotine content of cigarettes, as compared with U.S. Federal Trade Commission (FTC)-determined values (regression analysis). "Measured nicotine per cigarette" denotes total amount of nicotine in the length of cigarette tobacco rod smoked in the standard FTC smoking-machine assay. Fifteen popular commerical cigarette brands were assayed. (From Benowitz et al., 1983a, reprinted with permission.)

Although the major alkaloid in tobacco is nicotine, which is the topic of this review, there are several alkaloids that may be of pharmacologic importance. These include nornicotine, anabasine, myosmene, nicotyrine, and anatabine. These make up 8–12% of the total alkaloid content of tobacco products (Piade and Hoffmann, 1980). These minor alkaloids do have pharmacologic activity (Clark et al., 1965); however, little is known about the importance of these alkaloids in determining the effects of tobacco consumption in people.

## 3. PHARMACOKINETICS OF NICOTINE

### Absorption of Nicotine

Nicotine is distilled from burning tobacco and is carried proximally on tar droplets (0.1–0.4 $\mu$m), which are inhaled. Absorption of nicotine across biological membranes depends on pH (Armitage and Turner, 1970; Schievelbein et al., 1973). Nicotine is a weak base with a pKa of 7.9. In its ionized state, such as in acidic environments, nicotine does not rapidly cross membranes. The pH of smoke from flue-cured tobaccos, found in most cigarettes, is acidic (pH 5.5). At this pH, the nicotine is primarily ionized. As a consequence, there is little buccal absorption of nicotine from cigarette smoke, even when it is held in the mouth (Gori et al., 1985). The pH of smoke from air-cured tobaccos, such as in pipes, cigars and in a few European cigarettes, is alkaline (pH 8.5), and nicotine is primarily un-ionized. Smoke from these products is well absorbed through the mouth (Armitage et al., 1978; Russel et al., 1980c). Chewing tobacco, snuff, and nicotine gum are buffered to alkaline pH so as to facilitate nicotine absorption.

When tobacco smoke reaches the small airways and alveoli of the lung, the nicotine is rapidly absorbed independent of pH of the smoke. Armitage and co-workers (1975), measuring exhalation of radiolabeled nicotine, found that 82–92% of nicotine in mainstream smoke was absorbed by four habitual smokers, 29% by another habitual smoker who was presumed to be a noninhaler, and 30–66% by three nonsmokers (who were instructed to smoke as deeply as possible). The rapid absorption of nicotine from cigarette smoke through the lung is presumably because of the huge surface area of the alveoli and small airways and dissolution of nicotine into fluid of pH in the human physiologic range, which facilitates transfer across cell membranes.

Nicotine base can be absorbed through the skin and there have been cases of poisoning after skin contact with pesticides containing nicotine (Faulkner, 1933). Likewise, there is evidence of cutaneous absorption of nicotine in tobacco field workers (Gehlbach et al., 1975).

When taken orally, most typically as ingestion of tobacco by children, nicotine is usually innocuous even if a large dose is consumed. Nicotine is poorly absorbed from the stomach due to the acidity of gastric fluid (Travell, 1960), but is well absorbed in the small intestine (Jenner et al., 1973), which has a more alkaline pH and a large surface area. Despite good absorption, little nicotine reaches the systemic circulation, however, because of extensive first pass metabolism by the liver. Thus, the oral bioavailability of nicotine is low. Based on urinary excretion of nicotine, comparing oral and intravenous nicotine dosing, the data of Jenner et al. (1973) suggest a bioavailability of about 30% in humans.

Because of the complexity of the smoking process, the dose of nicotine consumed by the smoker cannot be predicted from the nicotine content of the tobacco or its absorption characteristics. To measure the dose, one needs to measure blood levels and know how fast an individual eliminates nicotine. This topic will be considered later in this article after discussion of the relevant pharmacokinetic issues.

## Distribution of Nicotine

After absorption, nicotine enters the bloodstream where, at pH 7.4, it is about 69% ionized and 31% un-ionized. Binding to plasma proteins is less than 5% (Benowitz et al., 1982a). The drug is distributed extensively to body tissues with a steady state volume of distribution averaging 2.6 times body weight (Fig. 2, Table 1). The pattern of tissue uptake cannot be studied in humans, but we have, by measuring concentrations of nicotine in various tissues after infusion of nicotine to steady state, examined tissue uptake in rabbits (Table 2). Spleen,

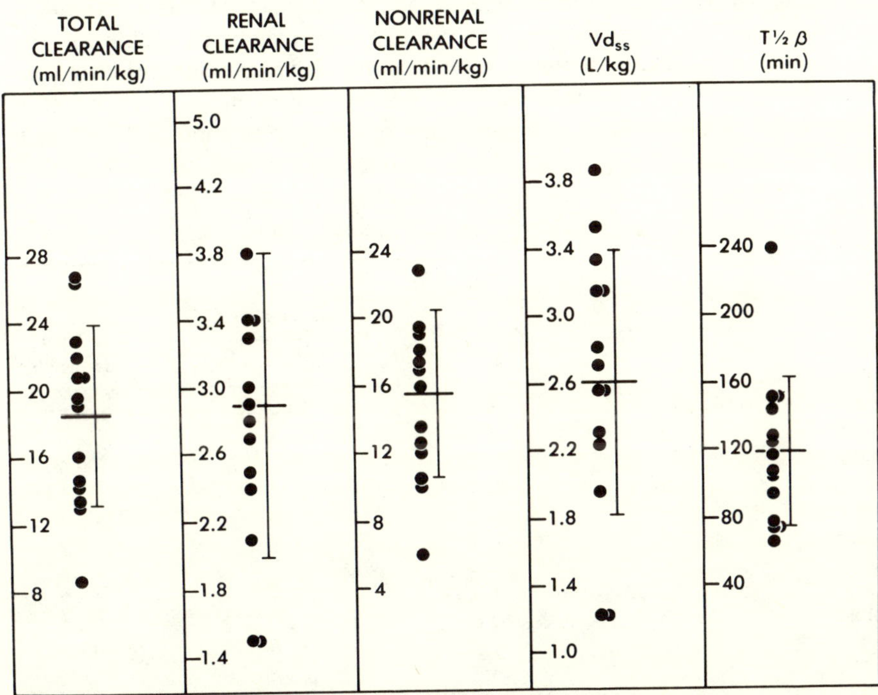

Figure 2.   Pharmacokinetics of nicotine in 14 healthy male cigarette smokers. Analysis was based on 30-min nicotine infusion, acid urine conditions. Vd$_{ss}$, steady state volume of distribution; t$_{1/2}\beta$ terminal half-life. (From Benowitz et al., 1983a, reprinted with permission.)

Table 1. Pharmacokinetics of Nicotine[a]

| Half-life | 120 min |
|---|---|
| Volume of distribution | 180 liters |
| Total clearance | 1300 ml/min |
| Renal clearance (acid urine) | 200 ml/min |
| Nonrenal clearance | 1100 ml/min |

[a] Average values, based on data from Benowitz et al. (1982a).

liver, lungs, and brain have high, and adipose tissue relatively low, affinity for nicotine.

After rapid intravenous injection, nicotine blood concentration falls rapidly, presumably due to tissue uptake of drug (Fig. 3). During this phase, blood concentrations are quite high while tissue concentrations would be expected to still be low. Distribution to brain and heart should be rapid because of relatively high rates of perfusion. Thus, direct effects on these organs are expected within 1 or 2 min. Distribution to muscle is by comparison predicted to be much slower, but muscle represents a storage reservoir of greater magnitude (Table 2). The consequence of uptake into muscle is that the blood concentration of drug continues to decline at a rate faster than can be explained by metabolism for 20 or 30 min after administration. Thereafter, the blood concentration declines much

Table 2. Distribution of Nicotine[a]

| Tissue | Rabbit Tissue : blood ratio (R) | Mass (M) (kg) | Human simulation M × R | % Total body nicotine |
|---|---|---|---|---|
| Blood | 1.0 | 5.4 | 5.4 | 4.8 |
| Brain | 3.0 | 1.5 | 4.5 | 4.0 |
| Heart | 2.0 | 0.3 | 0.6 | 0.5 |
| Muscle | 2.0 | 30.0 | 60.0 | 53.1 |
| Adipose | 0.5 | 10.0 | 5.0 | 4.4 |
| Kidney | 21.6 | 0.3 | 6.5 | 5.7 |
| Liver | 3.7 | 1.7 | 6.3 | 5.6 |
| Lung | 3.7 | 1.0 | 3.7 | 3.2 |
| GI tissue | 3.5 | 2.0 | 7.0 | 6.2 |
| Spleen | 9.3 | 1.5 | 14.0 | 12.4 |
| Totals | | 53.7 | 113.0 | |

$$V_D = \Sigma\, M \times R / \Sigma\, M = 2.1 \text{ liters/kg}$$

[a] Tissue to blood nicotine concentration ratio in rabbits based on measurements after 24 hr constant infusion of nicotine. Human simulation based on typical organ mass and partition ratios observed in rabbits. $V_D$ = volume of distribution predicted by simulation.

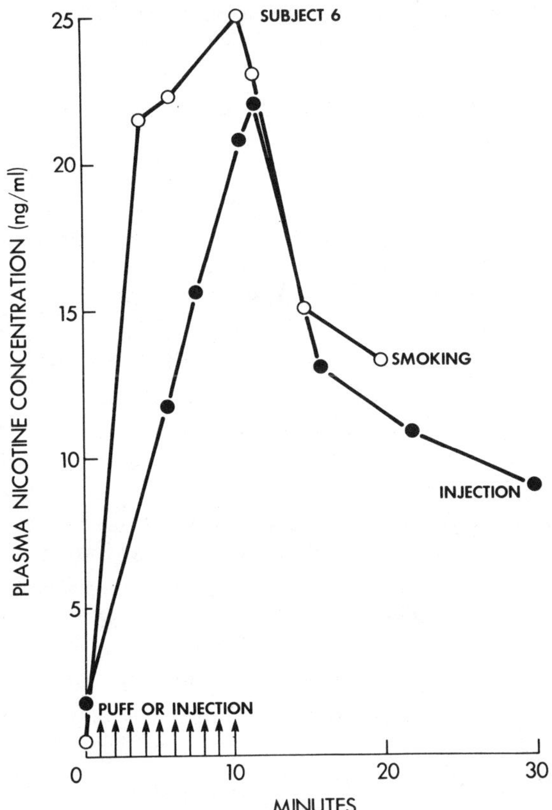


Figure 3.    Plasma nicotine concentrations in a subject given bolus intravenous injections of nicotine, 2 µg base/kg, or inhaling from a cigarette, once every minute for 10 min. There is a sharp decline in concentrations following the end of the series of nicotine injections versus smoking, due to distribution into body tissues. The terminal half-life (not shown in figure) becomes apparent 30–60 min following nicotine exposure.

more slowly, the rate being determined by rate of metabolism and rate of distribution out of tissues.

Smoking represents an exposure similar to that of rapid intraveous injection except that the point of entry into the circulation is through the pulmonary rather than the systemic venous circulation. As a consequence of delivery into the lung, the lag time between smoking and entry into the brain is shorter than that after intravenous injection. Thus, after smoking, nicotine will enter the brain quickly, but then brain levels will decline rapidly as nicotine is distributed to other tissues. The distribution half-life (about 8 min) rather than the elimination half-life (about 2 hr) determines the time course of central nervous system effects after smoking.

Nicotine is secreted into saliva (Russell and Feyerabend, 1978). Unfortunately, because of contamination of saliva from nicotine in tobacco smoke or chewing tobacco, the saliva is not useful as a way to monitor blood levels of nicotine. However, passage of saliva containing nicotine into the stomach, combined with a process of trapping of nicotine in the acidic gastric fluid and reabsorption from the small bowel, does provide a potential route for enteric recirculation of nicotine. This recirculation has been speculated to account for some of the oscillations in the terminal decline phase after intravenous infusion of nicotine or cessation of smoking (Russell, 1976).

Nicotine is excreted into urine and may, depending on urinary pH, be reabsorbed from the urinary bladder. When the urine is acidic (pH is less than 6), as is usually the case, nicotine is ionized and there is little reabsorption (Borzelleca, 1963). From alkaline urine, there may be significant reabsorption of nicotine, particularly where there is urine stasis. Alkaline urine as well as stasis may be seen in the presence of urinary tract infections with certain urea-splitting bacteria.

Nicotine freely crosses the placenta and has been found in amniotic fluid and in umbilical cord blood of neonates (Hibberd et al., 1978; Luck et al., 1982; Van Vunakis et al., 1974). Nicotine is also found in breast milk and breast fluid of nonlactating women (Petrakis et al., 1978; Hill and Wynder, 1979). The concentration of nicotine in breast milk is such that the dose of nicotine consumed by the infant is small and unlikely to have physiologic consequence. Relative concentrations of nicotine in various body fluids are shown in Table 3.

## Metabolism of Nicotine

*Major Metabolites.* Nicotine is extensively metabolized, primarily in the liver, but also to a small extent in the lung and kidney (Gorrod and Jenner, 1975; Turner et al., 1975). Renal excretion depends on urinary pH and urine flow, and accounts for from 2–35% of total elimination (Beckett et al., 1965; Benowitz et al., 1983*b*; Rosenberg et al., 1980). The major metabolites of nicotine are cotinine and nicotine-N-oxide (Fig. 4). The alpha carbon of the pyrrolidine ring is hydroxylated via cytochrome P450 to 5′-hydroxynicotine, which is in equilibrium with nicotine-$\Delta 1'(5')$-iminium ion (Brandange and Lindblom, 1979; Gorrod and Hibberd, 1982). The latter is metabolized by the cytoplasmic enzyme, aldehyde oxidase, to cotinine. Cotinine is also extensively metabolized, with only about 17% excreted unchanged in the urine (Bowman and McKennis, 1962; Benowitz et al., 1983*b*). The pharmacokinetics of cotinine will be described in a later section.

Nicotine-1′-N-oxide is a minor metabolite of nicotine. Oxidation of the nitrogen atom of the pyrrolidine ring is dependent upon a microsomal flavoprotein

Table 3. Nicotine and Cotinine Concentrations in Body Fluids of Smokers

| Body fluid | N | Nicotine concentration (ng/ml) | Nicotine ratio body fluid to plasma or blood | Cotinine concentration (ng/ml) | Cotinine ratio body fluid to plasma or blood |
|---|---|---|---|---|---|
| Saliva | | | | | |
| Feyerabend et al., 1982 | 82 | 567[a] (2–2000) | — | — | — |
| Haley et al., 1983 | 12 | — | — | 361 (±80) | 1.47 |
| Urine | | | | | |
| Benowitz (unpublished) | 26 | 2228 (159–6537) | — | 2799 (159–6537) | — |
| Feyerabend et al., 1982 | | 1278 (5–8000) | — | — | — |
| Amniotic fluid | | | | | |
| Luck and Nau, 1984[b] | 29 | (1.5–23) | 1.5 (±0.27) | (5–188) | 0.72 (±0.15) |
| Van Vunakis et al., 1974[c] | 37 | (ND–31) | — | (ND–129) | — |
| Umbilical venous blood | | | | | |
| Luck and Nau, 1984 | 27 | (0.5–25) | 1.12 (±0.30) | (15–233) | 1.05 (±0.27) |
| Breast milk | | | | | |
| Luck and Nau, 1984 | 23 | (2–62) | 2.9 (±1.1) | (12–222) | 0.8 (±0.21) |
| Breast fluid, nonlactating | | | | | |
| Petrakis et al., 1978 | 4 | (46–195) | — | — | — |
| Hill and Wynder, 1979 | 5 | 40.6 (20–56) | 1.05 (±0.31) | 169 (111–271) | 0.46 (±0.16) |
| Cervical fluid | | | | | |
| Sasson et al., 1985 | 10 | 740 (66–2620) | 67.2 (±91) | 316 (161–773) | 1.41 (±0.82) |

[a] Where available, mean values are given; numbers in parentheses are range or SD.
[b] 16–24 weeks gestation.
[c] 14–20 weeks gestation; ND = not detected.

system (Gorrod and Jenner, 1975). After intravenous injection, 100% of nicotine-N-oxide is excreted unchanged in the urine (Beckett et al., 1971a). After oral administration, only 30% is recovered unchanged; the remainder is recovered as nicotine and its metabolites. After rectal administration, less than 10% is recovered as nicotine oxide. These findings indicate extensive reduction of nic-

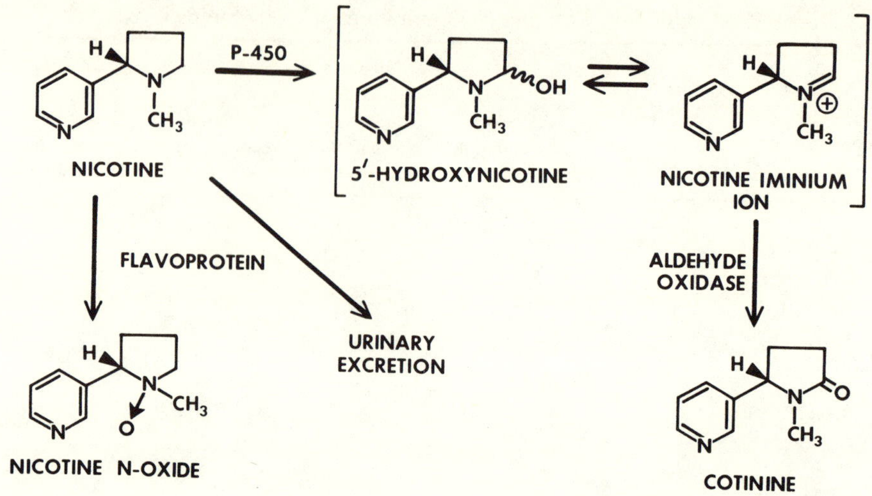

Figure 4.   Major pathways of nicotine metabolism. (From Benowitz et al., 1983*b*, reprinted with permission.)

otine oxide back to nicotine within the gastrointestinal tract, believed to be related to bacterial action.

Quantitative aspects of the conversion of nicotine to its metabolites have not been well defined. Studies of urinary excretion of cotinine in urine collected for 24 hr after intravenous injection of nicotine indicate a range from less than 10% in nonsmokers to an average of 25% in smokers (Beckett et al., 1971*b*). Another study, comparing 24-hr urinary excretion of cotinine to nicotine content of cigarette butts after smoking, indicated 46% recovery as cotinine (Schievelbein, 1982). However, both of these studies underestimate the conversion of nicotine to cotinine because the urine collection period, 24 hr, was too short. Cotinine has a half-life averaging 20 hr (see section on cotinine pharmacokinetics), so that in 24 hr only a little more than half of cotinine will be recovered. Urine collection for at least 72 hr is necessary to recover more than 90% of cotinine in most subjects. In addition, since only 17% of cotinine is excreted unchanged (Benowitz et al., 1983*b*), urinary recovery underestimates cotinine generation rate.

To investigate quantitative aspects of conversion of nicotine to cotinine, we measured blood levels of cotinine after nicotine infusion in abstinent smokers. At a different time, the same subjects were given an infusion of cotinine itself to determine its clearance. The clearance of cotinine, in conjunction with cotinine blood levels following nicotine infusion, can be used to estimate the generation of cotinine. Preliminary data in five subjects indicate that about 80% of nicotine is metabolized to cotinine.

Similar estimates can be obtained by review of published nicotine metabolite excretion data. In uncontrolled urinary pH conditions, about 10% of an intravenous dose of nicotine is excreted unchanged in the urine and 4% is excreted as nicotine-N-oxide (Beckett et al., 1971a). We have found in 26 habitual cigarette smokers the ratio in 24-hr urine collections of cotinine to nicotine-N-oxide averaged 2.98 ($\pm$1.08 SD). Assuming steady state, the rate of excretion of metabolites reflects generation rate. As noted above, 100% of nicotine-N-oxide but only 17% of cotinine is excreted unchanged in the urine. Therefore, the relative rate of generation of cotinine compared with nicotine-N-oxide is 17.5 to 1. If, as reported by Beckett et al. (1971a), 4% of nicotine is excreted as nicotine-N-oxide, then 70% of nicotine is converted to cotinine.

Beckett has shown in a small number of subjects that smokers generate more cotinine than do nonsmokers, possibly reflecting a state of accelerated metabolism in habitual smokers (Beckett et al., 1971b). Klein and Gorrod (1978a) found that the ratio of cotinine to nicotine-N-oxide tended to be lower in older compared with younger smokers, but found no differences between men and women. These authors also reported a reduction in nicotine-N-oxide excretion over the 29th through 32nd week of gestation in pregnant smokers (Klein and Gorrod, 1978b). The reason for and significance of this finding are unclear.

Gorrod et al. (1974) reported a higher ratio of cotinine to nicotine-N-oxide in patients with cancer of the bladder. The possibility of bacterial degradation of nicotine-N-oxide in the bladder was considered as an explanation for this observation. Bladder cancer patients frequently have urinary tract infections. However, the authors found no change in nicotine-N-oxide levels after the urine sat at room temperature for 24 hr. This finding suggests that bacterial degradation was not occurring. Whether the high ratio of cotinine/nicotine-N-oxide is due primarily to a decrease in nicotine oxide production or increased production or decreased rate of elimination of cotinine has not been determined.

Rate of Nicotine Metabolism. The rate of metabolism of nicotine can be determined by measuring blood levels after administration of a known dose of nicotine. We studied habitual cigarette smokers given intravenous infusions of nicotine for 30–60 min (Fig. 5) (Benowitz et al., 1982a). Total and renal clearance were computed directly, and the nonrenal or metabolic clearance computed as the difference between total and renal clearance. Nonrenal clearance averaged 1089 ml/min (Table 1). This represents about 70% of liver blood flow. Assuming most nicotine is metabolized by the liver (data in animals indicate only a small contribution by the lung), this means about 70% of the drug is extracted from the blood in each pass through the liver (Fig. 6). Based on an extraction ratio of 70%, one would predict an oral bioavailability of 30%, which is similar to that reported by Jenner et al. (1973).

An implication of the high degree of hepatic extraction is that clearance of nicotine should be dependent upon liver blood flow. That is, about 70% of

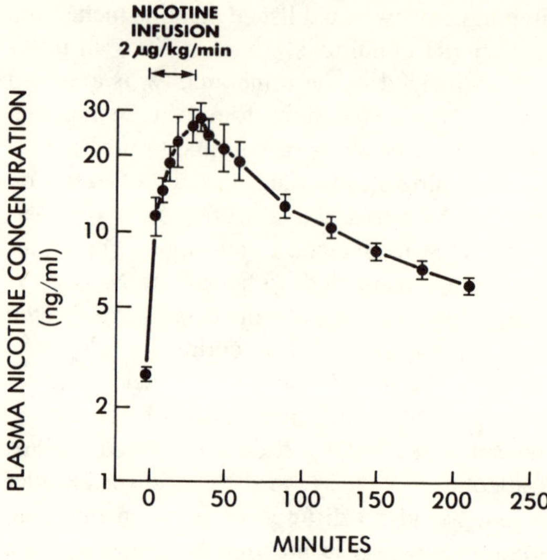



Figure 5. Plasma nicotine concentrations (± SEM) in five subjects during and after constant infusion for 30 min. (From Benowitz et al., 1982a, reprinted with permission.)

whatever volume of blood the circulation brings to the liver should be cleared of nicotine. Thus, physiologic events such as meals, posture, exercise, or other drugs which perturb hepatic blood flow are predicted to have significant effects on the rate of nicotine metabolism. The biological significance of this prediction remains to be determined.

The level of any drug in the body during repeated dosing depends on the rate of intake and the rate of elimination of the drug. There is considerable

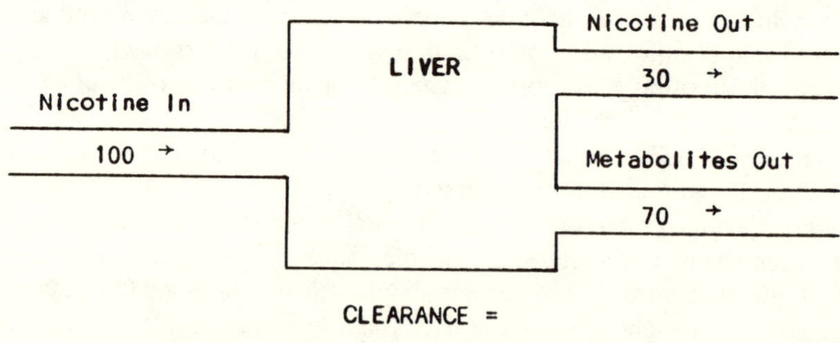



Figure 6. Liver clearance of nicotine. Nicotine enters the liver in portal venous and hepatic arterial blood. Seventy percent of nicotine entering the liver is metabolized; 30% leaves unchanged. Because the capacity of the liver to metabolize nicotine is high, the rate-determining factor for clearance is liver blood flow.

individual variability in rate of elimination of nicotine among individuals (Fig. 2). In pharmacokinetic terms, clearance is a number which predicts steady state blood level of a drug achieved at a particular level of intake. The relationship is expressed mathematically as $C_{ss} = R/CL$, where $C_{ss}$ is the steady state blood concentration, R is the rate of intake, and CL is clearance. Since there may be fourfold interindividual differences in clearance of nicotine, a person with an extremely high clearance will have blood and body nicotine levels one-fourth as high as a person consuming the same amount of nicotine but having a low clearance. If people smoke to achieve a particular body level of nicotine, then clearance could influence self-determined rate of cigarette smoking. Differences in clearance among people could result in different exposures to and risks from other combustion products.

The metabolism of cotinine is much slower than that of nicotine. Total clearance averages 72 ml/min and metabolic clearance 60 ml/min (Benowitz et al., 1983*b*). As it is slowly metabolized, the rate of elimination of cotinine is predicted not to be substantially influenced by changes in liver blood flow.

A number of other metabolites of nicotine have been characterized, but little is known about the pharmacologic activity of these metabolites. For a more detailed review of nicotine metabolism, the reader is referred to Gorrod and Jenner (1975).

## Renal Excretion

Nicotine is excreted by glomerular filtration and tubular secretion, with variable reabsorption depending on urinary pH (Table 4). In acid urine, nicotine is mostly ionized and tubular reabsorption is minimized; renal clearance may be

Table 4. Influence of Urinary pH on Renal Clearance of Cotinine and Nicotine[a]

| | Baseline[b]<br>(mean ± SD) | Acidified urine<br>(mean ± SD) | Alkalinized urine<br>(mean ± SD) |
|---|---|---|---|
| Urine pH | 5.8 ± 0.3 | 4.4 ± 0.4[c] | 6.7 ± 0.3[c] |
| Urine volume (ml) | 2350 ± 720 | 2738 ± 514 | 2466 ± 725 |
| Cotinine renal clearance (ml/min) | 12.3 ± 5.9 | 18.6 ± 10.0[c] | 14.0 ± 10.4 |
| Nicotine renal clearance (ml/min) | 102 ± 41 | 598 ± 238[c] | 43 ± 66[c] |

[a] From Benowitz et al. (1983*b*).
[b] Baseline smoking condition when urine pH was uncontrolled.
[c] $p < 0.05$.

as high as 600 ml/min (urinary pH 4.4), depending on urine flow rate (Benowitz et al., 1983*b*; Rosenberg et al., 1980). In alkaline urine, a larger fraction of nicotine is un-ionized, allowing net tubular reabsorption with renal clearances as low as 17 ml/min (urine pH 7.0). With uncontrolled urine pH, renal clearance averages about 100 ml/min, accounting for the elimination of 10–15% of the daily intake of nicotine. Because urinary pH is a determinant of elimination rate and, hence, body levels of nicotine, it has been suggested that urine pH may influence self-determined nicotine intake (Schachter, 1978). This issue will be addressed in a later section of this article.

Renal clearance of cotinine is much less than glomerular filtration rate (Benowitz et al., 1983*b*). Since cotinine is not appreciably protein bound, this indicates extensive tubular reabsorption. Renal clearance of cotinine can be enhanced by about 50% with extreme urinary acidification (Table 4). Cotinine excretion is less influenced by urinary pH than nicotine because it is less basic and, therefore, is primarily in the un-ionized form within the physiologic pH range. As for nicotine, the rate of excretion of cotinine is influenced by urinary flow rate. Renal excretion of cotinine is a minor route of elimination (averaging 17% of the total clearance); most cotinine is metabolized. In contrast, 100% of nicotine-N-oxide appears to be excreted unchanged in the urine (Beckett et al., 1971*a*). Since blood levels have not been determined in those studies, the clearance of nicotine oxide cannot be computed.

## 4.  NICOTINE AND COTININE BLOOD LEVELS DURING TOBACCO USE

### Nicotine Levels

Blood or plasma nicotine concentrations sampled in the afternoon in smokers generally range from 10–50 ng/ml, although we have observed one patient with a level of over 100 ng/ml. The increment in blood nicotine concentration after smoking a single cigarette ranges from 5 to 30 ng/ml, depending on how the cigarette is smoked (Armitage et al., 1975; Herning et al., 1983*b*; Isaac and Rand, 1972). Blood levels peak at the end of smoking a cigarette and decline rapidly over the next 20 min due to tissue distribution (Fig. 3).

Peak blood levels of nicotine are similar for cigar smokers and users of snuff and chewing tobacco, although the rate of rise of nicotine is slower (Armitage et al., 1978; Turner et al., 1977; Gritz et al., 1981; Russell et al., 1980*b*, 1981). Pipe smokers, particularly those who have previously smoked cigarettes, may have blood levels of nicotine as high as cigarette smokers (McCusker et al., 1982; Turner et al., 1977). Others who are primary pipe smokers, who have not previously smoked cigarettes, tend to have relatively low nicotine levels.

Blood levels of nicotine while chewing 4-mg gum are similar to those observed during cigarette smoking, while blood levels while chewing 2-mg gum are less than seen in most smokers (Russell et al., 1977).

The earliest published studies of the elimination kinetics of nicotine reported half-lives of 20–40 min (Armitage et al., 1975; Isaac and Rand, 1972). In those studies blood levels were followed only for 30 to 60 min, which is not long enough to determine the terminal half-life. Thus, half-lives were based on blood levels which included the distribution phase. As seen in Figure 5, when blood levels are followed for several hours after the end of nicotine infusion, there is a log linear decline phase with a half-life of about 2 hr.

The half-life of a drug is useful in predicting the rate of accumulation of that drug in the body with repetitive doses and the time course of decline after cessation of dosing. Based on a half-life of 2 hr for nicotine, one would predict accumulation over 6–8 hr (3–4 half-lives) of regular smoking and persistence of significant nicotine levels for 6–8 hr after cessation of smoking. If a smoker smokes until bedtime, significant levels should persist all night. Studies of blood levels in regular cigarette smokers confirm these predictions (Fig. 7) (Benowitz

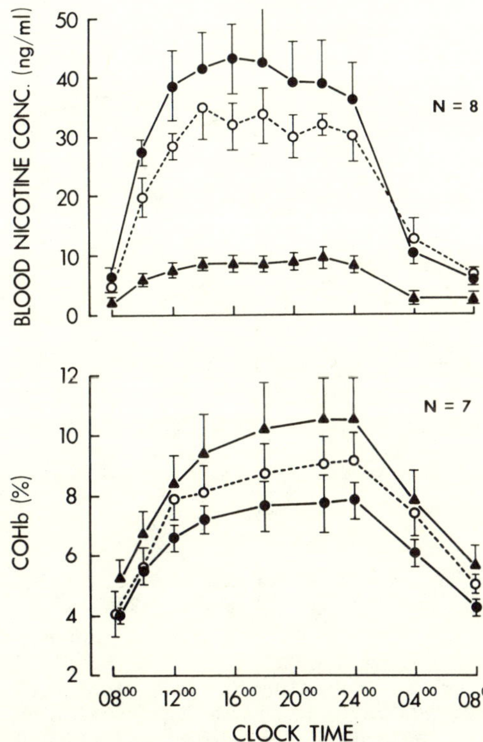


Figure 7. Blood nicotine and carboxyhemoglobin concentrations in subjects smoking high-nicotine (2.5 mg) (—●—) and low-nicotine (0.4 mg) (—▲—) Kentucky reference cigarettes and their usual brand (average nicotine yield, 1.2 mg) (--O--) of cigarettes. Subjects smoked on a fixed schedule of one cigarette every half hour from 8:30 AM to 11:00 PM for a total of 30 cigarettes per day. Blood samples were collected just before the next scheduled cigarette. (Figure taken from Benowitz et al., 1982b, reprinted with permission.)

et al., 1982*b*). Peaks and troughs follow each cigarette, but as the day progresses trough levels rise and the influence of peak levels becomes less important. Thus, nicotine is not a drug to which people are exposed intermittently and which is eliminated rapidly from the body. To the contrary, smoking represents a multiple-dosing situation with considerable accumulation while smoking and persistent levels for 24 hr of each day.

## Cotinine Levels

Cotinine is present in the blood of smokers in much higher concentrations than those of nicotine. Cotinine blood levels average about 250–300 ng/ml in groups of smokers (Benowitz et al., 1983*a*; Haley et al., 1983; Langone and Van Vunakis, 1975; Zeidenberg et al., 1977). We have seen levels ranging from 10 to 900 ng/ml. After stopping smoking, levels decline in a log linear fashion with a half-life averaging 20 hr (range 11–37 hr) (Fig. 8). Because of the long half-life, there is much less fluctuation in cotinine concentrations throughout the

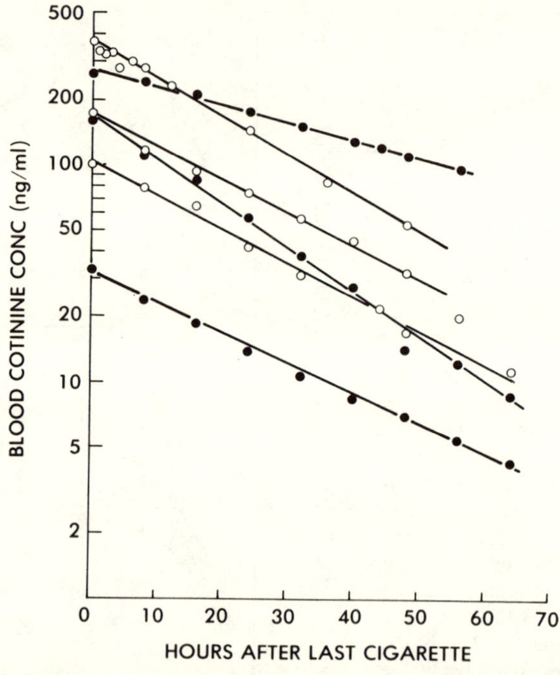


Figure 8.   Decline of blood concentrations of cotinine in six subjects, demonstrating log linearity of decline during abstinence from cigarette smoking. Symbols are intended to distinguish data from individual subjects. (Benowitz et al., 1983*b*, reprinted with permission.)

day than in nicotine concentrations (Fig. 9). As expected, there is a gradual rise in cotinine levels during the day, peaking at the end of smoking, and persisting in high concentrations overnight.

Because some studies have shown pharmacologic activity of cotinine in animals, we studied subjective and cardiovascular actions of cotinine in humans (Benowitz et al., 1982b). Cotinine was infused in abstinent smokers to achieve levels averaging 387 ng/ml, similar to those found in heavy cigarette smokers. No subjective or cardiovascular effects were noted. At levels to which smokers are generally exposed, therefore, cotinine appears to exert little, if any, direct pharmacologic effect.

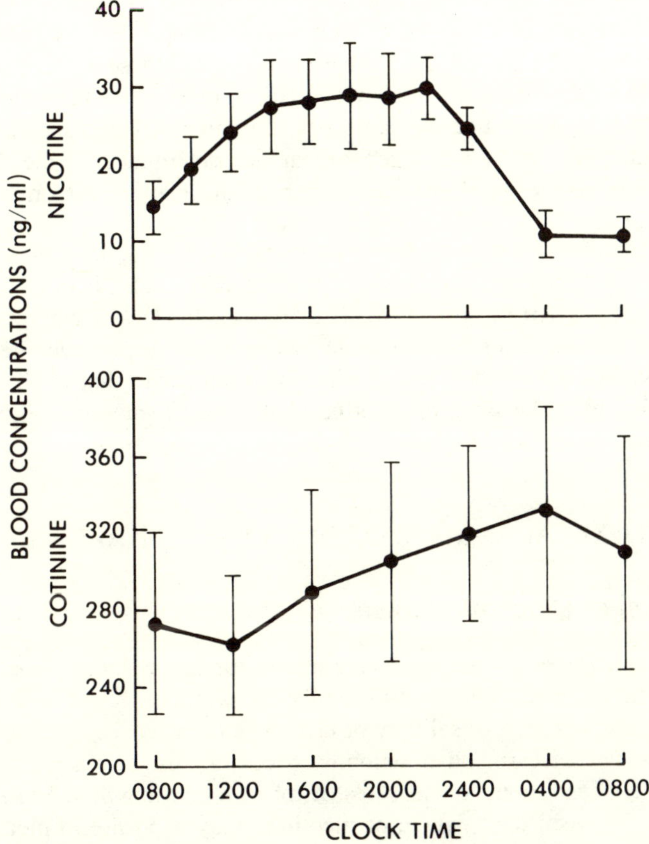

Figure 9.   Circadian blood concentrations of nicotine and cotinine during unrestricted smoking in the uncontrolled urine pH condition. There is much less relative fluctuation throughout the day in cotinine compared with nicotine concentrations. Data are shown as mean ± SE for eight subjects. (Benowitz et al., 1983b, reprinted with permission.)

## 5.  INTAKE OF NICOTINE DURING CIGARETTE SMOKING

Nicotine intake from single cigarettes has been measured by spiking cigarettes with radiolabeled nicotine (Armitage et al., 1975). That study of eight subjects, smoking a single filter-tipped cigarette, indicated an intake range of 0.36 to 2.62 mg. Intake was higher in smokers than in nonsmokers. This study was an important first step, but the method is not applicable to measuring nicotine consumption in natural smoking conditions. We estimated daily intake of nicotine by a method similar to that used in drug bioavailability studies (Benowitz and Jacob, 1984a). Metabolic clearance of nicotine was determined after intravenous injection. Metabolic clearance data were then used in conjunction with blood and urinary concentrations of nicotine measured during a 24-hr period of smoking to determine the daily intake of nicotine. This computation is based on the relationship $D = AUC_{nic} \times CL_{total}$, where D is the daily dose of nicotine, $AUC_{nic}$ is the area under the blood nicotine concentration time curve while smoking over 24 hr, and $CL_{total}$ is nicotine clearance based on nonrenal clearance determined from prior nicotine infusion and renal clearance determined on the day of the smoking study. An example of data and relevant analysis for one subject is shown in Fig. 10.

In 22 habitual cigarette smokers, we found an average daily intake of 37.6 mg nicotine, with a range from 10.5–78.6 mg (Fig. 11). Nicotine intake per cigarette averaged 1.0 mg (range 0.37–1.56 mg). Intake per cigarette did not correlate with smoking machine predicted yield. However, on average smoking machines predicted a yield of 1 mg which was similar to that observed in the smokers (Fig. 12). The issue of nicotine intake versus smoking machine yield will be discussed further in a later section.

## 6.  BIOCHEMICAL MARKERS OF NICOTINE INTAKE

### Comparison of Different Markers

Simple measures of nicotine intake are desirable for studies of smoking in the natural environment. Different investigators have used blood or urinary nicotine concentrations, blood, salivary or urinary cotinine concentrations, expired carbon monoxide or carboxyhemoglobin concentrations, or plasma or salivary thiocyanate concentrations as measures of tobacco consumption. We examined relationships between daily intake of nicotine, daily exposure to nicotine (that is, blood concentrations of nicotine integrated over 24 hr), various parameters of cigarette consumption, and different measures of nicotine intake (Benowitz and Jacob, 1984a).

The number of cigarettes smoked per day best correlated with nicotine

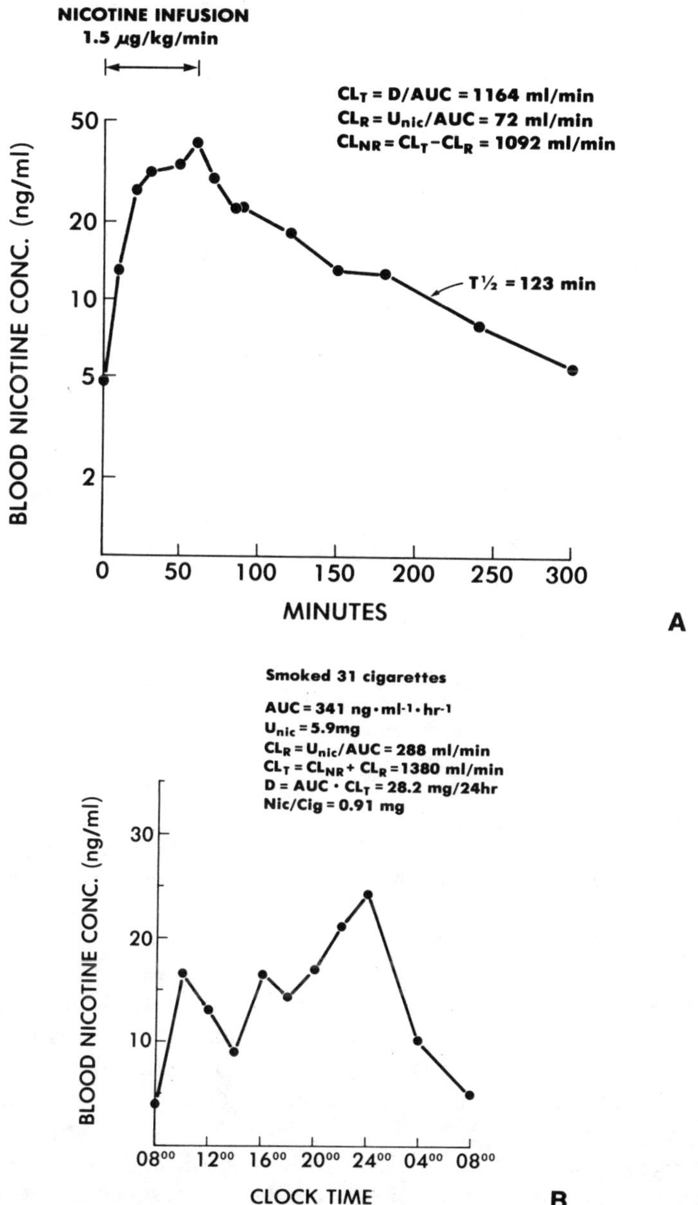


Figure 10.   Data and computations used to estimate daily intake of nicotine in one subject. (A) Blood concentrations of nicotine during and after nicotine infusion. (B) Blood concentrations of nicotine measured throughout the day during *ad lib.* cigarette smoking. (Benowitz and Jacob, 1984a, reprinted with permission.)

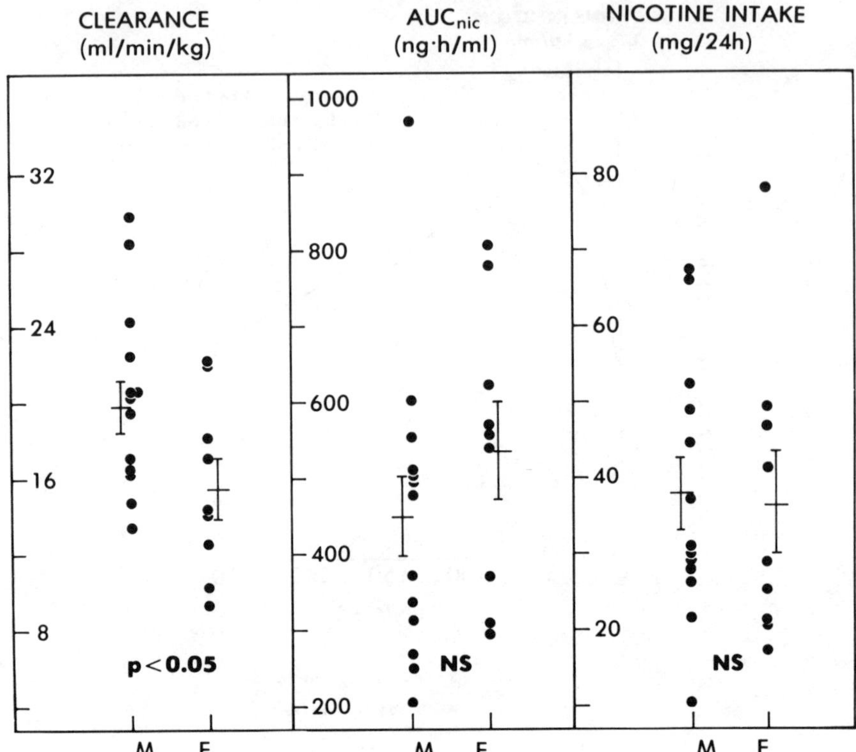



Figure 11.  Total nicotine clearance, area under the blood nicotine concentration–time curve (a measure of daily nicotine exposure), and estimated daily intake of nicotine in 22 subjects, separated according to sex (M, male; F, female). Bars indicate mean ± SE. Analysis of sex differences was by *t* test. (Benowitz and Jacob, 1984a, reprinted with permission.)

intake. Correlations between intake, exposure, and various biochemical markers are shown in Table 5. The best marker of nicotine intake and exposure was a random blood nicotine concentration measured at 4 PM. This level was drawn independently of when the last cigarette was smoked. This finding is consistent with the idea that nicotine levels accumulate throughout the day and plateau in the early afternoon (Fig. 7). At steady state, there should be a reasonably good correlation between nicotine concentrations and daily intake. Carboxyhemoglobin concentrations in the afternoon were the next best markers. Morning (8 AM) levels of nicotine and carboxyhemoglobin also correlated with intake, presumably reflecting persistence of nicotine and carboxyhemoglobin in the blood from exposure from the day before.

Although cotinine is a highly specific marker for nicotine exposure, blood

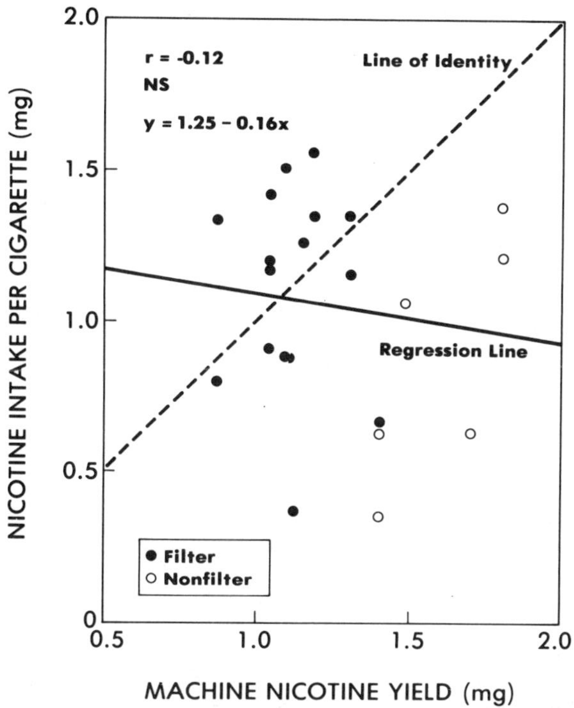


Figure 12.   Regression analysis of relationship between nicotine intake per cigarette and machine-determined nicotine yield. The solid line is the regression line. The dashed line is the line of identity, which indicates points at which measured nicotine intake per cigarette equals machine-determined nicotine yield. The data show no correlation between nicotine intake per cigarette and smoking-machine yield. Smokers of nonfiltered cigarettes consistently consumed less nicotine per cigarette than predicted by machine. (Benowitz and Jacob, 1984a, reprinted with permission.)

levels of cotinine were not as good a quantitative marker of nicotine intake as blood levels of nicotine or carboxyhemoglobin. This is presumably because of individual variability in fractional conversion of nicotine to cotinine and in the rate of elimination of cotinine itself. However, for longitudinal within-subject studies, cotinine level would be expected to be a good marker of changes in nicotine intake.

As expected by the known variation in renal clearance due to effects of urinary flow and pH, urinary concentrations of nicotine did not correlate well with nicotine intake. In contrast, urinary cotinine, which is less influenced by urinary flow of pH, was as good a marker as blood cotinine concentration.

Salivary cotinine concentrations correlate well with blood cotinine concen-

Table 5. Correlation of Various Markers of
Tobacco Smoke Intake with Daily Intake of
Nicotine and Nicotine Exposure[a]

| Measure[b] | Time | Intake | AUC |
|---|---|---|---|
| BNC | 4 PM | $0.81^c$ | $0.91^c$ |
| BNC | Noon | $0.76^c$ | $0.90^c$ |
| COHB | 4 PM | $0.69^c$ | $0.81^c$ |
| COHB | Noon | $0.65^c$ | $0.82^c$ |
| $U_{cot}$ | 24 hr | $0.62^c$ | $0.46^c$ |
| COHB | 8 AM | $0.61^c$ | $0.76^c$ |
| BNC | 8 AM | $0.56^c$ | $0.78^c$ |
| $B_{cot}$ | 4 PM | $0.53^d$ | $0.45^d$ |
| $B_{cot}$ | 8 AM | $0.51^d$ | 0.39 |
| $U_{nic}$ | 24 hr | 0.39 | 0.31 |

[a] From Benowitz and Jacob (1984a).
[b] BNC = blood nicotine concentration; COHB = carboxyhemoglobin level; $U_{cot}$ = 24-hr urinary excretion of cotinine; $B_{cot}$ = blood cotinine concentration.
[c] $p < 0.01$.
[d] $p < 0.05$.

trations (r = 0.90) (Haley et al., 1983). Therefore, salivary cotinine concentrations should be as useful as blood levels in indicating nicotine intake. However, cotinine is present in small amounts in tobacco and in tobacco smoke (Piade and Hoffmann, 1980). Contamination with cotinine from tobacco can potentially invalidate relationships between blood and salivary levels.

## Cotinine as a Quantitative Indicator

Because cotinine has a long half-life, time of sampling with respect to smoking the last cigarette or time of day is less critical than for nicotine. Cotinine levels are also easier for most laboratories to measure. It would be useful to be able to translate cotinine levels, at least on average, into daily intake of nicotine. Such a conversion can be computed using data presented earlier for average clearance of cotinine and fractional conversion of nicotine to cotinine. At steady state, which is assumed in persons with relatively stable daily smoking habits, the daily elimination of cotinine is the same as the daily conversion of nicotine to cotinine:

$$COT_{gen} = COT_{elim} = F \times D_{nic} \tag{1}$$

where $COT_{gen}$ and $COT_{elim}$ are the amounts of cotinine generated and eliminated per day, respectively; F is the fractional conversion of nicotine to cotinine; and

$D_{nic}$ is the daily intake of nicotine. Cotinine elimination rate can be estimated as the product of average concentration ($\overline{C_{COT}}$) and total body cotinine clearance:

$$COT_{elim} = \overline{C_{COT}} \times CL_{COT} \tag{2}$$

Combining and rearranging Equations 1 and 2,

$$D_{nic} = \frac{\overline{C_{COT}} \times CL_{COT}}{F} \tag{3}$$

Using our average data for cotinine clearance (72 ml/min) and F(0.80),

$$D_{nic} \text{ (mg/24 hr)} = (0.12) \times COT_C \text{ (ng/ml)} \tag{4}$$

Thus, an average blood concentration of cotinine of 100 ng/ml corresponds to an average 24-hr consumption of 12 mg nicotine. The typical smoker with a blood concentration of 300 ng/ml would be predicted to consume about 36 mg nicotine per day.

## 7. REGULATION OF NICOTINE INTAKE DURING CIGARETTE SMOKING

There is considerable evidence that smokers adjust their smoking behavior to try to regulate or maintain a particular level of nicotine in the body (Russell, 1976; Gritz, 1980). We have applied techniques for measuring daily intake of nicotine to study compensatory responses to accelerate elimination of nicotine and to switching to high- or low-yield brands of commercial cigarettes (Benowitz and Jacob, 1985).

### Influence of Nicotine Elimination Rate

Rate of renal elimination of nicotine was manipulated by administration of ammonium chloride or sodium bicarbonate to acidify or alkalinize the urine, respectively. Compared with daily excretion during placebo treatment (3.9 mg nicotine per day), acid loading increased (to 12 mg per day) and alkaline loading decreased (to 0.9 mg per day) daily excretion of nicotine (Fig. 13). Average blood nicotine concentrations were similar in placebo and bicarbonate treatment conditions, but were 15% lower during ammonium chloride treatment. Daily intake of nicotine was 18% higher during acid loading, indicating compensation for increased urinary loss (Fig. 14). The compensatory increase in nicotine consumption was only partial, replacing about half the excess urinary nicotine loss.

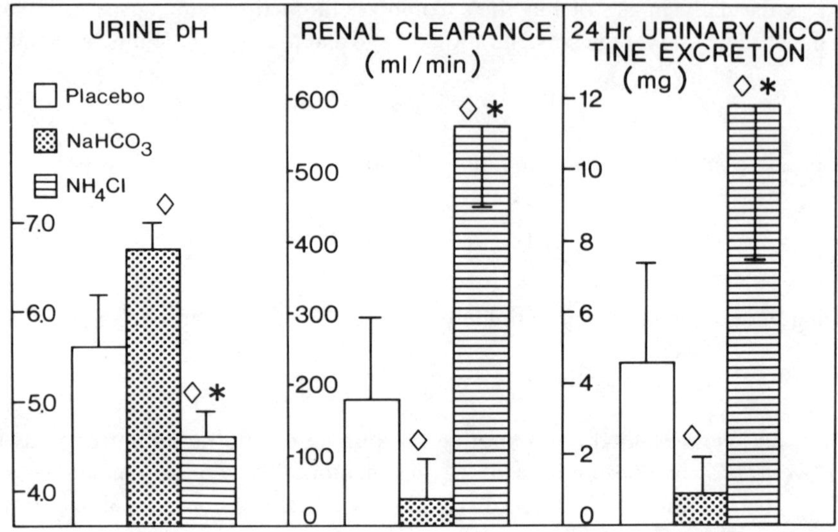



Figure 13.    Influence of urine pH on renal clearance of nicotine and 24-hr urinary excretion. The diamond indicates significant differences compared to placebo. The asterisk indicates significant differences, NaHCO3 versus NH4Cl treatment. (Benowitz and Jacob, 1985, reprinted with permission.)

Caution should be exercised in applying these findings to usual smoking situations. Our studies were performed under conditions of extreme urinary acidification or alkalinization so that the changes in renal clearance would be maximized. Even with extreme differences in urinary pH, differences in overall nicotine elimination rate and smoking behavior were modest. This is because renal excretion is a minor pathway for elimination of nicotine; most is metabolized. Smaller changes in urinary pH, such as occur spontaneously throughout the day or that might be related to stressful events, would not be expected to substantially influence nicotine elimination or smoking behavior. Alkali therapy has been advocated as a way to slow the rate of nicotine elimination and to reduce smoking (Fix et al., 1983). Our data show that there is relatively little excretion of nicotine in baseline conditions (averaging 4 mg per day) and a decrease to 0.9 mg per day with alkali treatment has very little effect on body levels or smoking behavior. Thus, bicarbonate therapy would be expected to have little, if any, effect on cigarette consumption. This prediction is supported by other experimental studies (Cherek et al., 1981).

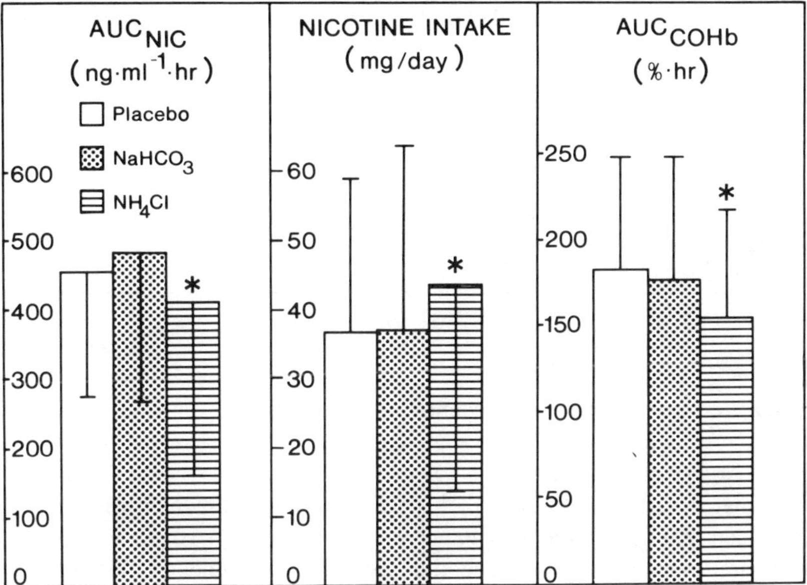


Figure 14.   Nicotine exposure (AUC$_{NIC}$) and intake and carbon monoxide exposure (AUC$_{COHb}$) in placebo, alkaline (NaHCO$_3$), and acid (NH$_4$Cl) treatment conditions. The asterisk indicates significant differences, NaHCO$_3$ versus NH$_4$Cl. (Benowitz and Jacob, 1985, reprinted with permission.)

## Switching to High- and Low-Yield Cigarettes

Several studies have reported that smokers, when switched from higher- to lower-yield cigarettes, consume more nicotine than predicted from machine-determined yields (Ashton et al., 1979; Hill and Marquardt, 1980; Russell et al., 1975, 1982; Sepkovic et al., 1984). Intake measures used in these studies include blood concentrations of nicotine, cotinine, and carboxyhemoglobin, usually on single samples, or urinary excretion of cotinine. We studied smokers switched for three or four days to commercial high-yield (Camel filter) cigarettes, similar in yield to the usual brand these smoker smoked, or to a popular low-yield cigarette (True filter) (Table 6). Daily intake of nicotine was less while smoking high-yield compared with usual cigarette (Fig. 15), presumably because the smokers enjoyed them less. However, average daily intake from the lower-yield cigarette was identical to that observed while smoking the higher-yield cigarette. Intake while smoking the two different brands was highly correlated.

Table 6. Nicotine, Carbon Monoxide, and Tar Yields of Cigarettes as Determined by the FTC[a]

|                          | FTC smoking machine yield | | |
|--------------------------|---------------|---------------------------|------------|
| Brand                    | Nicotine (mg) | Carbon monoxide (mg)      | Tar (mg)   |
| Usual (N = 11)           | 1.0 ± 0.15    | 15.1 ± 1.3                | 16.3 ± 2.7 |
|                          | (0.87–1.48)   | (12.3–16.5)               | (11.0–23.1) |
| Camel (85 mm, filtered)  | 1.18          | 16.0                      | 16.0       |
| True (85 mm, filtered)   | 0.43          | 4.3                       | 4.9        |

[a] Data are mean ± SD (range).

We interpret these data to be consistent with the idea of a lower limit of acceptable daily intake of nicotine (Herman and Kozlowski, 1979). When switched to a different high-yield cigarette or a lower-yield cigarette, smokers consume an amount that is sufficient to meet their needs, but not as much as when smoking their chosen brands. Analysis of intake per cigarette (Fig. 16) reveals that low-yield cigarettes are oversmoked and high-yield cigarettes undersmoked compared with machine-determined yields.

Studies of nicotine intake in individuals smoking their chosen brand also showed no correlation with machine yield (Fig. 12). Of note is that smokers of nonfilter cigarettes, which nominally have higher machine yields than most filter cigarettes, consumed less than predicted by machine. Their intake was similar to that of smokers of filter cigarettes.

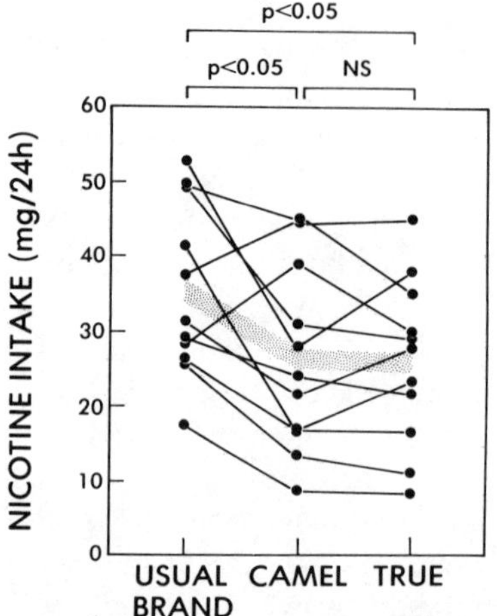


Figure 15. Daily intake of nicotine while smoking usual brand, Camel, or True cigarettes. Average smoking machine-determined yields are given in Table 6. Stippled areas indicate mean values. (Benowitz and Jacob, 1984a, reprinted with permission.)

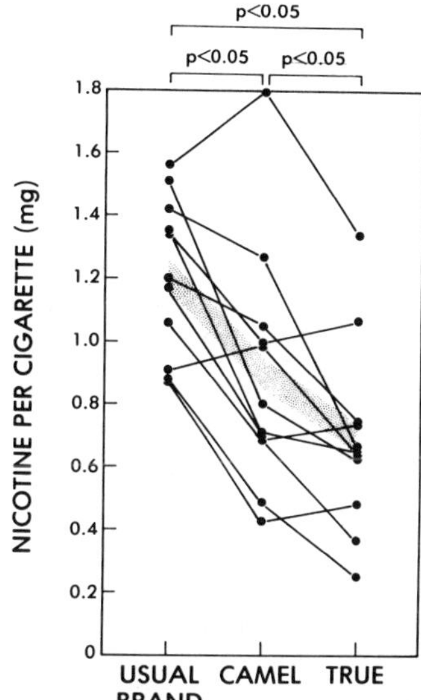


Figure 16. Nicotine intake per cigarette (averaged over all cigarettes smoked during the day) while smoking usual brand, Camel, or True cigarettes. Stippled areas indicate mean values. (Benowitz and Jacob, 1984*b*, reprinted with permission.)

Several studies of nicotine intake in people smoking their chosen brand of cigarettes have been reported (Benowitz et al., 1983*a*; Ebert et al., 1983; Hill et al., 1983; Rickert and Robinson, 1981; Russell et al., 1980*a*). The marker of intake of nicotine in these studies has been blood cotinine or nicotine concentrations. In most studies intake correlates weakly (although significantly) with yield. The slopes of the regression lines are quite shallow, indicating that only small differences in intake are seen with widely differing machine-determined yields. Considering the tendency for individuals to regulate intake of nicotine, as discussed above, and the fact that there is as much nicotine in the tobacco of low-yield as in higher-yield cigarettes, and that smokers can influence nicotine yield by smoking cigarettes differently than the machines, these findings are not unexpected.

Most of the above studies have not included ultra-low-yield cigarettes, that is, cigarettes yielding less than 0.1 mg nicotine and 1 mg tar. We have recently completed a study comparing intake from high- and ultra-low-yield cigarettes (Benowitz et al., 1986). Our data indicate an approximately 50% reduction in nicotine intake from the ultra-low-yield cigarettes compared with other low-yield cigarettes, although that intake was considerably higher than that predicted by the smoking machine. Analysis of population blood cotinine data by subgroups of machine-determined nicotine yields (Fig. 17) is consistent with the idea that smokers of ultra-low-yield cigarettes do consume less nicotine than other brands.

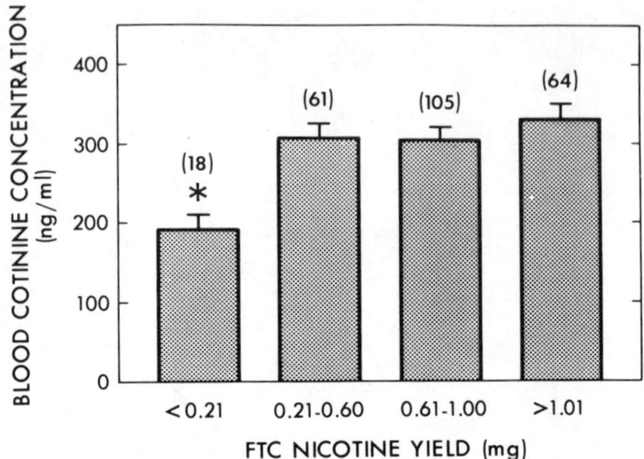


Figure 17.   Afternoon blood cotinine concentrations in smokers grouped according the U.S. Federal Trade Commission smoking-machine yield of nicotine. The asterisk indicates significant difference compared with all other groups. Bars indicate SEM. The parenthesis indicates number of subjects in each group. (Adapted from Benowitz et al., 1983a.)

## 8.   PHARMACOLOGY OF NICOTINE

### General Considerations

The basic pharmacology of nicotine is extremely complex. A detailed review of this pharmacology is beyond the scope of this chapter. Several fundamental pharmacologic issues are relevant to the human pharmacology of nicotine. First, the actions of nicotine in the intact organism are complex and difficult to predict from *in vitro* and animal studies. This is because nicotine can have effects on multiple neuroeffector systems, can affect many or all body organs, the effects of which depend on the prevalent autonomic tone of that particular organ. Effects on various organs may have opposing actions, the net effect depending upon the interaction of the two. For example, it is known that nicotine influences heart rate. Based on *in vitro* and animal studies, we know that nicotine can influence heart rate by action on peripheral chemoreceptors (via the central nervous system), direct actions on the medullary center of the brain, direct release of neuronal norepinephrine, facilitation of catecholamine release in response to neural activation, direct peripheral ganglionic stimulation with resultant cardiac sympathetic or parasympathetic (or both) nerve stimulation, release of epinephrine from the adrenal, vagal responses to nicotine-induced stimulation of the emetic chemoreceptor trigger zone in the medulla, and/or baroreceptor-mediated reflex responses to nicotine-induced blood pressure changes (Comroe, 1960; Su, 1982). It is obvious why it is difficult to predict from basic pharmacologic studies what the influence of nicotine on heart rate will be in the intact human.

A second basic pharmacologic issue is the nature of the dose–response relationship for nicotine. In classical pharmacology, nicotine is commonly discussed as an example of a drug which in low doses causes ganglionic stimulation and in high doses causes ganglionic blockade (following brief stimulation) (Comroe, 1960). Dose–response characteristics *in vivo* are often biphasic as well, although the mechanisms are far more complex. For example, at very low doses, similar to those seen during cigarette smoking, cardiovascular effects appear to be mediated by the central nervous system either via activation of chemoreceptor afferent pathways or via direct effects on the brain stem (Comroe, 1960; Su, 1982). The net result is sympathetic neural discharge with an increase in blood pressure and heart rate. At higher doses, there may be direct effects on the peripheral nervous system, such as ganglionic stimulation, with the release of adrenal catecholamines. With extremely high doses, there may be hypotension and heart rate slowing, mediated either by peripheral vagal activation or direct depressor effects mediated by effects on the brain (Ingenito et al., 1972; Porsius and Van Zwieten, 1978). The importance of considering dose–response relationships is illustrated in considering one of the widely quoted papers on the mechanism of release of vasopressin by nicotine. Studies were performed injecting doses of nicotine much higher than those relevant to humans into anesthetized dogs (Cadnapaphornchai et al., 1974). The net result was hypotension and renal effects consistent with release of vasopressin. The authors showed that blocking vagal afferents eliminated the vasopressin response and concluded that nicotine-mediated vasopressin release was a reflex response. Although this is undoubtedly true in that experimental model, humans respond to nicotine with tachycardia and an increase in blood pressure. Therefore, reflex activation such as would be seen following drug-induced hypotension does not occur. Before extrapolating pharmacologic observations from animals to humans, blood concentrations should be measured to insure that the effects are being studied in a portion of the dose–response curve related to smokers.

A third pharmacologic issue of importance is development of tolerance. Smokers know that tolerance develops to some effects of smoking. Smoking the first cigarette as a teenager is commonly associated with dizziness, nausea, and/or vomiting, effects to which the habitual smoker rapidly becomes tolerant. Likewise, in *in vitro* pharmacologic studies tolerance to various effects develops rapidly, although tolerance may not be complete (Loffelholz, 1970; Steinsland and Furchgott, 1975). Studies of the human pharmacology of nicotine necessarily take place in people who have a degree of tolerance prior to dosing with nicotine. The extent of tolerance depends on both the level of nicotine, the duration of exposure to a given level, and the rate of increase of nicotine exposure to a particular organ. In designing pharmacologic studies in man, it is essential to consider recent exposure to nicotine and the degree of tolerance that may be present at the time nicotine is administered. Individual differences in degree of tolerance may be important determinants of individual differences in pharmacologic responses observed in experimental studies.

In light of the above issues, it is important to consider not only the actions but also dose–response characteristics and tolerance, as well as pharmacokinetics and metabolism in interpreting studies of nicotine effects in man.

## Cardiovascular

Effects. The cardiovascular effects of cigarette smoking have been extensively studied. Smoking a cigarette activates the sympathetic nervous system and results, in healthy people, in an increase in heart rate and blood pressure (Cryer et al., 1976), cardiac stroke volume and output (Irving and Yamamoto, 1963), and coronary blood flow (Bargeron et al., 1957). In smokers with coronary heart disease, smoking may have different effects. It may reduce left ventricular contractility and cardiac output (Aronow et al., 1974; Pentecost and Shillingford, 1964), effects believed to be related to myocardial ischemia due to smoking-mediated tachycardia and/or effects of carbon monoxide. Coronary blood flow may also decrease after smoking, possibly related to nicotine-mediated increase in coronary vascular resistance (Klein et al., 1983; Reddy et al., 1983). Smoking or nicotine causes peripheral vascular changes including cutaneous vasoconstriction (Freund and Ward, 1960) associated with a decrease in skin temperature, systemic venoconstriction (Eckstein and Horsley, 1960), and increased muscle blood flow (Rottenstein et al., 1960). Smoking results in increased circulating concentrations of norepinephrine, consistent with neural adrenergic stimulation, and epinephrine, indicating adrenal medullary stimulation (Fig. 18) (Cryer et al., 1976). Circulating free fatty acids, glycerol, and lactate concentrations increase. Cardiovascular and metabolic effects are prevented by combined α- and β-adrenergic blockade, indicating that the cardiovascular effects of cigarette smoking are mediated by activation of the sympathetic nervous system.

Nicotine appears to be the substance in cigarette smoke responsible for activation of the sympathetic nervous system. Evidence for this comes from studies comparing the effects of nicotine-containing versus denicotinized cigarettes, comparing cigarettes of differing nicotine yield, and studying the effects of nicotine per se, either as intravenous nicotine or nicotine gum (Spohr et al., 1979; Aronow et al., 1971; Benowitz et al., 1982a; Nyberg et al.,1982).

Pharmacodynamics. Although there have been many studies of the actions of smoking or nicotine per se, until recently little has been known about the relationship between nicotine blood concentrations and cardiovascular effects. Such considerations are important to be able to relate knowledge of blood levels

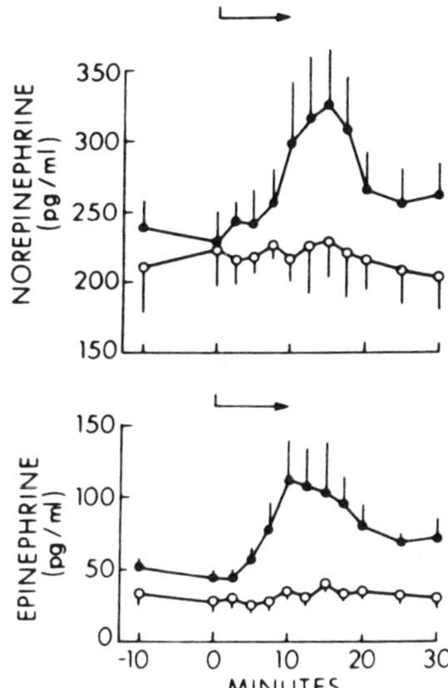


Figure 18. Plasma catecholamine concentrations after smoking two cigarettes. Mean values for 10 subjects (Cryer et al., 1976, reprinted with permission).

throughout the day while smoking to the expected cardiovascular consequences. We studied the effects of intravenous bolus injections of nicotine, given in a manner to simulate cigarette puffing (Rosenberg et al., 1980). A series of injections were given every 30 min for 3 hr. Blood levels (Fig. 19) and cardiovascular effects (Fig. 20) were similar to those observed after cigarette smoking. Heart rate and blood pressure increased sharply after the first series of nicotine injections. Values remained above baseline, but there was little increment with successive injections despite rising and falling nicotine blood levels. In contrast, skin temperature fell progressively during the period of nicotine dosing, gradually returing toward baseline at the end of the study. These data indicated rapid development of tolerance to heart rate and blood pressure responses, but tolerance was not complete in that heart rate and blood pressure remained above baseline.

To clarify these pharmacodynamic observations, a second group of subjects was studied with continuous infusion of nicotine. Heart rate and systolic blood pressure increased mostly in the first 5 to 10 min after infusion (Fig. 21) despite the progressive increase in nicotine blood levels over 30 min (Fig. 5). Skin temperature declined during the infusion and increased gradually after cessation of the infusion. Analysis of the blood concentration–response characteristics by hysteresis plots (Fig. 22) illustrates that heart rate increases at relatively low blood nicotine concentrations and plateaus despite rising levels. Heart rate was lower for a given blood nicotine concentration in the decline phase, indicating development of tolerance. Skin temperature declined and rose in association with changes in blood nicotine concentrations, showing no evidence of tolerance.

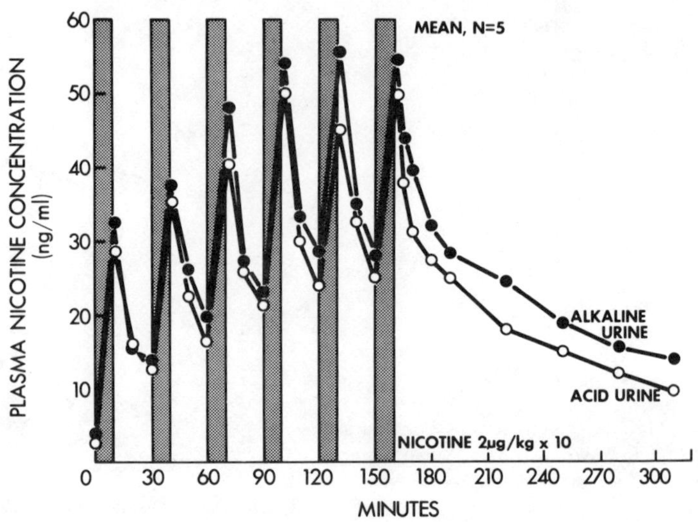


Figure 19.   Mean plasma nicotine concentrations for repetitive injections of nicotine with alkaline and acid urine. Hatched areas indicate 10-min intervals during which intravenous injections of nicotine, 2 µg/kg/min, were given. Data represent mean values for five subjects. (Rosenberg et al., 1980, reprinted with permission.)

These pharmacodynamic observations have potential clinical significance. That low concentrations of nicotine increase heart rate to a maximum suggests that heart rate will increase most with the first few cigarettes of the day, but subsequently will not vary in relation to the amount of nicotine consumed. That only partial tolerance develops to heart rate acceleration due to nicotine suggests that effects on heart rate may persist as long as significant levels of nicotine persist, including overnight. These predictions were tested in a study during which healthy volunteers smoked either high-or low-yield research cigarettes, or abstained from smoking (Benowitz et al., 1984). The research cigarettes were nonfilter cigarettes that were high or low yield because they *contained* more or less nicotine. Thus, full compensation for the low-yield cigarette was impossible. Resultant nicotine blood levels were fourfold different (Fig. 7). As predicted, heart rate, assessed by continuous ambulatory EKG monitoring, increased in the morning, more on smoking than nonsmoking days, and the increase occurred with the first few cigarettes for the day (Fig. 23). Subsequently, heart rate followed a normal circadian pattern, but was always higher when smoking compared with abstinence. Only at 6–7 AM prior to the next day's smoking, were heart rates during smoking similar to those in abstinence. Also, as predicted,

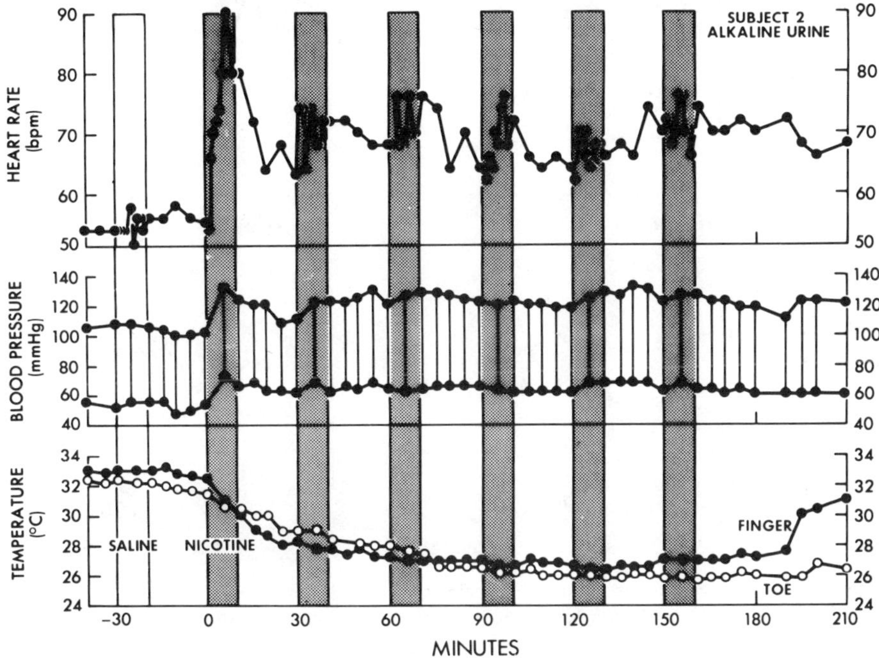


Figure 20.   Physiologic responses during repetitive series of injections of nicotine for a representative subject. Hatched areas indicate 10-min intervals during which intravenous injections of nicotine, 2 μg/kg/min, were given. Clear area indicates 10-min intervals during which intravenous injections of saline, 1 ml/min, were given. (Rosenberg et al., 1980, reprinted with permission.)

heart rate was no different when smoking low-yield compared with high-yield cigarettes, despite the fourfold difference in blood nicotine concentration.

If heart rate acceleration is, as we believe, a marker for sympathetic neural activation, then cigarette smoking can be viewed as producing a state of sympathetic activation 24 hr per day.

## Central Nervous System

Although smokers give different explanations for why they smoke, most agree that smoking produces arousal, particularly with the first few cigarettes of the day, as well as relaxation, particularly in stressful situations. Electroencephalographic desynchronization, decreased $\alpha$ and $\theta$ activity and increased $\alpha$ frequency consistent with arousal, is the usual response to cigarette smoking

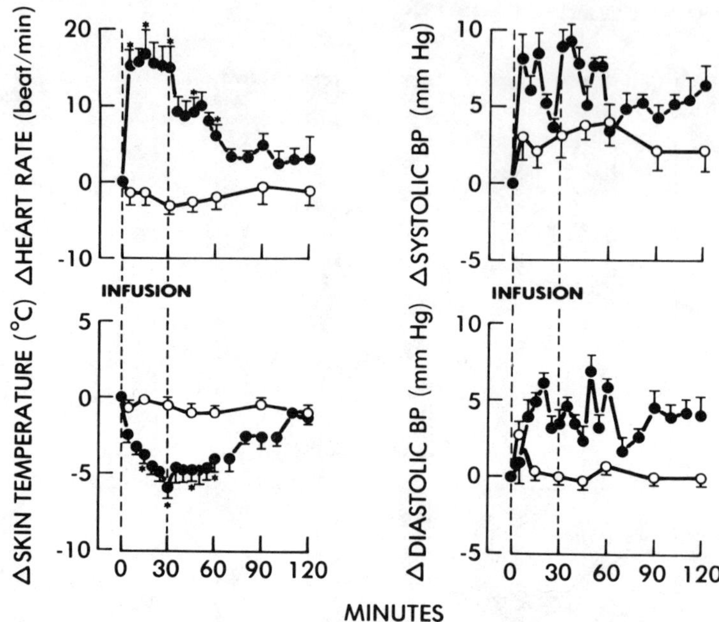


Figure 21.    Cardiovascular responses to nicotine ($N = 5$; ●) and saline ($N = 5$; ○) infusions. Asterisks indicate $p < 0.05$, comparing nicotine and saline conditions. (Benowitz et al., 1982a, reprinted with permission.)

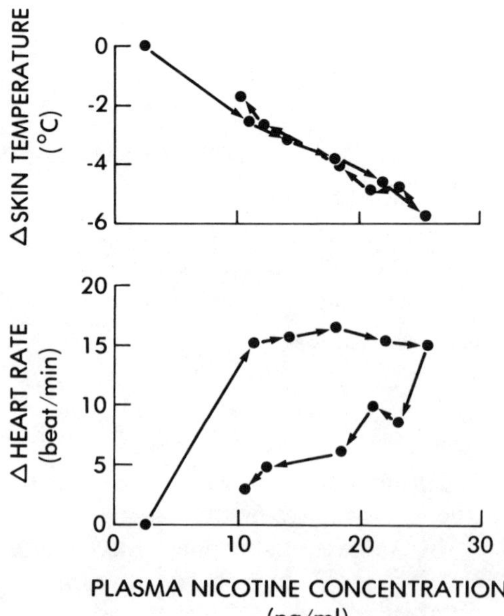


Figure 22.    Hysteresis loops for skin temperature and heart rate responses plotted against simultaneous plasma nicotine concentration (mean values $n = 5$). Arrows indicate progression of time during and after nicotine infusion. (Benowitz et al., 1982a, reprinted with permission.)

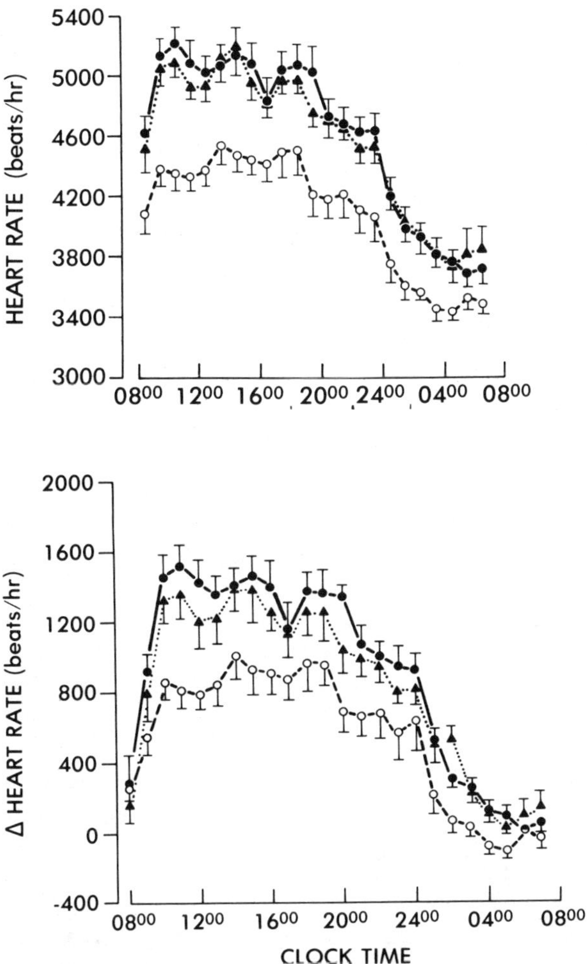



Figure 23.   (A) Hourly heart rate and (B) change in heart rate from 5 to 6 AM baseline while smoking low nicotine (0.4 mg; · · · ▲ · · ·) or high nicotine (2.5 mg; —●—) cigarettes and while abstaining (--○--) (mean ± $n$ = 10). (Benowitz et al., 1984, reprinted with permission.)

(Herning et al., 1983*a*; Knott and Venables, 1977). Electroencephalographic effects are blocked by mecamylamine, a centrally active nicotinic receptor antagonist, indicating a role for nicotinic cholinergic receptor activation (Domino, 1979). Tobacco abstinence is associated with effects opposite those of smoking; i.e., increased α power and reduced α frequency (Herning et al., 1983*a*; Ulett and Itil, 1969). Ashton and co-workers have found that the magnitude of another

EEG measure, contingent negative variation (CNV), a negative brain potential that occurs in the interval between an alerting stimulus and a second stimulus to which a response is required, responds in a biphasic, dose-dependent manner to intravenous injection of nicotine (Ashton et al., 1980). They studied doses ranging from 12.5 to 800 μg. Lower doses increased CNV magnitude, consistent with a stimulant effect, whereas higher doses decreased CNV magnitude, consistent with depression. Cigarette smoking was found to increase CNV in some and decrease CNV in other subjects. It was suggested that smokers titrate the dose of nicotine in smoking to achieve the desired level of alertness or relaxation, and that the level or effect which is sought depends on environmental needs. This model does not address the issue of development of tolerance, which would make titration of effect less meaningful as the day progressed.

The reinforcing role of nicotine as a drug of abuse has been investigated by Henningfield et al. (1983). They allowed abstinent cigarette smokers to self-administer intravenous nicotine in does of 0.75 , 1.0, or 3.0 mg per injection, each given over 10 sec. They found that the number of doses taken by subjects was inversely related to the dose of the drug per injection. Subjects described the subjective responses to nicotine as similar to the morphine groups of drugs. Subjects with previous drug abuse histories described the effects of nicotine as similar to those of cocaine. In some, the injections of nicotine produced dysphoric effects that limited the number of injections taken. However, three subjects continued to self-administer despite nausea. Thus, self-administration behavior for nicotine appears to be similar to that observed for other addicting drugs.

Predosing with mecamylamine diminished subjective responses to nicotine and the preferential selection of nicotine over saline (Henningfield, 1983). The above data support the idea that nicotine plays a central role in compulsive tobacco use.

With repeated intravenous nicotine injections, tolerance to the subjective effects develops rapidly (Henningfield, 1983; Jones et al., 1978). Similarly, in our studies with repeated injections or continuous infusion of nicotine, subjects reported a subjective feeling of being high or dizzy and a little euphoric for 5 or 10 min, but subsequently these responses abated (Benowitz et al., 1982a; Rosenberg et al., 1980). Thus, in contrast to the partial tolerance seen for heart rate effects, complete tolerance appears to develop for the euphoric and other psychological effects of nicotine. For a detailed recent review of the behavioral pharmacology of nicotine, the reader is referred to Henningfield (1984).


## Neuromuscular

One possible component of the relaxation response to cigarette smoking is the effect of nicotine on the neuromuscular system. The drug has been shown to stimulate Renshaw cell discharge, which inhibits motor anterior horn cell activity (Ueki et al., 1961). Tobacco smoking has been reported to decrease muscle tone in spastic patients (Webster, 1964) and has been shown to depress the amplitude of and EMG response to the patellar reflex (Domino and von Baumgarten, 1969). The depression was more prominent after high- compared with low-yield cigarettes and was absent after smoking lettuce leaf cigarettes, suggesting a specific role for nicotine. However, cigarette smoking has recently been shown to increase EMG activity and tonicity to the trapezius muscle (Fagerstrom and Gotestam, 1977), so that the net effect of relaxation versus muscular tension is difficult to predict.

Amplitude of hand tremor increases after smoking cigarettes or chewing nicotine gum (Shiffman et al., 1983; Frankenhaeuser et al., 1970). Whether the mechanism is systemic catecholamine release or a specific central nervous system-mediated action of nicotine has not been established.

## Endocrine

Cigarette smoking has been reported to increase circulating levels of catecholamines, vasopressin, growth hormone, cortisol, ACTH, and endorphins (Cryer et al, 1976; Husain et al., 1975; Pullan et al, 1979; Pomerleau et al, 1983; Rowe et al., 1980; Sandberg et al., 1973; Winternitz and Quillen, 1977; Karras and Kane, 1980). Effects of nicotine on catecholamines have been discussed previously (see Cardiovascular Effects).

It has been known for many years that nicotine can produce an antidiuresis in humans (Burn et al., 1945). Infusion of nicotine was once used as a diagnostic test for vasopressin release in evaluating patients with possible diabetes insipidus (Dingman et al., 1957). The endpoint for that test was the development of nausea. Nicotine may have a direct effect on the chemoreceptor trigger area of the brain stem to induce nausea. More recently, it has become known that nausea or emesis of any cause can produce vasopressin release (Rowe et al., 1979), raising a question about the specificity of the role of nicotine in causing vasopressin release. Many (but not all) smoking studies reporting vasopressin release do report nausea in smokers, possibly due to the use of rapid smoking paradigms. In one study, intravenous infusion of nicotine to habitual smokers did not increase vasopressin levels, and the authors raised the question as to whether the vasopressin releasing effects of nicotine act via pulmonary receptors (Rowe et al., 1980). However, in this study the doses of nicotine were low and might not be expected to achieve blood levels as high as those seen in many cigarette smokers. In another study, similar doses of nicotine given intravenously to nonsmokers did result in vasopressin release (Lightman et al., 1982). Many of the subjects were symptomatic.

We studied the effects of nicotine infusion on free water clearance and vasopressin levels. Infusion of nicotine was associated with antidiuresis. In some people, antidiuresis was associated with nausea and a sharp rise in vasopressin (Fig. 24) (Joseph et al., 1986). In other subjects, antidiuresis occurred without a measured rise in vasopressin level or development of nausea (Fig. 25). Conceivably, small amounts of vasopressin, sufficient to cause antidiuresis but below the sensitivity of the assay, could have been released. Animal studies have shown

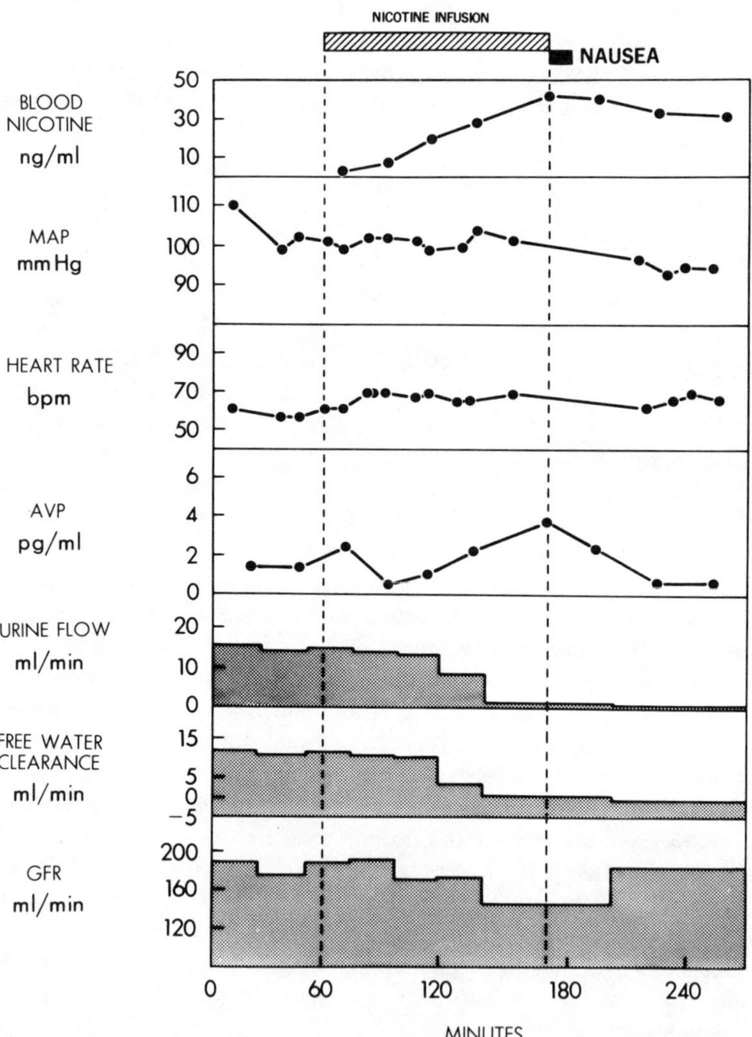


Figure 24.   Antidiuretic and renal hemodynamic responses to nicotine infusion in a water-loaded subject who developed nausea. MAP, mean arterial blood pressure; AVP, plasma arginine vasopressin; GFR, glomerular filtration rate. (From Joseph et al., 1986.)

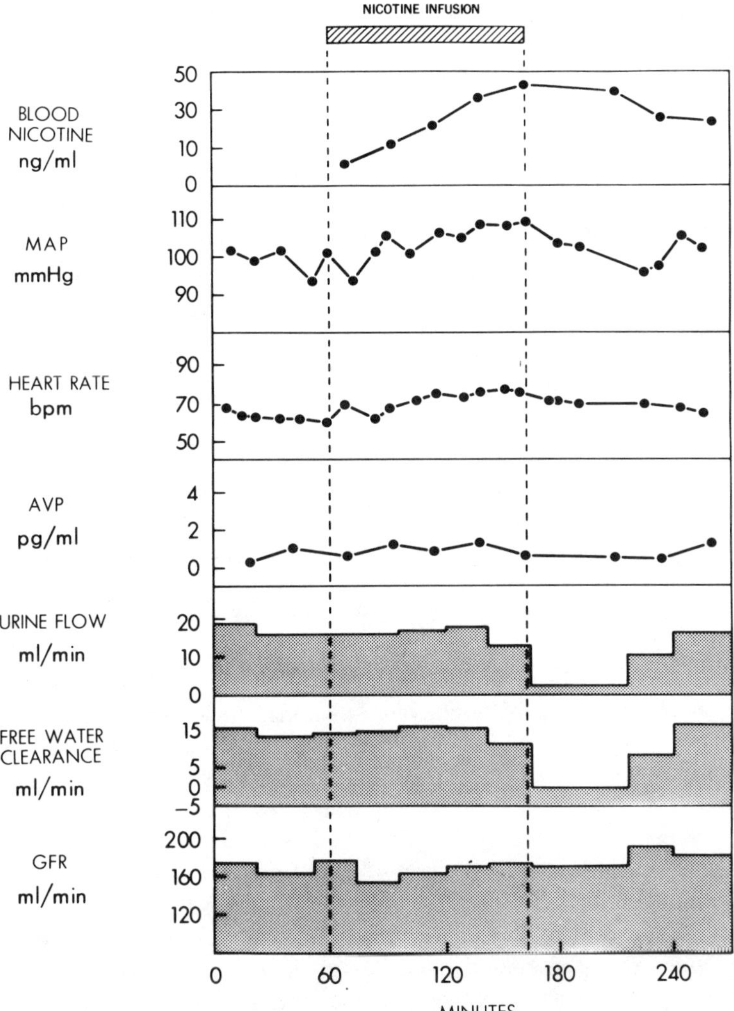


Figure 25.   Antidiuretic and renal hemodynamic response to nicotine infusion in a water-loaded subject who did not develop emesis. (From Joseph et al., 1986.)

vasopressin release after administration of nicotine into the brain (Bisset et al., 1975; Sladek and Joynt, 1979); however, these methods result in extremely high brain concentrations of nicotine. Another animal study showing nicotine-mediated vasopressin release was associated with systemic hypotension (Cadnapaphornchai et al., 1974). In the latter case, reflex responses to hypotension may mediate vasopressin release (see Section 8.1). Thus, the mechanism for nicotine-related antidiuresis in humans remains unresolved.

Tolerance is important in determining the antidiuretic effects of nicotine. Nausea, vasopressin release, and antidiuresis were most marked in our study in light smokers or those who recently quit. In heavy smokers, antidiuresis sometimes occurred but then disappeared despite continuing nicotine infusion, indicating the development of tolerance.

Several studies have shown that smoking increases plasma growth hormone and cortisol levels, effects which do not seem to be entirely mediated by adrenergic mechanisms in that they are not blocked by adrenergic blockade (Cryer et al., 1976). Often, hormonal responses to cigarette smoking are studied in smokers who, after having abstained for several hours or overnight, have smoked two or more cigarettes over a relatively short period of time. Rapid smoking of cigarettes, even in habitual cigarette smokers, is in itself stressful and may induce responses not seen with more ordinary cigarette smoking behavior (Miller et al., 1977).

With constant infusions of nicotine, we have confirmed an increase in growth hormone in some but not all subjects, but no increase in plasma cortisol. Circadian measurements of these hormones have shown no difference in levels when smoking compared with nonsmoking (Benowitz et al., 1984). Tucci and Sode (1972) have also shown that under normal smoking conditions plasma cortisol levels do not rise as a function of smoking. Thus, the plasma cortisol levels reported by others might have resulted from the stress of the experimental situation or possibly from the rapid smoking procedures, which induce effects different from smoking at normal rates. Alternatively, tolerance to effects of smoking may develop (we have seen this for growth hormone) such that hormonal changes are no longer apparent with repetitive smoking. In any case, the role of sustained adrenal cortical stimulation in inducing adverse health effects of cigarette smoking must be questioned.

Recently Pomerleau and co-workers (1983) reported that smoking increases β-endorphin levels and that increases are proportional to the increases in plasma nicotine. They speculate that endogenous opiate release may be a reason for which people smoke, particularly in stressful situations. In support of a role of endogenous opiate release in motivating smokers is a report that habitual smokers smoke less after naloxone pretreatment (Karras and Kane, 1980).

It has been shown in animal studies that nicotine inhibits the synthesis of prostacyclin in rabbit aorta and human peripheral veins and inhibits hypoxia-induced release of prostaglandin from rabbit hearts (Sonnenfeld and Wennmalm, 1980; Wennmalm, 1980). Prostacyclin is a prostaglandin that has antiaggregatory and vasodilating actions, believed to play a homeostatic role in preventing vascular thrombosis. Cigarette smoking has been reported to decrease the urinary excretion of prostacyclin metabolites in humans (Nadler et al., 1983), supporting the prediction from animal studies. The potential importance of nicotine effects on prostaglandin release in understanding human cardiovascular complications of smoking is discussed in the next section.

# 9.  IMPORTANCE OF NICOTINE IN HUMAN DISEASE

Smoking is a major risk factor for coronary and peripheral vascular disease, cancer, chronic obstructive lung disease, peptic ulcer disease, and reproductive disturbances, including prematurity. Tobacco smoke is a complex mixture of chemicals, many of which are felt to contribute to human disease, This review focuses on the potential role of nicotine per se in causing human disease. Nicotine could play a role in the pathogenesis of many smoking-related diseases, although at present there is no conclusive evidence that it contributes to any.

## Coronary and Peripheral Vascular Disease

Nicotine could contribute to atherosclerotic disease by actions on lipid metabolism, coagulation, and hemodynamic effects (Fig. 26). Cigarette smokers are known to have elevated low-density (LDL) and very-low-density lipoproteins (VLDL) and reduced high-density lipoprotein (HDL) levels compared with non-smokers (Stubbe et al., 1982; Wilson et al., 1983; Brischetto et al., 1983), a profile which is associated with an increased risk of atherosclerosis. It is hy-

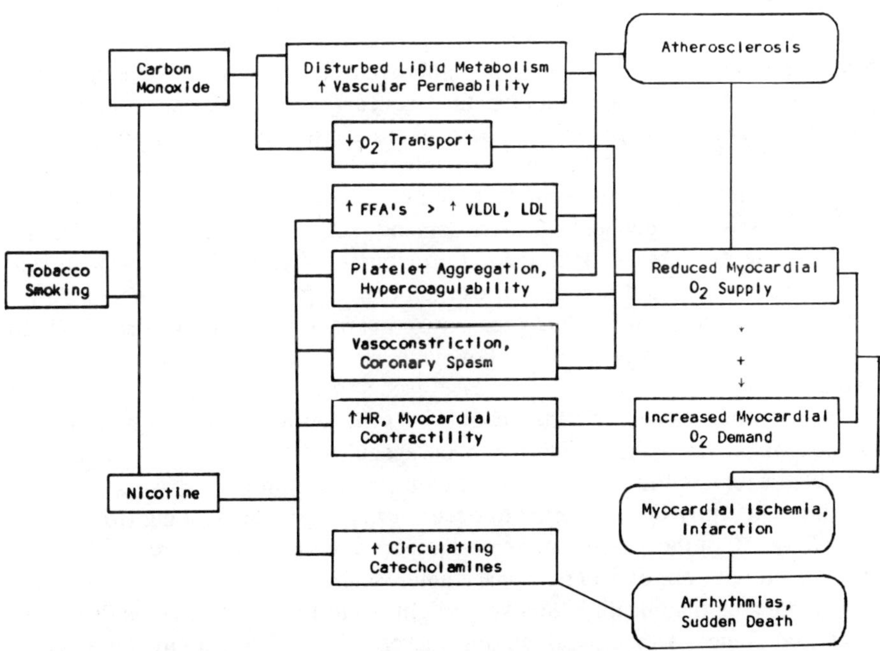


Figure 26.  Schematic diagram of mechanisms by which nicotine may promote coronary heart disease.

pothesized that nicotine, by releasing free fatty acids, increases the synthesis of triglycerides and VLDL by the liver, which in turn results in decreased HDL production (Brischetto et al., 1983).

The blood of smokers is known to coagulate more easily (Billimoria et al., 1975; Ring et al., 1983), platelets in many studies (but not all) are more reactive, and platelet survival is shortened in smokers compared with nonsmokers (Hawkins, 1972; Mustard and Murphy, 1963; Siess et al., 1983; Grignani et al., 1977; Levine, 1973). Thrombosis is believed to play an important role in atherogenesis (Mehta and Mehta, 1981). Platelets may release a growth factor that promotes the growth of vascular endothelial cells, contributing to the atherosclerotic plaque. The importance of nicotine as a determinant of platelet hyperreactivity has been suggested by a recent study showing an apparent relationship between nicotine concentrations after smoking different cigarettes and platelet aggregation response (Renaud et al., 1984). Nicotine could affect platelets by releasing epinephrine, which is known to enhance platelet reactivity; by inhibiting prostacyclin, an antiaggregatory hormone secreted by endothelial cells; or perhaps directly. Finally, by increasing heart rate and cardiac output, nicotine might increase blood turbulence and promote endothelial injury. A recent study in nonhuman primates has shown that lowering the heart rate (by electrical pacing after ablation of the sinoatrial node) retards the rate of development of coronary arteriosclerosis in animals on a hypercholesterolemic diet (Beere et al., 1984). It is possible that nicotine, by increasing heart rate throughout the day, would have a reverse effect (ie., accelerating atherogenesis) in humans. Although many potential mechanisms for promoting atherogenesis have been considered, nicotine has not been demonstrated to accelerate atherosclerosis in experimental animals.

It is likely that nicotine plays a role in causing acute coronary events. Myocardial infarction can occur with one or more of three precipitants: thrombosis, excessive oxygen and substrate demand, and/or coronary spasm. Nicotine can promote thrombosis, as discussed previously. Nicotine increases heart rate and blood pressure and myocardial oxygen consumption. Carbon monoxide inhaled in cigarette smoke reduces the oxygen-carrying capacity of the blood. In the healthy person, coronary blood flow increases to meet the increased demand. In the presence of coronary heart disease, ischemia may develop and myocardial dysfunction may occur. Nicotine could, by sympathomimetic actions or by inhibition of prostacyclin, which has been speculated to be a homeostatic local vasodilator released in response to ischemia, induce coronary spasm. Coronary spasm has recently been reported to occur during cigarette smoking (Maouad et al., 1984). All of the above could contribute to cause acute myocardial infarction in a person with pre-existing coronary atherosclerosis.

Sudden cardiac death in smokers might result from ischemia, as discussed above, combined with the arrhythmogenic effect of increased circulating catecholamines. It has been shown that cigarette smoking can decrease the ventricular fibrillation threshold after experimental myocardial infarction in dogs (Bellet et al., 1972). How much of this effect is due to nicotine and how much to carbon monoxide has not been established.

## Cancer

The importance of nicotine in causing cancer is more speculative than its role in causing cardiovascular disease. Nicotine is not in itself carcinogenic, but has been shown to be cocarcinogenic with benzo(a)pyrene in causing skin cancer in mice (Bock, 1980). The mechanism of cocarcinogenesis is not established. Nicotine can be nitrosated to form nitrosonornicotine and other related compounds in tobacco smoke (Hoffmann and Brunnemann, 1983). These nitrosamines are also found in snuff and in the saliva of snuff dippers (Hoffmann and Adams, 1981). Tobacco-specific N-nitrosamines are highly carcinogenic. Whether they are formed in humans in amounts adequate to contribute to cancer is unknown.

## Chronic Lung Disease

Cigarette smoking is the major risk factor for development of emphysema and chronic bronchitis. Chronic inflammation with migration of polymorphonuclear neutrophils into the lung occur with habitual cigarette smoke exposure. Neutrophils release elastase, an enzyme that destroys alveolar structure. Alpha-1-antitrypsin, an inhibitor of elastase, is partially inactivated by cigarette smoke, probably related to oxidant gases. Nicotine has been found to attract human neutrophils and to enhance the activity of other chemotactic factors (Hunninghake and Crystal, 1983; Totti et al., 1984). It is possible that nicotine in this way promotes the inflammatory response and contributes to development of chronic lung disease.

## Peptic Ulcer Disease

Smoking is strongly related to the prevalence of peptic ulcer disease and failure to stop smoking is the major predictor of failure to respond to ulcer therapy (Korman et al., 1981; Friedman et al., 1974). Smoking has been shown to decrease pancreatic fluid and bicarbonate secretion, resulting in greater and more prolonged acidity of gastric fluid of the duodenal bulb (Murthy et al., 1977, 1978). Similar effects after infusion of nicotine have been reported in animals (Konturek et al., 1972). Infusion of nicotine in humans has been reported to reduce gastric volume and acid secretion (Sonnenberg and Husmert, 1982). However, nicotine was infused in very low doses to nonsmokers, raising questions of relevance to habitual smokers. Smoking reverses the inhibition of nocturnal gastric secretion of acid by cimetidine or ranitidine (Boyd et al., 1983). If such effects are nicotine mediated, this observation can be explained by the pharmacokinetics of nicotine with persistence of nicotine overnight (Benowitz, 1983).

## Reproductive Disorders

Finally, smoking is a major risk factor for prematurity and, consequently, fetal morbidity and mortality (Abel, 1980). Tobacco smoke could influence the fetus either through alterations in maternal physiology, which would limit nutrient flow to the fetus, or by transplacental passage of components of smoke, which would have direct effects on the fetus. The factors considered most likely to affect the fetus are carbon monoxide and nicotine. Carbon monoxide inhalation has been shown to result in an increase in carboxyhemoglobin in both maternal and fetal blood and it has been suggested that this may limit oxygen supply to the fetus (Longo, 1977). However, it has been shown that although newborn infants of smoking mothers have higher concentrations of carboxyhemoglobin than neonates of nonsmokers, there are only trivial differences in hemoglobin concentrations, hematocrits, and various characteristics of hemoglobin (Burea et al., 1984). Thus, it is difficult to explain an adverse effect based on chronic hypoxia due to carbon monoxide in tobacco smoke. It is likely that nicotine is important in causing adverse effects.

Effects of nicotine on the fetus could include reduction of uterine blood flow or a direct effect on fetal function (Ayromlooi et al., 1981; Resnik et al., 1979; Suzuki et al., 1971). The presence of nicotine and its principal metabolites has been demonstrated in umbilical cord blood and urine of newborn infants of smoking mothers, as well as in amniotic fluid, indicating transplacental passage (Hibberd et al., 1978). It is not known whether nicotine can be metabolized by the fetus because the metabolites in fetal tissues could be a result of transfer of metabolites from mother to fetus.

Most of all, nicotine is important in causing human disease because it is the primary cause of tobacco dependence.

ACKNOWLEDGMENTS

I wish to acknowledge the scientific contributions of my colleagues—Peyton Jacob, Reese Jones, Jon Rosenberg, Florence Kuyt, Sharon Hall, and Ronald Herning—without whose involvement the research presented in this paper would not have been possible.

Research supported in part by United States Public Health Service Grants No. DA02277, CA32389, HL29476, and DA01696.

# REFERENCES


Abel, E. L., 1980, Smoking during pregnancy: A review of effects on growth and development of offspring, *Human Biol.* **52:**593–625.

Armitage, A. K., and Turner, D. M., 1970, Absorption of nicotine in cigarette and cigar smoke through the oral mucosa, *Nature* **226:**1231–1232.

Armitage, A. K., Dollery, C. T., George, C. F., Houseman, T. H., Lewis, P. J., and Turner, D. M., 1975, Absorption and metabolism of nicotine from cigarettes, *Br. Med. J.* **4:**313–316.

Armitage, A., Dollery, C., Houseman, T., Kohner, E., Lewis, P. J., and Turner, D., 1978, Absorption of nicotine from small cigars, *Clin. Pharmacol. Ther.* **23:**143–150.

Aronow, W. S., Dendlinger, B. S., and Rokaw, S. N., 1971, Heart rate and carbon monoxide level after smoking high, low and non-nicotine cigarettes, *Ann. Intern. Med.* **746:**697–702.

Aronow, W. S., Cassidy, J., Vangrow, J. S., March, H., Kern, J. C., Goldsmith, J. R., Khemka, M., Pagano, J., and Vawter, M., 1974, Effect of cigarette smoking and breathing carbon monoxide on cardiovascular hemodynamics in anginal patients, *Circulation* **50:**340–347.

Ashton, H., Stepney, R., and Thompson, J. W., 1979, Self-titration by cigarette smokers, *Br. Med. J.* **2:**357–360.

Ashton, H., Marsh, V. R., Millman, J. E., Rawlins, M. D., Telford, R., and Thompson, J. W., 1980, Biphasic dose-related responses of the CNV (contingent negative variation) to I.V. nicotine in man, *Br. J. Clin. Pharmacol.* **10:**579–589.

Ayromlooi, J., Desiderio, D., and Tobias, M., 1981, Effect of nicotine sulfate on the hemodynamics and acid base balance of chronically instrumented pregnant sheep, *Dev. Pharmacol. Ther.* **3:**205–213.

Bargeron, L. M., Ehmke, D., Gonlubol, F., Castellanos, A., Siegel, A., and Bing, R. J., 1957, Effect of cigarette smoking on coronary blood flow and myocardial metabolism, *Circulation* **15:**251–257.

Beckett, A. H., Rowland, M., and Triggs, E. B., 1965, Significance of smoking in investigations of urinary excretion rates of amines in man, *Nature* **207:**200–201.

Beckett, A. H., Gorrod, J. W., and Jenner, P., 1971a, The analysis of nicotine-1′-oxide in urine, in the presence of nicotine and cotinine, and its application to the study of *in vivo* nicotine metabolism in man, *J. Pharm. Pharmacol.* **23:**55S–61S.

Beckett, A. H., Gorrod, J. W., and Jenner, P., 1971b, The effect of smoking on nicotine metabolism *in vivo* in man, *J. Pharm. Pharmacol.* **23:**62S–67S.

Beere, P. A., Glagov, S., and Zarins, C. K., 1984, Retarding effect of lowered heart rate on coronary atherosclerosis, *Science* **226:**180–182.

Bellet, S., DeGuzman, N. T., Kostis, J. B., Roman, L., and Fleischmann, D., 1972, The effect of inhalation of cigarette smoke on ventricular fibrillation threshold in normal dogs and dogs with acute myocardial infarction, *Am. Heart J.* **83:**67–76.

Benowitz, N. L., 1983, Smoking and gastric inhibition by $H_2$ antagonists (Letter), *Lancet* **1:**584.

Benowitz, N. L., and Jacob, P., III, 1984a, Daily intake of nicotine during cigarette smoking. *Clin. Pharmacol. Ther.* **35:**499–504.

Benowitz, N. L., and Jacob, P., III, 1984b, Nicotine and carbon monoxide intake from high- and low-yield commercial cigarettes, *Clin. Pharmacol. Ther.* **36:**265–270.

Benowitz, N. L., and Jacob, P., III, 1985, Nicotine renal excretion rate influences nicotine intake during cigarette smoking, *J. Pharmacol. Exp. Ther.* **234:**153–155.

Benowitz, N. L., Jacob, P., III, Jones, R. T., and Rosenberg, J., 1982a, Interindividual variability in the metabolism and cardiovascular effects of nicotine in man, *J. Pharmacol. Exp. Ther.* **221:**368–372.

Benowitz, N. L., Kuyt, F., and Jacob, P., 1982b, Circadian blood nicotine concentrations during cigarette smoking, *Clin. Pharmacol. Ther.* **32:**758–764.

Benowitz, N. L., Hall, S. M., Herning, R. I., Jacob, P., III, Jones, R. T., and Osman, A-L., 1983a, Smokers of low yield cigarettes do not consume less nicotine, *N. Engl. J. Med.* **309:**139–142.

Benowitz, N. L., Kuyt, F., Jacob, P., Jones, R. T., and Osman, A-L., 1983b, Cotinine disposition and effects, *Clin. Pharmacol. Ther.* **309:**139–142.

Benowitz, N. L., Kuyt, F., Jacob, P., III, 1984, Influence of nicotine on cardiovascular and hormonal effects of cigarette smoking, *Clin. Pharmacol. Ther.* **36:**74–81.

Benowitz, N. L., Jacob, P., III, Yu, L., and Talcott, R., 1986, Reduced tar, nicotine and carbon monoxide exposure while smoking ultra-low but not low-yield cigarettes, *J. Am. Med. Assoc.* **256:**241–246.

Billimoria, J. D., Pozner, H., Metselaar, B., Best, F. W., and James D. C. O., 1975, Effect of cigarette smoking on lipids, lipoproteins, blood coagulation, fibrinolysis and cellular components of human blood, *Atherosclerosis* **21:**61–76.

Bisset, G. W., Feldberg, W., Guertzenstein, P. B., and Silva, M. R. E., 1975, Vasopressin release by nicotine: The site of action, *Br. J. Pharmacol.* **54:**463.

Bock, F. G., 1980, Cocarcinogenic properties of nicotine in: *Banbury Report—A Safe Cigarette?* (G. B. Gori and F. G. Bock, eds.), pp. 129–139, Cold Harbor Spring Laboratory.

Borzelleca, J. F., 1963, Drug movement from the isolated urinary bladder of the rabbit (abst.), *Fed. Proc.* **22:**661.

Bowman, E. R., and McKennis, H., Jr., 1962, Studies on the metabolism of (-)-cotinine in the human, *J. Pharmacol. Exp. Ther.* **135:**306–311.

Boyd, E. J. S., Wilson, J. A., and Wormsley, K. G., 1983, Smoking impairs therapeutic gastric inhibition, *Lancet,* **1:**95–97.

Brandange, S., and Lindblom, L., 1979, The enzyme "aldehyde oxidase" is an iminium oxidase. Reaction with nicotine delta-1-(5')iminium ion, *Biochem. Biophys. Res. Commun.,* **91:**991–996.

Brischetto, C. S., Connor, W. E., Connor, S. L., and Matarazzo, J. D., 1983, Plasma lipid and lipoprotein profiles of cigarette smokers from randomly selected families: Enhancement of hyperlipidemia and depression of high-density lipoprotein, *Am. J. Cardiol.* **52:**675–680.

Bureau, M. A., Shapcott, D., Berthiaume,Y., Monette, J., Blovin, D., Blanchard, P., and Begin, R., 1984, A study of P50,2,3-diphosphoglycerate, total hemoglobin, hematocrit and type F hemoglobin in fetal blood, *Pediatrics* **73:**22.

Burn, J. H., Truelove, L. H., and Burn, I., 1945, The antidiuretic action of nicotine and of smoking, *Br. Med. J.* **1:**403.

Cadnapaphornchai, P., Boykin, J. L., Berl, T., McDonald, K. M., and Schrier, R. W., 1974, Mechanism of effect of nicotine on renal water excretion, *Am. J. Physiol.* **227:**1216–1220.

Cherek, D. R., Lowe, W. C., and Friedman, T. T., 1981, Effects of ammonium chloride on urinary pH and cigarette smoking behavior, *Clin. Pharmacol. Ther.* **29:**762–770.

Clark, M. S. G., Rand, M. J., and Vanov, S., 1965, Comparison of pharmacological activity of nicotine and related alkaloids occurring in cigarette smoke, *Arch. Int. Pharmacodyn.* **1965:**363–379.

Comroe, J. H., 1960, The pharmacological actions of nicotine, *Ann. N.Y. Acad. Sci.* **90:**48–51.

Cryer, P. E., Haymond, M. W., Santiago, J. V., and Shah, S. D., 1976, Norepinephrine and epinephrine release and adrenergic mediation of smoking-associated hemodynamic and meta-bolic events, *N. Engl. J. Med.* **295:**573–577.

Dingman, J. F., Benirschke, K., and Thorn, G. W., 1957, Studies of neurohypophyseal function in man, *Am. J. Med.* **23:**226.

Domino, E. F., 1979, Behavioral, electrophysiological, endocrine, and skeletal muscle actions of nicotine and tobacco smoking, in: *Electrophysiological Effects of Nicotine* (A. Remond and C. Izard, eds.), pp. 133–146, Elsevier/North-Holland Biomedical Press, Amsterdam.

Domino, E. F., and von Baumgarten, A. M., 1969, Tobacco cigarette smoking and patellar reflex depression, *Clin. Pharmacol. Ther.* **10:**72–79.

Ebert, R. V., McNabb, M. E., McCusker, K. T., and Snow, S. L., 1983, Amount of nicotine and carbon monoxide inhaled by smokers of low-tar, low-nicotine cigarettes, *J. Am. Med. Assoc.* **20**:2840–2842.

Eckstein, J. W., and Horsley, A. W., 1960, Responses of the peripheral veins in man to the intravenous administration of nicotine, *Ann. N.Y. Acad. Sci.* **90**:133–137.

Fagerstrom, K. O., and Gotestam, K. G., 1977, Increase in muscle tonus after tobacco smoking, *Addict. Behav.* **2**:203–206.

Faulkner, J. M., 1933, Nicotine poisoning by absorption through the skin, *J. Am. Med. Assoc.* **100**:1664–1665.

Feyerabend, C., Higenbottam, T., and Russell, M. A. H., 1982, Nicotine concentrations in urine and saliva of smokers and non-smokers, *Br. Med. J.* **284**:1002–1004.

Fix, A. J., Daughton, D., Kass, I., Smith, J. L., Wickiser, A., Golden, C. J., and Wass, A. R., 1983, Urinary alkalinization and smoking cessation, *J. Clin. Psychol.* **39**:617–623.

Frankenhaeuser, T., Myrsten, A. L., and Pot, B., 1970, Psychophysiological reactions to cigarette smoking, *Scand. J. Psychol.* **11**:237.

Freund, J., and Ward, C., 1960, The acute effect of cigarette smoking on the digital circulation in health and disease, *Ann. N.Y. Acad. Sci.* **90**:85–101.

Friedman, G. D., Siegelaub, A. B., and Seltzer, C. C., 1974, Cigarettes, alcohol, coffee and peptic ulcer. *N. Engl. J. Med.* **209**:469–472.

Gehlbach, S. H., Perry, L. D., Williams, W. A., Freeman, J. I., Langone, J. J., Peta. L. V., and Van Vunakis, H., 1975, Nicotine absorption by workers harvesting green tobacco, *Lancet* **1**:478–480.

Gori, G. B., Benowitz, N. L., and Lynch, C. J., 1986, Mouth and deep airways absorption of nicotine in cigarette smokers, in press, *Clin. Pharmacol. Therap.*

Gorrod, J. W., and Hibberd, A. R., 1982, The metabolism of nicotine-delta-1'(5')-iminium ion, *in vivo* and *in vitro, Eur. J. Drug Metab. Pharmacokinet.* **7**:293–298.

Gorrod, J. W., and Jenner, P., 1975, The metabolism of tobacco alkaloids, in: *Essays in Toxicology,* Vol. 6, pp. 35–78, Academic Press, New York.

Gorrod, J. W., Jenner, P., Keysell, G. R., and Mikhael, B. R., 1974, Oxidative metabolism of nicotine by cigarette smokers with cancer of the urinary bladder, *J. Natl. Cancer Inst.* **52**:1421–1424.

Grignani, G., Gamba, G., and Ascari, E., 1977, Cigarette smoking effect on platelet function, *Thromb. Haemostasis.* **37**:422–428.

Gritz, E. R., 1980, Smoking behavior and tobacco abuse, in: *Advances in Substance Abuse,* Vol. 1 (N. Mello, ed.), pp. 91–158, JAI Press, Greenwich, Connecticut.

Gritz, E. R., Baer-Weiss, V., Benowitz, N. L., Van Vunakis, H., and Jarvik, M. E., 1981, Plasma nicotine and cotinine concentrations in habitual smokeless tobacco users, *Clin. Pharmacol. Ther.* **30**:201–209.

Haley, N. J., Axelrad, C. M., and Tilton, K. A., 1983, Validation of self-reported smoking behavior: Biochemical analyses of cotinine and thiocyanate, *Am. J. Publ. Health* **73**:1204–1207.

Hawkins, R. I., 1972, Smoking, platelets and thrombosis, *Nature* **236**:450–452.

Henningfield, J. E., 1983, Measurement issues in cigarette smoking research: Basic behavioral and physiological effects and patterns of nicotine self-administration, in *Measurement in the Analysis and Treatment of Smoking Behavior,* National Institute on Drug Abuse Research Monograph No. 48 (J. Grabowski and C. Bell, eds.), pp. 27–38, U.S. Government Printing Office, Washington, D.C.

Henningfield, J. E., 1984, Behavioral pharmacology of cigarette smoking, in: *Advances in Behavioral Pharmacology, Vol. 4 (T. Thompson, P. B. Dews, and J. E. Barrett, eds.), pp. 131–210, Academic Press, New York.*

Henningfield, J. E., Miyasato, K., and Jasinski, D. R., 1983, Cigarette smokers self-administer intravenous nicotine, *Pharmacol. Biochem. Behav.* **19**:887–890.

Herman, C. P., and Kozlowski, L. T., 1979, Indulgence, excess, and restraint: Perspectives on consummatory behavior in everyday life, *J. Drug Issues* Spring:185–196.

Herning, R. I., Jones, R. T., and Bachman, J., 1983a, EEG changes during tobacco withdrawal, *Psychophysiology* **20**:507–512.

Herning, R. I., Jones, R. T., Benowitz, N. L., and Mines, A. H., 1983b, How a cigarette is smoked determines nicotine blood levels, *Clin. Pharmacol. Ther.* **33**:84–90.

Hibberd, A. R., O'Connor, V., and Gorrod, J. W., 1978, Detection of nicotine, nicotine-1'-N-oxide and cotinine in maternal and fetal body fluids, in: *Biological Oxidation of Nitrogen* (J. W. Gorrod, ed.), pp. 353–361, Elsevier, Amsterdam.

Hill, P., and Marquardt, H., 1980, Plasma and urine changes after smoking different brands of cigarettes, *Clin. Pharmacol. Ther.* **27**:652–658.

Hill, P., and Wynder, E. L., 1979, Nicotine and cotinine in breast fluid, *Cancer Lett.* **6**:251–254.

Hill, P., Haley, N. J., and Wynder, E. L., 1983, Cigarette smoking: Carboxyhemoglobin, plasma nicotine, cotinine and thiocyanate versus self-reported smoking data and cardiovascular disease, *J. Chronic Dis.* **36**:439–449.

Hoffmann, D., and Adams, J. D., 1981, Carcinogenic tobacco-specific N-nitrosamines in snuff and in the saliva of snuff-dippers, *Cancer Res.* **42**:4305–4308.

Hoffmann, D., and Brunnemann, K. D., 1983, Endogenous formation of N-nitrosoproline in cigarette smokers, *Cancer Res.*, **43**:5570–5574.

Hoffmann, D., Adams, J. D., and Haley, N. J., 1983, Reported cigarette smoke values: A closer look, *Am. J. Pub. Health* **73**:1050–1053.

Hunninghake, G. W., and Crystal, R. G., 1983, Cigarette smoking and lung destruction: Accumulation of neutrophils in the lungs of cigarette smokers, *Am. Rev. Respir. Dis.* **128**:833–838.

Husain, M. K, Frantz, A. G., Ciarochi, F., and Robinson, A. G., 1975, Nicotine-stimulated release of neurophysin and vasopressin in humans, *J. Clin. Endocrinol. Metab.* **41**:1113.

Ingenito, A. J., Barrett, J. P., and Procita, L., 1972, Direct central and reflexly mediated effects of nicotine on the peripheral circulation, *Eur. J. Pharmacol.* **17**:375–385.

Irving, D. W., and Yamamoto, T., 1963, Cigarette smoking and cardiac output, *Br. Heart J.* **25**:126–132.

Isaac, P. F., and Rand, M. J., 1972, Cigarette smoking and plasma levels of nicotine, *Nature* **236**:308–310.

Jenner, P., Gorrod, J. W., and Beckett, A. H., 1973, The absorption of nicotine-1'-N-oxide and its reduction in the gastrointestinal tract in man, *Xenobiotica* **3**:341–349.

Jones, R. T., Farrell, T. R., and Herning, R. I., 1978, Tobacco smoking and nicotine tolerance, in: Self-Administration of Abused Substances, National Institute on Drug Abuse Research Monograph Series No. 18, (N. A. Krasnegor, ed.), pp. 202–208, U.S. Government Printing Office, Washington, D.C.

Joseph, M., Benowitz, N., and Keil, L., 1986, Antidiuretic and renal hemodynamic actions of intravenous nicotine in man (submitted for publication).

Karras, A., and Kane, J. M., 1980, Naloxone reduces cigarette smoking, *Life Sci.* **27**:1541–1545.

Klein, A. E., and Gorrod, J. W., 1978a, The metabolism of nicotine in cigarette smokers during pregnancy, *Eur. J. Drug Metab. Pharmacokinet.* **2**:87–93.

Klein, A. E., and Gorrod, J. W., 1978b, Age as a factor in the metabolism of nicotine, *Eur. J. Drug Metab. Pharmacokinet.* **1**:51–58.

Klein, L. W., Ambrose, J., Pichard, A., Holt, J., Gorlin, R., and Teichholz, L. E., 1983, Acute effects of cigarette smoking on coronary vascular dynamics, *Circulations* (abst.) **68**:165.

Knott, V., and Venables, P., 1977, EEG alpha correlates of nonsmokers, smokers, smoking and smoking deprivation, *Psychophysiology* **14**:150–156.

Konturek, S. J., Dale, J., Jacobson, E. D., and Johnson, L. R., 1972, Mechanisms of nicotine-induced inhibition of pancreatic secretion of bicarbonate in the dog. *Gastroenterology* **62**:425–429.

Korman, M. G., Shaw, R. G., Hansky, J., Schmidt, G. T., and Stern, A. I., 1981, Influence of smoking on healing rate of duodenal ulcer in response to cimetidine or high-dose antacid, *Gastroenterology* **80**:1451–1453.

Kozlowski, L. T., Rickert, W. S., Pope, M. A., Robinson, J. C., and Frecker, R. C., 1982*a*, Estimating the yield to smokers of tar, nicotine and carbon monoxide from the "lowest yield" ventilated filter cigarettes, *Br. J. Addict.* **77**:159–165.

Kozlowski, L. T., Appel, C.-P., Frecker, R. C., and Khouw, V., 1982*b*, Nicotine, a prescribable drug available without prescription, *Lancet* **1**:334.

Langone, J. J., and Van Vunakis, H., 1975, Quantitation of cotinine in sera of smokers, *Res. Commun. Chem. Pathol. Pharmacol.* **10**:21–28.

Levine, P. H., 1973, An acute effect of cigarette smoking on platelet function, *Circulation* **48**:619–623.

Lightman, S., Landgon, N., Todd, K., and Forsling, M., 1982, Naloxone increases the nicotine-stimulated rise of vasopressin secretion in man, *Clin. Endocrinol.* **16**:353–358.

Loffelholz, K., 1970, Autoinhibition of nicotinic release of noradrenaline from postaganglionic sympathetic nerves, *Naunyn-Schmied. Arch Pharmacol. Bd.* **267**:49–63.

Longo, L. D., 1977, The biological effects of carbon monoxide on the pregnant woman, fetus and newborn infant, *Am. J. Obst. Gynecol.* **129**:69.

Luck, W., and Nau, H., 1984, Exposure of the fetus, neonate, and nursed infant to nicotine and cotinine from maternal smoking, *N. Engl. J. Med.* **311**:672.

Luck, W., Hansen, R., Steldinger, R., and Nau, H., 1982, Nicotine and cotinine—Two pharmacologically active substances as parameters for the strain on fetuses and babies of mothers who smoke, *J. Perinatal Med.* **10**:107–108.

Maouad, J., Fernandez, F., Barrillon, A., Gerbaux, A., and Gay, J., 1984, Diffuse or segmental narrowing (spasm) of the coronary arteries during smoking demonstrated on angiography, *Am. J. Cardio.* **53**:354–355.

McCusker, K., McNabb, E., and Bone, R., 1982, Plasma nicotine levels in pipe smokers, *J. Am. Med. Assoc.* **248**:577–578.

Mehta, J., and Mehta, P., 1981, Role of blood platelets and prostaglandins in coronary artery disease, *Am. J. Cardiol.* **48**:366–373.

Miller, L. C., Schilling, A. F., Logan, D. L., and Johnson, R. L., 1977, Potential hazards of rapid smoking as a technic for the modification of smoking behavior, *N. Engl. J. Med.* **297**:590–592.

Murthy, S. N. S., Dinoso, V. P., Clearfield, H. R., and Chey, W. Y., 1977, Simultaneous measurement of basal pancreatic, gastric acid secretion, plasma gastrin, and secretin during smoking, *Gastroenterology* **73**:758–761.

Murthy, S. N. S., Dinoso, V. P., Clearfield, H. R., and Chey, W. Y., 1978, Serial pH changes in the duodenal bulb during smoking, *Gastroenterology* **75**:1–4.

Mustard, J. F., and Murphy, E. A., 1963, Effect of smoking on blood coagulation and platelet survival in man, *Br. Med. J.* **1**:846–849.

Nadler, J. L., Velasco, J. S., and Horton, R., 1983, Cigarette smoking inhibits prostacyclin formation, *Lancet* **1**:1248–1250.

Nyberg, G., Panfilov, V., Sivertsson, R., and Wilhelmsen, L., 1982, Cardiovascular effects of nicotine chewing gum in healthy nonsmokers, *Eur. J. Clin. Pharmacol.* **23**:303–307.

Pentecost, B., and Shillingford, J., 1964, The acute effects of smoking on myocardial performance in patients with coronary arterial disease, *Br. Heart J.* **26**:422–429.

Petrakis, N. L., Gruenke, L. D., Beelen, T. C., Castagnoli, N., and Craig, J. C., 1978, Nicotine in breast fluid of nonlactating women, *Science* **199**:303–305.

Piade, J. J., and Hoffmann, D., 1980, Chemical studies on tobacco smoke LXVII. Quantitative determination of alkaloids in tobacco by liquid chromatography, *J. Liq. Chromatogr.* **3**:1505–1515.

Pomerleau, O. F., Fertig, J. B., Seyler, L. E., and Jaffe, J., 1983, Neuroendocrine reactivity to nicotine in smokers, *Psychopharmacology* **81**:61–67.

Porsius, A. J., and Van Zwieten, P. A., 1978, The central actions of nicotine on blood pressure and heart rate after administration via the left vertebral artery of anaesthetized cats, *Arzneim.-Forsch.* **28:**1628–1631.

Pullan, P. T., Clappison, B. H., and Johnston, C. I., 1979, Plasma vasopressin and human neurophysins in physiological and pathological states associated with changes in vasopressin secretion, *J. Clin. Endocrinol. Metab.* **49:**580.

Reddy, C. V. R., Khan, R. G., Feit, A., Chowdry, I. H., and El Sherif, N., 1983, Effects of cigarette smoking on coronary hemodynamics in coronary artery disease (abst.), *Circulation* **68:**165.

Renaud, S., Blache, D., Dumont, E., Thevenon, C., and Wissendanger, T., 1984, Platelet function after cigarette smoking in relation to nicotine and carbon monoxide, *Clin. Pharmacol. Ther.* **36:**389–395.

Resnik, R., Brink, G. W., and Wilkes, M., 1979, Catecholamine-mediated reduction in uterine blood flow after nicotine infusion in the pregnant ewe, *J. Clin. Invest.* **63:**1133–1136.

Rickert, W. S., and Robinson, J. C., 1981, Estimating the hazards of less hazardous cigarettes. II. Study of cigarette yields of nicotine, carbon monoxide and hydrogen cyanide in relation to levels of cotinine, carboxyhemoglobin, and tiocyanate in smokers, *J. Toxicol. Environ. Health* **7:**391–403.

Ring, T., Kristensen, S. D., Jensen, P. N., Mourits-Andersen, T., Madsen, H., and Dyerberg, J., 1983, Cigarette smoking shortens the bleeding time, *Thromb. Res.* **32:**531–536.

Rosenberg, J., Benowitz, N. L., Jacob, P., and Wilson, K. M., 1980, Disposition kinetics and effects of intravenous nicotine, *Clin. Pharmacol. Ther.* **28:**516–522.

Rottenstein, H., Pierce, G., Russ, E., Felder, D., and Montgomery, H., 1960, Influence of nicotine on the blood flow of resting skeletal muscle and of the digits in normal subjects, *Ann. N.Y. Acad. Sci.* **90:**102–113.

Rowe, J. W., Shelton, R. L., Helderman, J. H., Vestal, R. E., and Robertson, G. L., 1979, Influence of the emetic reflex on vasopressin release in man, *Kidney Int.* **16:**729.

Rowe, J. W., Kilgore, A., and Robertson, G. L., 1980, Evidence in man that cigarette smoking induces vasopressin release via an airway-specific mechanism, *J. Clin. Endocrinol. Metab.* **51:**170.

Russell, M. A. H., 1976, Tobacco smoking and nicotine dependence, in: *Research Advances in Alcohol and Drug Problems,* Vol. 3, (R. J. Gibbins, Y. Israel, H. Kalant, R. E. Popham, W. Schmidt, and R. E. Smart, eds.), pp. 1–48, John Wiley, New York.

Russell, M. A. H., and Feyerabend, C., 1978, Cigarette smoking: A dependence or high-nicotine boli, *Drug Metab. Rev.* **8:**29–57.

Russell, M. A. H., Wilson, C., Patel, U. A., Feyerabend, C., and Cole, P. V., 1975, Plasma nicotine levels after smoking cigarettes with high, medium, and low nicotine yields, *Br. Med. J.* **2:**414–416.

Russell, M. A. H., Sutton, S. R., Feyerabend, C., Cole, P. V., and Saloojee, Y., 1977, Nicotine chewing gum as a substitute for smoking, *Br. Med. J.* **1:**1060–1064.

Russell, M. A. H., Jarvis, M., and Feyerabend, C., 1980a, Relation of nicotine yield of cigarettes to blood nicotine concentration in smokers, *Br. Med. J.* **280:**972–976.

Russell, M. A. H., Jarvis, M. J., and Feyerabend, C., 1980b, A new age for snuff? *Lancet* **1:**474–475.

Russell, M. A. H., Raw, M., and Jarvis, M. J., 1980c, Clinical use of nicotine chewing gum, *Br. Med. J.* **280:**1599–1602.

Russell, M. A. H., Jarvis, M. J., Devitt, G., and Feyerabend, C., 1981, Nicotine intake by snuff users, *Br. Med. J.* **283:**814–817.

Russell, M. A. H., Sutton, S. R., Iyer, R., Feyerabend, C., and Vesey, C. J., 1982, Long-term switching to low-tar low-nicotine cigarettes, *Br. J. Addict.* **77:**145–158.

Sandberg, H., Roman, L., Zavodnick, J., and Kupers, N., 1973, The effect of smoking on serum somatotropin, immunoreactive insulin and blood glucose levels of young adult males, *J. Pharmacol. Exp. Ther.* **184:**787–791.

Sasson, I. M., Haley, N. J., Hoffmann, D., Wynder, E. L., Hellberg, D., and Nilsson, S., 1985, Cigarette smoking and neoplasia of the uterine cervix: Smoke constituents in cervical mucus, *N. Engl. J. Med.* **312:**315–316.

Schachter, S., 1978, Pharmacological and psychological determinants of smoking, *Ann. Intern. Med.* **88:**104–114.

Schievelbein, H., 1982, Nicotine. Resorption and fate, *Pharmacol. Ther.* **18:**233–248.

Schievelbein, H., Eberhardt, R., Loschenkohl, K., Rahlfs, V., and Bedall, F. K., 1973, Absorption of nicotine through the oral mucosa. I. Measurement of nicotine concentration in the blood after application of nicotine and total particulate matter, *Agents Actions* **3/4:**254–258.

Sepkovic, D. W., Parker, K., Axelrad, C. M., Haley, N. J., and Wynder, E. L., 1984, Cigarette smoking as a risk for cardiovascular disease. V. Biochemical parameters with increased and decreased nicotine content cigarettes, *Addict. Behav.* **9:**255–263.

Shiffman, S. M., Gritz, E. R., Maltese, J., Lee, M. A., Schneider, N. G., and Jarvik, M. E., 1983, Effects of cigarette smoking and oral nicotine on hand tremor, *Clin. Pharmacol. Ther.* **33:**800–805.

Siess, W., Lorenz, R., Roth, P., and Weber, P. C., 1982, Plasma catecholamines, platelet aggregation and associated thromboxane formation after physical exercise, smoking or norepinephrine infusion, *Circulation* **66:**44–48.

Sladek, C. D., and Joynt, R. J., 1979, Characterization of cholinergic control of vasopressin release by the organ-cultured rat hypothalamoneurohypophyseal system, *Endocrinology* **104:**657.

Sonnenberg, A., and Husmert, N., 1982, Effect of nicotine on gastric mucosal blood flow and acid secretion, *Gut* **23:**532–535.

Sonnenfeld, T., and Wennmaim, A., 1980, Inhibition by nicotine of the formation of prostacyclin-like activity in rabbit and human vascular tissue, *Br. J. Pharmacol.* **71:**609–613.

Sophr, U., Hoffmann, K., Steck, W., Harenberg, J., Walter, E., Hengen, N., Augustin, J., Morl, H., Koch, A., Horsch, A., and Weber, E., 1979, Evaluation of smoking-induced effects on sympathetic, hemodynamic and metabolic variables with respect to plasma nicotine and COHb levels, *Atherosclerosis* **33:**271–283.

Steinsland, O. S., and Furchgott, R. F., 1975, Desensitization of the adrenergic neurons of the isolated rabbit ear artery to nicotinic agonists, *J. Pharmacol. Exp. Ther.* **193:**138–148.

Stubbe, I., Eskilsson, J., and Nilsson-Ehle, P., 1982, High-density lipoprotein concentrations increase after stopping smoking, *Br. Med. J.* **284:**1511–1513.

Su, C., 1982, Actions of nicotine and smoking on circulation, *Pharmacol. Ther.* **17:**129–141.

Suzuki, K., Horiguchi, T., Comas-Urrutia, A. C., Mueller-Heubach, E., Morishima, H. O., and Adamson, K., 1971, Pharmacologic effects of nicotine upon the fetus and mother in the rhesus monkey, *Am. J. Obstet. Gynecol.* **111:**1092–1101.

Totti, N., McCusker, K. T., Campbell, E. J., Griffin, G. L., and Senior, R. M., 1984, Nicotine is chemotactic for neutrophils and enhances neutrophil responsiveness to chemotactic peptides, *Science* **223:**169–172.

Travell, J., 1960, Absorption of nicotine from various sites, *Ann. N.Y. Acad. Sci.* **90:**13–30.

Tucci, J. R., and Sode, J., 1972, Chronic cigarette smoking: Effect on adrenocortical and sympathoadrenomedullary activity in man, *J. Am. Med. Assoc.* **221:**282–285.

Turner, D. M., Armitage, A. K., Briant, R. H., and Dollery, C. T., 1975, Metabolism of nicotine by the isolated perfused dog lung, *Xenobiotica* **5:**539–551.

Turner, J. A. M., Sillett, R. W., and McNicol, M. W., 1977, Effect of cigar smoking on carboxyhemoglobin and plasma nicotine concentrations in primary pipe and cigar smokers and ex-cigarette smokers, *Br. Med. J.* **2:**1387–1389.

Ueki, S., Koketsu, K., and Domino, E. F., 1961, Effects of mecamylamine on the Golgi recurrent collateral-Renshaw cell synapse in the spinal cord, *Exp. Neurol.* **3:**141–148.

Ulett, J., and Itil, T., 1969, Quantitative electroencephalogram in smoking and smoking deprivation, *Science* **164:**969–970.

Van Vunakis, H., Langone, J. J., and Milunsky, A., 1974, Nicotine and cotinine in the amniotic fluid of smokers in the second trimester of pregnancy, *Am. J. Obstet. Gynecol.* **120:**64–66.

Webster, D. D., 1964, The dynamic quantitation of spasticity with automated integrals of passive motion resistance, *Clin. Pharmacol. Ther.* **5:**900–908.

Wennmalm, A., 1980, Nicotine inhibits hypoxia- and arachidonate-induced release of prostacyclin-like activity in rabbit hearts, *Br. J. Pharmacol.* **69:**545–549.

Wilson, P. W. F., Garrison, R. J., Abbott, R. D., and Castelli, W. P., 1983, Factors associated with lipoprotein cholesterol levels. The Framingham Study, *Arteriosclerosis* **3:**273–281.

Winternitz, W. W., and Quillen, D., 1977, Acute hormonal response to cigarette smoking, *J. Clin. Pharmacol.* **17:**389–397.

Zeidenberg, P., Jaffe, J. H., Kanzler, M., Levitt, M. D., Langone, J. J., and Van Vunakis, H., 1977, Nicotine: Cotinine levels in blood during cessation of smoking, *Compr. Psychiatry* **18:**93–101.

2

# Benzodiazepines as Drugs of Abuse and Dependence

HOWARD D. CAPPELL, EDWARD M. SELLERS, and USOA BUSTO

## 1. INTRODUCTION

One of the most difficult problems in discussing drug abuse is definitional. Where benzodiazepines are concerned the definitional problems are particularly complex, since most of their use occurs within a legitimate medical/therapeutic context, and decisions must be made as to where appropriate therapy ends and abuse or misuse begins. Moreover, although a dispassionate appraisal of the abuse problems associated with benzodiazepines is now in process in the scientific literature, information available to the general population of actual and potential consumers of benzodiazepines is not always conveyed with comparable objectivity (cf. Cohen, 1983).

The Third Edition of the *Diagnostic and Statistical Manual of Mental Disorders* (*DSM-III*) (American Psychiatric Association, 1980) specifies criteria for sedative or hypnotic abuse and dependence that provide the basis for reasonable working definitions upon which to organize this chapter. The diagnostic criteria for *abuse* of benzodiazepines are:

1. Pattern of pathological use: inability to cut down or stop use; intoxication throughout the day; frequent use of 60 mg or more of diazepam; amnesic periods for events that occurred while intoxicated.

HOWARD D. CAPPELL, EDWARD M. SELLERS, and USOA BUSTO ● Addiction Research Foundation, Toronto, Ontario, M5S 2S1, Canada.

2. Impairment in social or occupational functioning due to substance use: e.g., fights, loss of friends, absence from work, loss of job, or legal difficulties (other than arrests due to possession, purchase, or sale of the substance).
3. Duration of disturbance of at least one month.

The diagnostic criteria for *dependence* are:

1. Tolerance: Need for markedly increased amounts of the substance to achieve the desired effect or markedly diminished effect with regular use of the same amount.
2. Withdrawal: Development of characteristic withdrawal syndrome after cessation or reduction in substance use.

In subsequent sections, we will attempt to provide an appraisal of what is known about benzodiazepines as drugs of abuse and dependence, using the criteria in *DSM-III* as an organizational guide. Before doing this, however, we will attempt to place the issue of benzodiazepine abuse and dependence in a general social and medical context.

## The Social Context of Benzodiazepine Use and Abuse

Perhaps more than is the case for other abusable drugs, the literature on benzodiazepines contains numerous discourses that combine a review of research findings with some appraisal of the costs and benefits of benzodiazepines in the general social context of their use. Although such discourses do not necessarily provide new scientific information about benzodiazepine abuse, they do provide a useful picture of the value judgments that become evident when such scientific information is processed.

Some authors have emphasized the importance of cultural values, particularly as expressed through public information media, as colorants of issues that could otherwise be decided by a more-or-less scientific appraisal. Cohen (1983) recounts the brief history of benzodiazepines using diazepam (Valium) as an example. According to Cohen, important new drugs, and in particular those developed as anxiolytics, seem to go through a sequence of three phases of judgment. The first he called "euphoric ignorance, in which the new agent outperforms the older drugs and any of its long-term problems are unapparent." However, this is succeeded by the stage of "paranoid uncertainty" which is engendered because "as experience is gained, some of the adverse consequences that were not evident when the drug was introduced present themselves, and there is disappointment and perhaps disillusionment about the agent." The third stage, called "judicious evaluation," eventually "emerges when the drug's advantages and disadvantages become thoroughly clear, and it has been learned how to use the agent properly and more skillfully."

In Cohen's view, the paranoid uncertainty phase was seized by the various media as a newsworthy opportunity. An example of the kind of journalistic response referred to by Cohen can be found in a book entitled *The Tranquillizing of America* (Hughes and Brewin, 1979). This book is not about benzodiazepines exclusively, but the titles of some of its chapters are revealing. These include: Sedativism; The New Health Crisis; Women as Victims; Poisoning Our Future; Chemical Solitary Confinement; and Pacification of the Elderly.

The appropriateness of a good deal of medically sanctioned benzodiazepine use has also been challenged from a sociological perspective. Such an analysis begins with the argument that there are a variety of social and health costs that have been unnecessarily high because of the dominance of a "biomedical model" of anxiety. The essence of such a critique (cf. Cooperstock and Hill, 1982; Cooperstock and Lennard, 1979) is that in their role as prescribers of benzodiazepines, physicians have become treaters of a variety of "problems of living" (e.g., anxiety, stressful interpersonal problems) which may not be appropriate for the health care system to address in the first instance. The inappropriate prescribing that this entails engenders gratuitous visits to physicians. This is an economic cost apart from "the physical and psychosocial costs incurred through toxic reactions, adverse reactions and overdose." The solution proposed to reduce these costs is a change in health care delivery that might include "the involvement of patients themselves as well as other agencies and professions as sources of support in dealing with social and psychological predicaments" (Blackwell and Cooperstock, 1983). This was described as a "biopsychosocial" model, and Blackwell and Cooperstock observed that some shift in this direction has already occurred.

Current medical opinion appears very sensitive to the possiblility of inappropriate prescribing and insufficient monitoring of patients' use of benzodiazepines. Moreover, there are some recurrent arguments within the medical opinion literature that have been used to legitimize the use of benzodiazepines and to rebut criticisms of casualness in their prescription. Generally, such arguments take the form of a cost–benefit analysis in which the problems associated with benzodiazepines are balanced against their utility and safety. One such analysis was provided by Clare (1983), who considered benzodiazepine use in the context of psychotropic drug use in general. The assumption implicit in his approach is that some form of psychotropic drug use is inevitable, and that a *comparative* analysis is necessary to gain perspective on the costs and benefits of benzodiazepines. To this end, Clare compared statistics on various aspects of benzodiazepine use with those on alcohol and tobacco use in the United Kingdom. His conclusion was that compared with alcohol and tobacco, the use of minor tranquilizers does not constitute an unreasonable burden on the economic cost of health care, nor does it exact a comparable toll in physical and social consequences (e.g., traffic accidents, psychological and physical pathology). On the benefit side, he suggested that benzodiazepines can be used to treat conditions for which less costly alternatives do not exist. In effect, Clare was arguing that with simple guidelines to physicians to prevent misuse, benzodiazepines have an appropriate role to play in the management of psychologically distressed individuals. Caution in prescribing for him was a more reasonable stance to adopt than "fussing about their cost, moralizing over their effects or demanding their prohibition."

Other analyses in a medical context (Lader, 1978; Marks, 1981) make common arguments if not always arriving at identical conclusions. Both of these authors urge that information on the extent of prescriptions not be interpreted in a vacuum but in the context of the prevalence of anxiety and stress conditions. Marks (1981) concluded from various studies that physicians almost always prescribe benzodiazepines only in the event of an appropriate indication, and are if anything conservative prescribers. Mindful of a frequent criticism, Marks cast the fact that prescriptions for women are roughly twice as frequent as those for men in the context of a correspondingly greater prevalence of psychological distress in women, combined with their greater willingness than men to discuss psychic distress with physicians and receive "appropriate treatment" for this distress (cf. also Rosser, 1982). The coping response of men to such distress is quite different: "They drink, gamble and smoke tobacco and marihuana, for example, far more than women, and there is a greater tendency on the part of men towards absenteeism from work during periods of emotional stress. Indeed, it can be argued that there is not too much use of benzodiazepines in women but too little use in men." Indeed, Marks argued that the social and economic costs of alcoholism, smoking, and absenteeism might be reduced by *more frequent* appropriate prescription of benzodiazepines to men.

Lader (1978) accepted that "the benzodiazepines provide worthwhile but by no means total symptomatic relief" from anxiety. However, he wondered whether "stress reactions to social problems" should be dealt with by tranquilizers as the treatment of choice, and questioned the "medicalization" of such social problems. Moreover, he expressed the fear that benzodiazepines could become the "opium of the masses," with physicians in the unenviable role of persuading individuals to tolerate social conditions that perhaps should be addressed through social reform (cf. also Cooperstock and Lennard, 1979, for a similar view from nonphysicians).

What is worth noting is that Lader and Marks begin from the same base of objective data and do not disagree about what it contains. Any differences in how to respond to the facts of benzodiazepine use are value-laden. Whereas Lader emphasizes that physicians "must be re-educated in the use of drugs, in realizing that counselling is not really more time-consuming than repeat prescriptions," Marks asserts that physicians' "prescriptions of these drugs for the emotionally sick may be held to be rather on the conservative side." Arguments in this arena have been based on indirect inferences from data. Recently, a group of investigators in Britain (Catalan et al., 1984) reported some data from a controlled investigation that bear directly on the alternative methods by which physicians might deal with the presenting problems that give rise to most prescriptions for benzodiazepines.

No matter what value orientation may exist with regard to pharmacologic versus nonpharmacologic therapies for psychic distress, there is less divergence of opinion that long-term use (e.g., more than 6 months) warrants more concern than short-term use. There may be a number of reasons for this. One is that longer-term use may lead to physical dependence. This could make cessation of use more difficult should the patient elect to stop. Also, there is a small risk of medical complications beyond a generalized discomfort associated with withdrawal (Greenblatt et al., 1983*a,b*). At the same time, Greenblatt et al. argued that any judgment on this point must take into account the clinical circumstances of a particular patient, with particular sensitivity to the chronicity of the indication for the prescription: "For some patients with chronic anxiety, long-term anxiolytic pharmacotherapy will be necessary and appropriate in much the same way that insulin is needed for diabetes. Such chronically anxious patients are seldom physiologically addicted to benzodiazepines, nor should they be considered drug abusers." Greenblatt et al. concluded in this prestigiously placed (in the *New England Journal of Medicine*) article that the "sensationalistic and terrifying depictions of benzodiazepine use and addiction . . . in the lay media" are totally unwarranted in the light of scientific studies.

Another eminent researcher in this field (Rickels, 1981) has raised some interesting points about the interpretation of prescription audits as bases for judging whether benzodiazepines are misused. Rickels notes that although the use of antianxiety drugs as measured by prescription audits rose rapidly from the early 1960s, this peaked in 1973, since which time prescriptions have become less frequent. He was as much concerned with the *under*use that could be construed in the post-1973 data as the *over*use that was often seen in the suveys before 1973. To critics who suggested that benzodiazepines were overused and abused, Rickels continued: "It has been suggested that male physicians often prescribe dangerous, ineffective and addictive drugs primarily to females. Such tales regarding the use of minor tranquilizers for the treatment of anxiety are very disturbing, in that the major result of such publicity may be that those patients for whom anxiolytic medication is definitely indicated will take these medications either not at all or only very reluctantly, and thus will either continue to suffer from anxiety with its many consequences or will turn to other, frequently less appropriate, coping mechanisms. Some of these alternatives—including use of alcohol or marihuana, smoking and overeating—are certainly much less appealing from the point of view of public health than is the use of benzodiazepines" (1981, p. 1).

The medical debate on benzodiazepines has yet to reach a resolution. In a recent editorial in the *Journal of Clinical Psychopharmacology*, Shader and Greenblatt (1984) once more decried "the negative view of benodiazepines on the part of the media because we are encountering more and more patients who are reluctant to use these drugs even when they could be beneficial." Along with other medical observers on this scene, they suggest that viewing benzodiazepine use in contrast to the use of other substances, such as alcohol and tobacco, is the key to a reasonable perspective. Survey data on the relationship between tranquilizer and alcohol use as *alternative* coping mechanisms for dealing with psychological distress (e.g., Uhlenhuth et al., 1978) appear to support their view.

To the extent that the available data can be interpreted on either side of the debate, the role of an appraisal of scientific evidence concerning benzodiazepines can be influential but perhaps not decisive. In any event, the purpose of this chapter is not to take a posture pro or con. Instead, we attempt to review categories of empirical literature that appear relevant to a consideration of this topic and to identify the conclusions that seem warranted by the evidence. In Cohen's (1983) terms, the intent is to contribute to the process of "judicious evaluation." Our emphasis will be on clinical and experimental research. First, however, we will present a brief overview of the general pharmacology of benzodiazepines as it relates to the questions of dependence and abuse.

## 2.  PHARMACOLOGY OF BENZODIAZEPINES

Although it is beyond the scope of this chapter to include a detailed review of the pharmacology of the benzodiazepines, there is some basic information about pharmacology, especially as it relates to the issue of abuse, that is important to give as background. This is especially relevant in the present context because benzodiazepines are marketed in so many different preparations that vary in potency, rate of disappearance from the body, and perhaps in the relative balance of various effects such as reinforcing efficacy and behavioral impairment (Roache and Griffiths, 1985).

In 1960, chlordiazepoxide was the first benzodiazepine to be marketed. This was followed by diazepam in 1963 and oxazepam in 1965. By 1984 there were more than 40 benzodiazepines marketed world-wide. The popularity of these agents rests on their pharmacological actions, their relative safety, and on their ability to meet the needs of both patients and their physicians.

### Mechanism of Action

In 1977 a specific receptor for the benzodiazepines was discovered (Braestrup and Squires, 1977; Mohler and Okada, 1977). Receptors are highest in density in the cerebral and cerebellar cortices, and their pattern of distribution is compatible with the sites of anticonvulsant action of these drugs (Braestrup et al., 1977). Recently, two subclasses of receptor related to different pharmacological actions have been postulated (Klepner et al., 1979). "Type I" receptors are predominant in the cerebellum and are associated primarily with the anxiolytic effects of benzodiazepines. "Type II" receptors are concentrated in the cerebral cortex and appear to be involved in the sedating action of these drugs. The structural distinctions between the two types are not clear; however, it appears that drugs such as diazepam have equal affinity for both "types," whereas some new benzodiazepines under investigation seem to interact relatively selectively with one or the other receptor types. If the dependence liability of benzodiazepines is predicated primarily on their sedative properties rather than their anxiolytic action, this distinction could have important implications for the development of safer benzodiazepines (Sellers and Busto, 1982a). In principle, it should be possible to design an anxiolytic agent with a reduced capacity to produce physical dependence and for which the dynamic interaction with ethanol is reduced.

In recent years, knowledge concerning the details of the mechanism of action of benzodiazepines has increased rapidly. Of particular importance has been the development of the selective and specific receptor antagonist Ro 15-1788, which blocks or reverses diazepam effects. It can therefore be used clinically and experimentally to evaluate the capacity of benzodiazepines to produce physical dependence. Originally thought to be devoid of pharmacological action on its own, Ro 15-1788 now appears to possess intrinsic activity (e.g., De Vry and Slangen, 1985), the extent and significance of which await further investigation. An endogenous transmitter "diazepam-binding inhibitor" has been isolated from rat and human brain with selectivity and specificity for the benzodiazepine receptor (Ferrero et al., 1984).

## Pharmacokinetics and Biotransformation

There are significant variations in the kinetic properties of benzodiazepines that have implications for problems of abuse and misuse. Among these are the large variations in *absorption* among benzodiazepines. After oral administration, the absorption of chlordiazepoxide and diazepam is rapid. Plasma concentration peaks approximately 1 hr following ingestion, and systemic bioavailability is 100%. In contrast, oxazepam is absorbed more slowly, with plasma concentration peaking 2–3 hr after ingestion; its bioavailability when taken orally is approximately 50–70%. Clorazepate is converted to its active form (desmethyldiazepam) in the stomach; its conversion can be slowed by antacids and its peak effects can thereby be decreased. The relatively rapid absorption of diazepam accounts for the acute subjective "high," drowsiness, and motor impairment following drug ingestion, and may be related to its differential reinforcing efficacy compared with other benzodiazepines (cf. Section 3.1).

A second pharmacokinetic consideration is *binding*. Most benzodiazepines are extensively (e.g., 98% for diazepam) bound to serum albumin. This decreases the concentration of free active drug in equilibrium with sites of action and elimination. The importance of extensive binding is that effects are prolonged and elimination is slowed.

Yet another factor is *biotransformation*. All benzodiazepines undergo transformation in the liver by microsomal enzymes. The rates and patterns of biotransformation vary greatly both in healthy and sick humans. Despite the large number of benzodiazepines, there are only a few patterns of biotransformation, which are summarized in Figure 1. In Table 1 many of the benzodiazepines have been grouped according to their half-lives and pattern of disposition.

The prototypic benzodiazepines are chlordiazepoxide and diazepam. They are highly lipid soluble and widely distributed in the body. Diazepam and many other benzodiazepines are metabolized to *N*-desmethyldiazepam or nordiazepam (ND). The metabolite is pharmacologically active. Clorazepate and prazepam are two benzodiazepines that are rapidly converted to ND and owe their clinical effects to the metabolite. Triazolam, on the other hand, shares with other benzodiazepines a rather rapid biotransformation without active metabolites.

Chlordiazepoxide and diazepam each have two major active metabolites. Chlordiazepoxide is converted to desmethylchlordiazepoxide and then to demoxepam, whereas diazepam is converted to desmethyldiazepam and then to

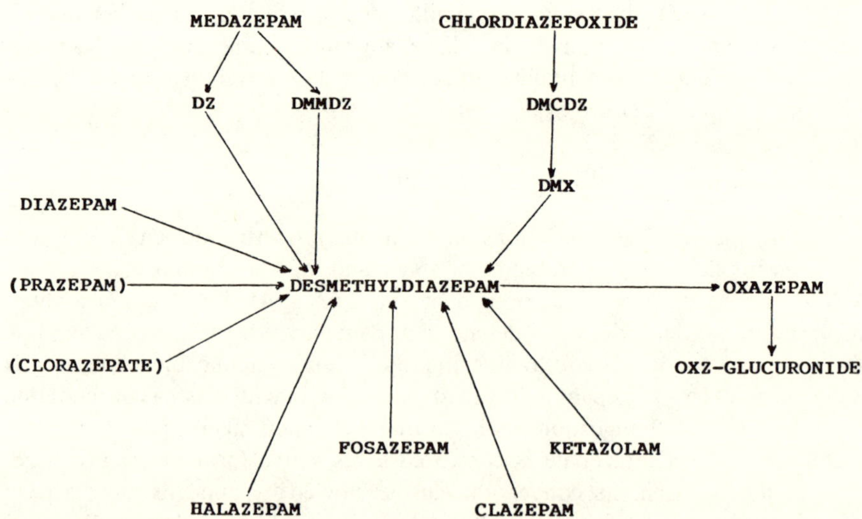


Figure 1.   Patterns of biotransformation of benzodiazepines.

Table 1. Kinetic Classification of Benzodiazepines

| Type | Half-life (hours) |
|---|---|
| Type I.   Long half-life | |
| A.   Rapid pro-drug conversion | |
| Clorazepate ⟶ (desmethyldiazepam) | 30–120 |
| Flurazepam ⟶ (desalkylflurazepam) | 40–250 |
| B.   Slow  pro-drug conversion | |
| Cloradiazepoxide ⟶ (desmethyldiazepam) | 5–30 |
| Desmethyldiazepam ⟶ (oxazepam) | 30–200 |
| Diazepam ⟶ (desmethyldiazepam) | 30–200 |
| Flunitrazepam ⟶ (desmethylflunitrazepam) | 15–35 |
| Prazepam ⟶ (desmethyldiazepam) | 30–200 |
| Type II.   Intermediate half-life | |
| Alprazolam | 6–20 |
| Clotiazepam | 5–15 |
| Lorazepam | 10–20 |
| Nitrazepam | 15–40 |
| Oxazepam | 5–15 |
| Temazepam | 10–20 |
| Type III.   Short half-life | |
| Brotizolam | 2.0–7 |
| Midazolam | 1.5–5 |
| Triazolam | 1.5–5 |

oxazepam. Unlike chlordiazepoxide and diazepam, oxazepam and lorazepam are biotransformed to an inactive metabolite. Since the mean half-life of oxazepam is 7 hr, cumulation is minimized in therapeutic use and excessive sedation can be averted while a therapeutic effect is achieved. Knowledge of the details of biotransformation can be extremely important to an understanding of both the appropriate therapeutic use of benzodiazepines and to an understanding of abuse and problems. Clearly, it is important to know whether the parent is transformed to a metabolite that is active and has a long half-life (as is the case with diazepam and chlordiazepoxide) or is rapidly converted to an inactive form (as is the case with triazolam, oxazepam, and lorazepam).

A number of factors are known to affect the half-life of diazepam. For example, a rule of thumb is based on the coincidence that age in years is on average equal to half-life in hours. However, this rule often does not apply and there is wide variation among individuals in the rates of biotransformation of benzodiazepines.

The occurrence of long half-lives and wide interindividual variations make prediction of the time of maximum clinical effect or toxicity very difficult, but during long-term administration cumulative and long-lasting effects can be expected for those benzodiazepines with long half-lives. Generally, the time for cumulation to peak concentrations during oral administration is five times the half-life; for example, the time to peak concentration is 3, 7, and 10 days for chlordiazepoxide, diazepam, and desmethyldiazepam, respectively. The slow cumulation means that the maximal effect may be delayed for some time, with the result that the onset of toxic effects can occur with unpredictable delays as concentration continues to increase. At the same time, it may take a prolonged period for detoxification to occur when the elimination of the drug is slow. Slow detoxification can be advantageous in minimizing the expression of withdrawal signs, and is probably responsible for the relative lack of intensity of such signs clinically (cf. *Dependence on Benzodiazepines—Physical Dependence*).

Problems stemming from kinetic properties can emerge when a drug is inappropriately assumed to have special efficacy for a particular disorder. For example, though flurazepam is marketed as a hypnotic, it has few actual properties to qualify it as such. It possesses no unique efficacy in inducing sleep; however, the cumulation of its slowly eliminated active metabolite (desalkylflurazepam) over 7 to 10 days results in a slow onset of the hypnotic effect and the side effect of daytime drowsiness during long-term use. This can lead to toxic results, especially in older persons in whom there may be greater sensitivity as well as less efficient biotransformation of the drug (Greenblatt et al., 1982).

## Pharmacokinetic Factors and Benzodiazepine Abuse

The kinetic properties of benzodiazepines that may contribute to abuse liability and persistent self-administration can be grouped into those that relate to their potency as positive reinforcers and those that contribute to the potential for negative reinforcement through the development of physical dependence and the occurrence of withdrawal. These are summarized in Table 2. *Absorption rate and bioavailability* are pharmacokinetic factors that may be of some importance in relation to reinforcement. In principle, rapid delivery of the drug to the CNS (e.g., when absorption is rapid) should provide the optimal kinetic conditions for reinforcement and hence for repeated use. This certainly appears to be the case for drugs such as cocaine and heroin that can be taken by methods that produce rapid entry into the brain. Conversely, reinforcement should be diminished when absorption is slow and subjective effects are dampened or eliminated. There are few empirical data relating kinetic factors to reinforcement by benzodiazepines, but there is some suggestive evidence. For example, diazepam is more rapidly absorbed than oxazepam. Although other factors such as ease of access are critical determinants of illicit use, the rapid absorption rate and high bioavailability of diazepam may contribute to its preferred status among drug abusers (Busto et al., 1986; Griffiths et al., 1984*b*; Stitzer et al., 1981). Another illustrative drug is halazepam. The structure of halazepam differs slightly from that of diazepam, but it is a "pro-drug" for the same active metabolite, *N-*

Table 2. Kinetics and Abuse Potential

1.  Properties that increase potency as a reinforcer
    High intrinsic pharmacological activity of drug
    Rapid absorption
    Rapid entry into specific brain regions
    High oral bioavailability
    Low protein binding
    Short half-life
    Small volume of distribution
    High clearance

2.  Factors that promote physical dependence
    High intrinsic pharmacological activity of drug
    Cumulative drug load (dose, frequency, duration of treatment)
    Small volume of distribution
    Long half-life
    Low clearance

3.  Factors that promote appearance of the withdrawal syndrome
    High intrinsic pharmacological activity of drug
    Short half-life
    High clearance
    Rapid exit from specific brain regions
    Small volume of distribution

desmethyldiazepam, which is slowly (half-life 50 hr) eliminated from the body. These kinetic features are associated with a slow onset of action and with persistent effects. In a comparative study of halazepam, diazepam, and placebo (Jaffe et al., 1983; compare also p. 67), a variety of subjective effects of halazepam appeared 2 hr after oral ingestion, whereas the same effects were already detected within 30 min after diazepam was taken. Moreover, the peak effect of halazepam across all scales was less intense for halazepam than diazepam. The authors suggested that such differences should confer a lower abuse potential on halazepam than diazepam. Such data can be influential, as this study provided the basis for recommending the exclusion of halazepam from Schedule IV of the Psychotropic Convention 1971 by the World Health Organization Expert Committee for the 6th Review of Psychoactive Substances (World Health Organization, 1982).

To possess addiction liability, a drug must gain access to a site or sites of action in the brain. Therefore, *distribution of drugs into the brain and other tissues* can also be of importance in reinforcement. Entry into the brain is determined by blood flow, lipid solubility, molecular size, and the presence of active transport processes (Brodie et al., 1960). Rate of distribution into the brain appears to be associated with abuse potential for barbiturates and opiates. In the case of benzodiazepines, higher lipid solubility is associated with more rapid entry into the brain (Arendt et al., 1983). For example, the half-life of the appearance of diazepam and midazolam in the cerebrospinal fluid of cats is 0.44 and 0.28 min, respectively. Clinically, diazepam is characterized by rapid onset of anxiolytic and sedating effects when compared with other benzodiazepines that enter the brain more slowly. These relationships comment only indirectly on the relative potency of benzodiazepines as reinforcers, but it appears likely that rate of distribution into the brain can be used as a predictor of reinforcement.

The *elimination* kinetics of some addictive drugs are related to self-administration rate (a direct measure of reinforcing potency) and to the appearance of withdrawal symptoms (which may provide a basis for reinforcement). The average clinical effect of a drug during chronic administration will depend on its accumulation in the body. This in turn depends on the rate of free clearance, with low rates of clearance resulting in accumulation. Elimination kinetics are correlated with the rate of self-administration and the appearance of withdrawal signs, but the data cannot be interpreted unequivocally. For example, Griffiths et al. (1981) found that baboons self-administered six different benzodiazepines intravenously. The highest rate of self-administration occurred with midazolam, which is also rapidly eliminated. Its pattern of self-administration resembled that obtained with barbiturates. The benzodiazepines that were slowly eliminated or had active metabolites with long half-lives were less readily self-administered. Although such data are suggestive, they are not without problems of interpretation. First, as the authors noted cautiously, there are no pharmacokinetic data available from baboons, and the kinetics may not be the same as for humans. Moreover, there is the alternative possibility that midazolam was relatively more reinforcing because of its high lipid solubility that permitted relatively rapid entry into the brain (Arendt et al., 1983).

Elimination kinetics are clearly related to the severity of withdrawal in dependent organisms. For example, long-acting barbiturates are less liable to produce physical dependence in animals as measured by withdrawal severity than shorter-acting ones (Boisse and Okamoto, 1978a). These data correspond well to clinical observations that individuals dependent on barbiturates with relatively short half-lives such as secobarbital or amobarbital (half-life of 25 hr) can develop severe withdrawal symptoms, whereas phenobarbital (half-life of 86 hr) is seldom or never associated with important withdrawal symptoms. Whether a similar phenomenon applies to the benzodiazepines has not yet been systematically determined either clinically or in animal studies. There are some clinical reports that patients dependent on rapidly eliminated benzodiazepines are at higher risk of experiencing withdrawal symptoms (Einarson, 1980; Hanna, 1972; Hollister, 1980). However, no systematic study of the influence of elimination rates on the appearance of withdrawal has been reported; in any event, interpretation would be clouded because elimination half-life is only one of many variables that can influence the appearance of a withdrawal syndrome. Nor is the relationship between physical dependence, withdrawal, and drug-taking behavior as a dependent variable a simple one (Cappell and LeBlanc, 1981).

The clearest application of pharmacokinetic principles is in the clinical management of withdrawal. An understanding of these principles has helped in the development of simplified detoxification procedures in which patients who have been using excessive amounts of benzodiazepines with short or intermediate half-lives are switched to equivalent doses of diazepam. A subsequent gradual tapering regimen may then be instituted, capitalizing on the half-life of diazepam and its metabolites to protect against the emergence of symptoms that would occur in the case of rapidly eliminated benzodiazepines (Harrison et al., 1984).

A consideration of the pharmacokinetics of benzodiazepines could easily constitute a chapter in its own right. This brief overview has been meant only to provide a rudimentary acquaintance with the issues in pharmacokinetics and biotransformation of benzodiazepines as they may apply to problems of abuse and dependence. These should be borne in mind as context for the findings that are reviewed in the sections that follow.


## 3.   BENZODIAZEPINES AS REINFORCERS

Drugs with abuse potential are typically self-administered in a variety of experimental models in both animal and human subjects. Voluntary self-administration is virtually a defining characteristic of a drug with abuse potential; conversely, drugs that are avoided or not preferred over placebo in self-administration models are unlikely to be abused. The literature concerning reinforcement by benzodiazepines is not large, but there has been a good deal of recent work that is of special interest because the subjects have been humans. That a drug is reinforcing does not necessitate that it will be abused; however, it would be perplexing if a drug of abuse were not reinforcing.

### Human Studies

Much of the experimental work on reinforcement by benzodiazepines has been generated in the human self-administration laboratory of the Baltimore City Hospital. An early study in this series (Griffiths et al., 1976) demonstrated that diazepam was clearly reinforcing to subjects with a history of sedative abuse and that self-administration responded to various experimental manipulations similarly to other established reinforcing drugs. Reinforcing efficacy increased monotonically over a dose range of 2, 5, and 10 mg per ingestion. In a study of two subjects (Bigelow et al., 1976) it was found that as the "cost" (in points earned by riding an exercise bicycle) of diazepam (10 mg) increased, the rates of self-administration decreased. In this respect, diazepam was similar to pentobarbital. Diazepam could be consumed during a 7-hr session, and a cumulative dose of diazepam as high as 200 mg was often consumed. Interestingly, doses of this magnitude produced only limited gross effects on physical and social functioning. This should be kept in mind in considering the daily dosage levels typically achieved in long-term therapeutic use of diazepam.

A more extensive study (Griffiths et al., 1979) included a contrast of two doses each of pentobarbital, diazepam, and chlorpromazine, as well as a placebo condition. As in the previous work, these subjects all had histories of sedative drug abuse. Currency to purchase drugs could be earned by pedaling a stationary bicycle. Generally, neither placebo nor chlorpromazine was reinforcing, and subjects did not purchase these drugs. However, both diazepam and pentobarbital were purchased frequently, and higher doses of each appeared to be preferred. Pentobarbital appeared to be more reinforcing, but the authors were unable to determine why.

In a subsequent study, diazepam and pentobarbital were compared with a view to establishing a possible qualitative basis for the difference in reinforcing efficacy (Griffiths et al., 1980). The most important comparison for present purposes was one in which subjects (all with a history of sedative and other drug abuse) could choose between pentobarbital (400 mg) and diazepam (200 mg). In a previous study, these doses were shown to be roughly comparable in quantitative magnitude of effect according to both subject and observer rating scales. Despite this quantitative equivalence, subjects consistently preferred pentobarbital over diazepam when given the choice (although both active drugs were preferred over placebo). A possible mechanism for the difference was explored by reviewing nursing notes containing observations of the subjects' general mood and behavior. These observations indicated a higher frequency of "negative" consequences of diazepam ingestion over pentobarbital ingestion. Associated with diazepam consumption were more frequent general complaints from the subjects, greater irritability and mood swings, more frequent hostility, confusion, and depressed mood. The interpretation of these interesting differences, however, remains unclear. When the subjects were asked to give "liking" ratings for the two drugs, diazepam (200 mg) and pentobarbital (400 mg) were not discriminated from each other although both were more liked than placebo. Thus, although "negative" consequences of drug self-administration may affect the actual relative preference for it, they need not affect a general appraisal of how much it is liked. These differences in the effects of pentobarbital and diazepam on mood and behavior have been described in more detail by Griffiths et al. (1983). Perhaps the most important observation for present purposes was that liking for diazepam decreased over 5 days of administration, and staff and self-reports of deterioration in mood occurred over the same period. This did not happen with pentobarbital, and may help to explain why diazepam was less preferred in the choice experiment than pentobarbital (cf. Griffiths et al., 1980).

This important strategy of *comparing* agents has been extended to a study of two benzodiazepines (Griffiths et al., 1984*b*). The purpose of this comparison was to determine potential pharmacological mechanisms underlying the apparently greater abuse liability of diazepam over oxazepam. Diazepam was studied across a range of 10–160 mg per ingestion, and oxazepam over a range of 30–480 mg per ingestion. Measurements of relative liking were obtained as well as a variety of subjective and objective drug effects. The effects of diazepam were liked more than those of oxazepam, and a number of possible reasons for the difference emerged. Diazepam had a more rapid onset of action, a shorter latency to peak effect, and was more potent, especially on the dimension of "liking." Subjectively, diazepam was judged to be like a barbiturate more often than oxazepam, whereas oxazepam was judged to be like a placebo more often than diazepam. Liking for diazepam increased with repeated administration, but this was not so for oxazepam. In a second study, (Griffiths et al., 1984*a*), diazepam and oxazepam were compared for reinforcing efficacy in a self-administration choice test. Diazepam was a much more effective reinforcer than oxazepam; its relative potency as a reinforcer was estimated as 8 : 1, whereas its relative potency in producing psychomotor effects was only 3 : 1. Subjects assigned diazepam a higher "street" monetary value than oxazepam. Finally, various sources of epidemiological data were offered to show that diazepam is in actuality more abused than oxazepam. Some aspects of this methodology have recently been applied to an appraisal of the abuse liability of triazolam in a comparison with pentobarbital (Roache and Griffiths, 1985).

Comparisons such as this give clear guidelines concerning the features of benzodiazepines that make them subject to abuse. What remains to be seen is whether abuse potential and anxiolytic activity are separable pharmacologic actions. Some evidence on the latter question has been produced by Jaffe et al. (1983). The subjects in this study were male hospital inpatients who had recently completed an alcohol detoxification program. Many of them had a history of abusing a variety of other drugs. The drugs compared were diazepam (20 and 40 mg), halazepam (160 and 320 mg), and placebo. The assumption was made from (undocumented) clinical experience that the anxiolytic potency of diazepam is eight times that of halazepam. Each subject was tested on three occasions at weekly intervals, thus providing data on three of five possible treatments. Subjects ingested the drug and gave various subjective ratings at 0.5, 2, 3, 4, 6, and 8 hr postingestion. The results on a scale of euphoria were the most impressive. The onset of euphoria from diazepam was much more rapid than halazepam and reached a higher peak. Indeed, as a euphoriant halazepam was not much more effective than placebo. Halazepam's relative potency in producing sedation was greater than as a euphoriant but still less than that of diazepam. Its effects were delayed and its peaks were lower. It was also liked less than diazepam. From these data it would appear that there is some dissociation between the anxiolytic potency of a benzodiazepine and those of its actions that give rise to abuse potential. They also support the suggestion that pharmacological properties of these agents (e.g., those related to rapidity of onset and peak effect) can be exploited to minimize abuse potential without necessarily compromising anxiolytic activity.

Preston et al. (1984) studied the self-administration of oxazepam among opiate abusers while they were being detoxified from a regimen of methadone maintenance. Patients were first given experience with oxazepam in no-choice trials, during which 30 mg was given twice daily. A series of preference trials ensued during which the six patients could select up to 30 mg of oxazepam or a placebo. Although some individual subjects showed a consistent preference for oxazepam over placebo, there was no difference on average. This was despite the fact that the active drug produced significant changes on a variety of subjective and objective measures (although not on a measure of opiate withdrawal symptomatology).

The foregoing studies all employed populations of drug abusers as subjects. de Wit et al. (1984*a,b*) compared self-administration of lorazepam (0.5, 1.0, and 2.0 mg) with diazepam (5.0 mg) and placebo. The subjects were normal college students with no history of drug abuse. Subjects had no preference between lorazepam and placebo at doses of 0.5 and 1.0 mg, and actually preferred placebo over 2.0 mg. When offered a choice between lorazepam 1.0 mg and diazepam 5.0 mg there was no preference. From the latter result it might be concluded by inference that diazepam would not have been chosen more often than placebo. The active drugs did exert some pharmacologic effects in comparison to placebo, so inertness was not a likely reason for the results. The highest dose of lorazepam was less "liked" than the lowest, possibly explaining its reduced choice compared with placebo. However, there were no reports of statistically significant differences in liking between placebo and any drug dose. This is of course not what has been observed with higher doses and subjects with a history of drug abuse.

Johanson and Uhlenhuth (1980) have provided important data to show that the failure to find reinforcement by benzodiazepines in normal subjects is not due to an inadequate methodology. In a preference procedure, subjects chose to self-administer *d*-amphetamine (5 mg) over diazepam (2 mg), but were indifferent between diazepam at 2 mg and placebo. Placebo was actually preferred over diazepam at 5 and 10 mg. The higher doses of diazepam produced self-reports of increased fatigue and confusion compared with placebo, and it is possible that anticipation of these effects was responsible for avoidance of diazepam when it occurred. Such effects might be especially problematic for the normal volunteers who served as subjects in this study, since it was conducted in the morning prior to taking on the everyday tasks of school or work (cf. also Orzack et al., 1982).

The preference technique has also been used to study other variables that might affect reinforcement by benzodiazepines (de Wit et al., 1982). The variables were time of day (9–10 AM and 4–5 PM), subject characteristics (control or anxious), and drug (*d, l*-amphetamine 5 mg, diazepam 5 and 10 mg, or placebo). As before, amphetamine was chosen significantly more often than placebo. The lower dose of diazepam was chosen about as often as placebo, but the higher dose significantly less often. The anxious and control groups did not differ in their choice of diazepam at either dose. Nor did time of day affect the choice of diazepam, which was less than placebo in both morning and afternoon sessions. The data concerning anxiety are especially interesting because the subjects in the group in question were diagnosed as having a Generalized Anxiety Disorder by *DSM-III* criteria and had much higher baseline anxiety scores than controls in the Anxiety subscale of the Profile of Mood States inventory.

Some interesting observations on the anxiolytic (and thus perhaps the reinforcing) potential of diazepam in normal subjects have been reported by Ghoneim et al. (1984). They administered doses of 0.1, 0.2, and 0.3 mg/kg of diazepam to a large number of normal volunteers. This is an appreciable dosage range as the highest dose was equivalent to 20 mg in the prototypical 70 kg person. Although there were substantial dose-dependent effects on self-reports of general sedation (e.g., scales of alert/drowsy, fuzzy/clear-headed), diazepam had no effect on self-report of calmness and relaxation. Ghoneim et al. concluded that in "healthy" subjects the anxiolytic actions of diazepam cannot be studied because they do not occur, except perhaps in anxiety-provoking circumstances. This suggests a possible dissociation between the anxiolytic and other effects of diazepam, and perhaps thereby between *reinforcing* and other effects.

The inclination of normal subjects to avoid relatively high doses of benzodiazepines has been further confirmed in a study by de Wit et al. (1984c). College students were given a choice of self-administering placebo or flurazepam (15 or 30 mg) in a series of choice trials. The lower dose could not be distinguished from placebo and was chosen equally often in preference tests. The higher dose of flurazepam was distinguishable on the basis of subjective effects, and it was chosen *less* often than placebo. This is consistent with results obtained with normal subjects and relatively higher doses of other benzodiazepines. Clearly, benzodiazepines are less reinforcing to subjects with no history of sedative abuse than to those with such a history.

There are a few studies of actual clinical populations that comment on the reinforcing action of diazepam in particular. In one of these (Winstead et al., 1974), inpatients in a psychiatric service were allowed diazepam (10 mg) on demand. A Drug-Seeking Index (DSI) was devised by dividing each patient's number of drug requests by the total days on the ward. One of the strongest predictors of drug-seeking behavior by this index was level of anxiety as measured by a standard test inventory; the higher the anxiety the greater the DSI. Thus not surprisingly, anxiety may make diazepam a more potent reinforcer. However, mere availability did not necessarily lead to diazepam use even in this population, since 27% of patients never once requested diazepam. Also not surprisingly, the presence of anxiety and the experience of relief by using tranquilizers is by far the most common reason given by persistent therapeutic users of the drugs for their continued reliance upon them (e.g., Murray, 1981). It should be borne in mind that these observations with a clinical population are somewhat at odds with the findings by de Wit et al. (1982) considered immediately above.

Another interesting test of diazepam as a reinforcer in a clinical setting was conducted by Balmer et al. (1981). Two groups of "neurotic" patients received diazepam ($n = 21$) or placebo ($n = 23$) in a double-blind procedure over a 6-month period. This was combined with 1-hr sessions of psychotherapy every two weeks. Subjects were instructed to take their medication on a PRN basis, but not to exceed 6 doses (5 mg) per day. One dependent variable was the number of patients who stopped (by self-report) taking medication altogether. In the placebo group, 73% of patients chose this course of action as compared with 45% who received diazepam, an interesting difference if not significant statistically. The mean number of pills consumed over the study was approximately 220 per patient for those who received diazepam and 150 for those given the placebo. This difference was significant, and appeared due to the more extended duration of use in the diazepam group rather than a higher daily intake of tablets. Indeed, the mean daily intake for the heaviest user of diazepam was 17.5 mg/day. Higher tablet intake, *irrespective of whether the tablets were diazepam or placebo,* was positively associated with an index of "psychic irritability." Thus, although the pharmacologically active medication appeared more reinforcing than placebo, the simple ingestion of a *potentially* anxiolytic medication, regardless of its pharmacologic potency, was most reinforcing to individuals with the greatest degree of psychological "need" for it. As a general subjective appraisal of diazepam as a therapeutic agent, twice as many subjects in the diazepam group (52.8%) reported a "good" effect of the drug than those who received placebo (26.1%).

We (Cappell et al., 1985) have completed a study of "unauthorized" self-administration of benzodiazepines in patients who volunteered to participate in a study of outpatient treatment of long-term low-dose benzodiazepine dependence. Half the patients were tapered on active diazepam ($n = 21$), while the remainder ($n = 19$) were tapered on a placebo. Neither the patient nor the therapists were aware of the actual drug assignment. The data related to self-administration were self-report and weekly plasma benzodiazepine determinations from which experts judged whether patients had supplemented beyond the agreed dosage levels. According to the plasma determinations, 84% of the patients in the placebo condition used a supplementary benzodiazepine at least once, whereas this occurred among only 33% of patients on active drug ($p < .01$).

Moreover, seven patients assigned to placebo spontaneously requested a switch to their own prescribed supply as compared with only one who did so in the diazepam condition ($p < .05$). Repeated assessments showed that withdrawal symptoms were more frequent and more intense among patients in the placebo condition. These data showed that low doses of benzodiazepines were clearly reinforcing to this population; the effect of the placebo manipulation shows that the reinforcement is based on a *bona fide pharmacological property* of benzodiazepines even at the low self-administered doses that prevailed in this population.

## Animal Studies

Animal studies have been confined to a limited number of issues, including the very fundamental one of whether various benzodiazepines are or are not reinforcing. This literature, which was reviewed recently by Griffiths and Ator (1980), offers much less information than the literature on humans and will therefore be covered here very selectively. One of the most thorough studies in animals (Griffiths et al., 1981) compared a selection of barbiturates, benzodiazepines, and chlorpromazine using intravenous self-administration technology in baboons. The major conclusion was that the avidity with which the various compounds were self-administered (barbiturates > benzodiazepines > chlorpromazine) corresponded well to the analogous behavior in humans. All of the benzodiazepines tested (lorazepam, clorazepate, diazepam, flurazepam, medazepam, midazolam) were moderately effective reinforcers. One observation in this study tended to confirm the importance of the relationship between pharmacokinetics and reinforcement that was apparent in human studies. Midazolam was the benzodiazepine that maintained the highest rate of self-administration, and it is also the most rapidly eliminated of this group. However, the intravenous method clearly does not allow the study of some of the kinetic variations among drugs that have been apparent in oral self-administration studies in humans.

In a review of research on the self-administration of various benzodiazepines, Yanagita (1981) pointed out a pattern rarely observed with other drugs. This was true with diazepam self-administered intravenously and nitrazepam intragastrically in monkeys. Early during a course of 10 weeks, the monkeys self-administered increasingly large daily doses. Subsequently, however, they reduced their intake substantially and maintained these reduced levels for weeks without change. Yanagita took this to indicate that the reinforcing efficacy of benzodiazepines is less than that of pentobarbital, which does not diminish in self-administration over time. Considering the same discrepancy in their study of self-administration in humans, Griffiths et al. (1983) conjectured that there may be progressively increasing disruptive effects of benzodiazepines on mood and behavior with persistent self-administration that eventually limit the amount of drug taken. It is an important question whether this occurs at therapeutic doses or only the high doses for which the phenomenon seems definitely to occur. Self-managed lowering of daily dose has been reported in therapeutic use (e.g., Bowden and Fisher, 1980), but the reasons for this have not been described.

There have been a limited number of studies on the relationship between alcohol and benzodiazepines as reinforcers. Amit et al. (1973) induced rats to prefer alcohol by applying hypothalamic stimulation. Subsequently, when alcohol was removed the rats displayed a preference for a solution of diazepam but not for a solution of hashish. Amit et al. concluded that the rats ingested diazepam solution either to self-medicate for alcohol withdrawal effects, the occurrence of which was not assessed, or simply to replace the reinforcing effect of ethanol. The fact that the diazepam solution contained 10% ethanol unfortunately compromises interpretation of these results.

Bergman and Johnson (1985) were interested in the effect of self-administration of other reinforcing drugs on the reinforcing efficacy of intravenous diazepam. Rhesus monkeys were maintained on intravenous self-administration of cocaine or pentobarbital, both of which were clearly reinforcing. Diazepam in various doses per infusion was made available. Prior experience with cocaine had little effect on the self-administration of diazepam, which generally produced no more responding than a saline vehicle. However, prior self-administration of pentobarbital did increase the probability that diazepam would be reinforcing. Bergman and Johanson did not determine what mechanism may have been operating to distinguish the effectiveness of cocaine and pentobarbital, but they did conclude that their data illustrated the importance of prior drug history as a determinant of the reinforcing potential of other agents.

Although the reinforcing efficacy of diazepam could be increased by prior experience with other drugs, it should still be recalled that diazepam was a relatively weak reinforcer in the study by Bergman and Johanson (1985). Pilotto et al. (1984) found that rats would not self-administer diazepam at a greater rate than an inert control solution unless they were also maintained on a particular schedule of food reinforcement. No mechanistic explanation for this result could be offered, but it was similar to those obtained with other drugs that were relatively weak reinforcers; amphetamine, cocaine, and heroin, on the other hand, were self-administered without the need for a manipulation of the schedule of food reinforcement.

An exception to the usual result with diazepam was reported by Fuchs et al. (1984), who found that 9 of 20 naive rats showed a strong preference for a diazepam solution over tap water in an oral preference test. In fact, the preference appeared comparable to that achieved with etonitazene, a potent opioid, and much greater than that for barbital. This is interesting because barbiturates and opioids are usually much better reinforcers than diazepam. However, the basis for this atypical result awaits further exploration.

In a brief report, McClearn and Erwin (1980) reported that injections of diazepam reduced oral ethanol self-administration in mice in a dose-dependent fashion. This was taken as evidence that the pharmacological feedback signal governing ethanol self-administration could be mimicked to some degree by diazepam. Roehrs et al. (1984) have reported comparable findings with chlordiazepoxide in rats with two qualifications. First, their effect was not dose-related, and second, it occurred only when ethanol consumption was relatively high to begin with. More recently, Samson and Grant (1985) have reported a dose-dependent decrease in ethanol self-administration in rats by chlordiazepoxide.

## Overview

In summary, it appears that benzodiazepines are weak reinforcers compared with other drugs of abuse. This may rest in part on the constellation of *pharmacodynamic* effects produced (or perhaps not produced) by the class of drugs. However, *pharmacokinetic* profiles also appear to be an important determinant of relative reinforcing efficacy within the benzodiazepine class. A history of drug abuse, including alcohol abuse, may enhance reinforcement by benzodiazepines in humans. Anxiety also appears to be a substrate for reinforcement in humans with no drug abuse history, but this is not sufficient by itself to make diazepam reinforcing in a laboratory trial. There has been limited study of reinforcement by benzodiazepines in "normal" humans, but what evidence there is suggests that these drugs have at most modest potency as reinforcers compared with many other drugs. A reasonable speculation from these studies is that there are individual differences in the behavioral or biological substrates that make for reinforcement by benzodiazepines. This appears to distinguish them from amphetamine, which is more indiscriminately reinforcing even to subjects with limited previous experience with drugs. Animal studies reveal little beyond confirmation of some of the findings of the human research. However, there has been a suggestion that diazepam reduces reinforcement by ethanol in mice. This gives credence to the suggestion that benzodiazepine substitution therapy for excessive alcohol consumption (cf. Greenblatt and Shader, 1974) may have a bona fide pharmacologic basis, although it appears likely that such a proposal would be suspect to many clinicians on other grounds.

## 4. DEPENDENCE ON BENZODIAZEPINES

Earlier we referred to *DSM-III* criteria for the diagnosis of benzodiazepine dependence. These included both tolerance and withdrawal. However, one should not blindly review these features of benzodiazepine use without comment on how tolerance and the induction of a withdrawal syndrome constitute problems. For a discussion of the general conceptual issues involved here, readers are referred to recent reviews (Cappell and LeBlanc, 1981, 1983).

## Tolerance

In this section we will selectively review the literature on tolerance to benzodiazepines in animals and humans. The purpose will not be to draw conclusions about the underlying mechanisms (i.e., the neurochemical or metabolic basis) of tolerance. Instead, we will attempt to illustrate and summarize the literature in a more descriptive sense and to relate this as far as possible to questions of clinical importance.

In speaking of tolerance to benzodiazepines, it is important to keep particular *effects* in mind. Two major classes of benzodiazepine action are "sedative/depressant" and "anxiolytic." These actions can be measured in various ways, and it may be that the categories are not totally independent from a functional or mechanistic point of view. However, it is useful to use these two broad working categories of effect in an appraisal of the relationship of tolerance to dependence and abuse.

In a clinical context, the single most important question is whether tolerance to the *therapeutic* effects of benzodiazepines is acquired. If so, an increase in dose would be required to restore and maintain therapeutic eficacy if it were decided that the patient should continue on the drug. A need to increase dosage due to tolerance should not be reflexively rejected as a necessarily bad thing. However, it could be problematic for several reasons:

1. The cost of treatment would increase.
2. The development of physical dependence would increase in likelihood and severity. This could complicate any attempt to terminate the drug treatment should this be desired and at least would need to be taken into account in planning a regime of withdrawal.
3. If there is some chronic toxicity associated with the drug (organ damage in particular), the risk would likely be enhanced by increasing total exposure through a long-term dosage increase.
4. If there are some acute problematic effects of the drug to which tolerance does *not* develop (e.g., on performance), these would become exaggerated by an increase in dose to accommodate tolerance to the therapeutic effect.

It should be borne in mind that tolerance might be beneficial in some instances. For example, tolerance to impairment not accompanied by tolerance to the therapeutic effect would be considered desirable.

Human Studies. One of the earliest reports of apparent tolerance in a clinical setting was made by Warner (1965). Although detailed data were not presented, Warner reported that most of his patients receiving oxazepam experienced drowsiness at the outset of treatment, but that this typically subsided after three or four doses (see also Hillestad et al., 1974, *re* diazepam; Scharf and Jacoby, 1982, *re* lorazepam; Bond et al., 1983). A well-controlled demonstration of tolerance in a clinical setting was reported by Giles et al. (1978) in patients about to undergo endoscopic examination. They determined the dose required to attain a "relaxation point" at which subjects would respond to verbal prompts but would drift into sleep in the absence of verbal stimulation. Subjects were classified as "users" or "nonusers" of benzodiazepines based on self-report and blood tests positive for diazepam or metabolites. The users required a median dose of intravenous diazepam 2.5 times greater than the nonusers to achieve the relaxation point. Whether or not this reflects tolerance to the anxiolytic action of diazepam, as opposed to a separate sedative/hypnotic action, cannot be determined from this study.

Some interesting additional data on this point were reported recently by Cook et al. (1984). This was a study of patients who required intravenous diazepam for dental or endoscopic procedures. The endpoint of the injection was a clinical judgment of sedation based on relaxation of grip, ataxia, and lowering of the forearm. "Pronounced" tolerance was observed in "patients taking benzodiazepines regularly, even those taking low doses, or those taking regular alcohol." What this suggests is that a low dose may remain adequate for some therapeutic effects notwithstanding that tolerance to a more generalized sedative effect can occur. A similar dissociation of tolerance to different effects is evident elsewhere in this review.

There have been a substantial number of experimental studies of tolerance in humans. Tolerance to diazepam has occurred quite rapidly in such studies. Seppala et al. (1980) administered 10 mg of diazepam orally to healthy volunteers followed by a battery of behavioral tests. The following day the test battery was administered once more after the same dose. Diminished effects were seen in tests of eye–hand coordination and reaction time, but not in the case of critical flicker fusion (CFF), a test of exophoria (diazepam caused the eyes to diverge), or a number of subjective estimates of drug effect described by the authors as "euphoria." Using a similar battery of tests, Aranko et al. (1983) studied tolerance to lorazepam and cross-tolerance to diazepam in normal volunteers. After one week of daily treatment with lorazepam (2 mg), the effect of a 3-mg test dose of lorazepam was significantly reduced in tests of reaction time, attention, hand–eye coordination, and body sway, but not CFF. One week of daily treatment with diazepam (5 mg BID did not diminish all effects of lorazepam, but it did produce cross-tolerance to its effects on CFF, body sway, and an attentional task.

File and Lister (1983) studied tolerance to single doses of lorazepam (2.5 mg) administered at weekly intervals. Even with this infrequent dosing interval there was tolerance to effects on digit–symbol substitution and tapping-rate. There was no tolerance to the increase in heart rate that lorazepam produced nor to its effect on recall for nonsense syllables. There was tolerance to the increase in self-rated dizziness, but not sedation. Interestingly, the tolerance observed was substantial after only a single previous exposure to lorazepam. Thus for this drug with these indexes of impairment, tolerance was swift and occurred with a remarkably infrequent dosage schedule.

Although File and Lister (1983) found no tolerance to the effect of lorazepam on recall, Ghoneim et al. (1981) did report such a result with diazepam. The tolerance was described as "incomplete," but it should be recalled that in daily dosing studies, any effect of tolerance to diazepam may be offset by cumulation of the drug.

Earlier it was noted (File and Lister, 1983) that tolerance to lorazepam did not require extensive exposure. This may not be surprising in view of the phenomenon of acute tolerance. Such tolerance is manifest as a significant diminution of the functional effect of a single dose of a drug at a faster rate than would be expected given the rate of decline in the plasma level. Acute tolerance to diazepam (Ellinwood et al., 1981, 1983) and clorazepate (Greenblatt et al., 1979) have been reported.

One somewhat paradoxical result was reported in a study (Lader et al., 1980) in which normal volunteers took clorazepate daily for 15 days. Although there was tolerance as measured by some indices by the 15th day, for other parameters including digit–symbol substitution and self-rating of subjective effect, the drug exerted a *greater* effect on day 15 than day 1. This "sensitization" cannot be explained as a reflection of drug accumulation, since if this was so it should occur with all indices. As interesting as this result is however, there has never been any other suggestion of sensitization in the literature. Another study suggestive of an increased effect with repeated exposure was reported by Griffiths et al. (1984*a*). However, the data did not permit comment on whether this reflected sensitization or drug cumulation.

There have been a few useful experimental studies of tolerance in patient populations. Lader and Petursson (1983) compared eight patients recently withdrawn from long-term therapeutic benzodiazepine use with eight normal controls. The patients were tested at 4 and 21 days following total withdrawal. Various responses to intravenous diazepam (10 mg/75 kg) were measured. The patients displayed comparative tolerance to the increase in pulse rate produced by diazepam, but this tolerance appeared to be lost 21 days after withdrawal, when if anything, patients were *more* responsive to the drug than normals. Patients were also less affected in a self-rating of alertness–drowsiness. Although no specific data were reported on this important point, the authors claimed that tolerance appeared to develop to the sedative but not the "euphoriant" actions of diazepam.

Swift et al. (1983) administered a brief battery of tests to elderly (over 65 years) patients who had been using nitrazepam or flurazepam for anywhere from 1 month to 15 years. Most had been using them "regularly" at night for the hypnotic effect. The authors characterized the population as normal in postural sway and on performance in an "abbreviated mental test." Reports of drowsiness and other potentially drug-related symptoms were obtained. The authors concluded that their "findings are suggestive of CNS adaptation."

Lucki et al. (1985) compared chronic users of benzodiazepines still on medication with a control group that had been free of benzodiazepines for at least 3 weeks. The comparison included tests of psychomotor performance, mood, and memory. There were no differences between the groups except on a scale of "tranquilization"; patients still taking a benzodiazepine reported greater tranquilization, but were no different in "mental" or "physical sedation." When patients still taking a benzodiazepine were tested shortly after consuming their usual therapeutic dose, their behavior was not much affected on any of these tests, although there was a measurable decrement in recall of a word list and an actual reduction in physical sedation. Although the data were far from compelling, the authors concluded their findings supported "the clinical impression that tolerance develops rapidly to the subjective feelings of sedation but not to the anxiety-reducing effects of (benzodiazepines) medications."

The research considered to date has concentrated on the effects to which tolerance *does* develop. Where benzodiazepines are concerned, however, it is equally important to consider whether tolerance does *not* develop to some effects, especially the therapeutic effects. Fortunately, there are several studies relevant to this question involving anxiety and insomnia. Rickels et al. (1983) reported a prospective study in which patients were assigned to receive diazepam (range 15–40 mg/day, mean 25 mg/day) for 6, 14, or 24 weeks. Scores on several standardized anxiety rating scales indicated that the anxiolytic effect of diazepam found upon initiation of treatment was retained for up to 22 weeks. Plasma determinations indicated that this was achieved without patients' increasing their dose. This seems to suggest, as the authors claimed, that tolerance to the anxiolytic effect of diazepam did not develop. At the same time, half of the patients shifted to *placebo* after 6 weeks on diazepam maintained whatever improvement had been achieved on the drug. Thus, if one is dealing with a substantial proportion of placebo responders, it is inappropriate in these cases to speak of lack of tolerance to a *drug* effect.

Fontaine et al. (1984) compared anxious patients treated with diazepam (15 mg/day), bromazepam (18 mg/day), and placebo over a 4-week period. Both benzodiazepines produced a reduction in anxiety compared with placebo as measured by the Hamilton Rating Scale for Anxiety. The improvement was maintained over 4 weeks of treatment with no evidence of tolerance. Such findings are important in the light of evidence that tolerance to sedation and drowsiness occurs within a few days.

Not all investigations show that diazepam is efficacious for more than a transient period. For example, Shapiro et al. (1983) found that diazepam was more effective than placebo on only the first week of 6 weeks of monitored treatment, and that the effect was not large to begin with. Patients improved and maintained their improvement over time with both diazepam (dose range 5–40 mg/day, mean dose 21.4 mg/day) and placebo. Why there should be such a large discrepancy in efficacy in this study compared with those of Rickels et al. (1983) and Fontaine et al. (1984) is not evident (although an explanation based on differences in mean daily dose seems unlikely).

The mere consideration of tolerance to an effect of a drug becomes moot if the effect cannot be demonstrated. A study by Laughren et al. (1982b) appears to complicate this issue even more. They collected information on diazepam use and anxiety symptoms from a group of psychiatric outpatients attending a Mental Hygiene Clinic in a Veterans Administration hospital. Duration of use of diazepam ranged from 1 to 12 years. Daily dose ranged from 5–40 mg with a mean of 17.7 mg. Clinician ratings using the Hamilton Anxiety Scale were that these patients were not getting an adequate anxiolytic effect. However, the authors stated that most of the patients were satisfied with the anxiolytic effect they were getting. It seems clear from such data that although the question of whether benzodiazepines retain anxiolytic efficacy over the long term is related in part to tolerance, tolerance is only a part of a much more complex issue.

The question of long-term efficacy has been addressed in conjunction with the use of benzodiazepines as hypnotics (Oswald et al., 1982). Over a 24-week period, subjects who had not used hypnotics previously took placebo, nitrazepam (5 mg), or lormetazepam (2 mg) at bedtime. All had referred themselves to the study as poor sleepers. Both benzodiazepines were effective early in treatment. Compared with placebo, both produced improvement in self-rated quality of sleep and a reduction in latency to fall asleep. The improvement in sleep quality waned over a 24-week period, but the reduction in sleep latency was still present after 24 weeks of treatment. Efficacy of this duration is not universal, however. Kales and Scharf (1973) reported that with the exception of flurazepam, tolerance developed to the sleep-facilitating effect of benzodiazepines within weeks of nightly administration. The long-term efficacy of flurazepam as well as nitrazepam has also been reported by Swift et al. (1983). More than 250 elderly patients were prescribed nitrazepam ($n = 220$) or flurazepam ($n = 33$) as hypnotics. In the large majority of cases efficacy was maintained over a very long time, with little evidence of tolerance to the therapeutic (hypnotic) effect; at the same time, there was little evidence of "problems of unwanted sedation, unsteadiness, or mental confusion." Approximately 90% of patients did not escalate dosage. Clearly then, one should be circumspect in drawing general conclusions about tolerance to benzodiazepines as an undifferentiated class of drugs.

The relationship of tolerance to consumption of benzodiazepines in actual clinical use is not extensively described. It is evident that dosages of hundreds of milligrams of diazepam per day can be ingested without gross toxicity (e.g., Griffiths et al., 1984*b*) in a laboratory setting. Subjects in these investigations have not been therapeutic users. Such consumption is indicative of a relatively high *capacity* for diazepam ingestion, but is not evidence that capacity of this magnitude is likely to be used. Indeed, studies of chronic benzodiazepine consumption in the context of medical use suggest that escalation to an extremely high daily dose is not common. Khan et al. (1981) described the use of benzodiazepines in 87 patients drawn from a larger sample of general practice patients. These patients were current users of benzodiazepines who had been doing so for shorter (less than 2 years) or longer (more than 6 years) periods. Of the sample, 34 displayed an increase in dose over time and 53 displayed no change or a decrease. An increase in dosage was significantly associated with the longest duration of use, which may indicate that tolerance led to the increased dosage. Nonetheless, dosage was always within the recommended therapeutic range for each benzodiazepine. It should be noted, however, that patients with a history of substance abuse disorder, including alcohol abuse, were excluded. Other investigators (Bowden and Fisher, 1980) have commented on the difficulty in finding patients exceeding or even attaining 30 mg/day of diazepam in a mental health clinic population. Indeed, their experience was that self-determined *reduction* of dosage was a common behavior in their clinic population. Thus a cycle of tolerance, dosage escalation, and more tolerance is apparently not a common feature of clinical use of benzodiazepines, even when the use is over long periods (cf. also Tessler et al., 1978). Escalation to very high doses probably has tolerance as a prerequisite, but this may only occur in a limited subpopulation of all of those persons exposed to benzodiazepines. In this, benzodiazepines are probably no different in principle from alcohol.

Animal Studies. There have been numerous studies on tolerance to effects of various benzodiazepines in different animal species. Because of the large number, here we will consider only studies that appear to make a point of particular importance, or with particular clarity.

Earlier it was suggested that one of the most important questions concerning tolerance was whether benzodiazepines lose their anxiolytic efficacy with chronic use. Data from animals are instructive but not unanimous on this point. Perhaps the most frequently cited study on this subject is one by Margules and Stein (1968). The work incorporated a version of the anticonflict or antipunishment test of anxiolytic activity. In this task, an animal is trained to perform a response (usually a lever press) to obtain a reinforcer (usually food). Punishment is subsequently administered by delivering an electric shock when the lever press response is made. Suppression of responding which varies with shock intensity is the result of the punishment. Agents that are anxiolytic in man will restore responding suppressed by such punishment, and indeed this is a standard screening test for anxiolytic activity of new drugs. Segments of unpunished responding may be interspersed with segments of punished responding to assess nonspecific (mainly depressant) effects of a drug. Margules and Stein (1968) tested the effect of chronic treatment with oxazepam in rats using this methodology. On the first few administrations, unpunished responding was severely depressed and punished responding (which was at zero levels following saline) rose slightly above baseline levels. Within a few drug trials, the suppressive effect of oxazepam on unpunished responding virtually disappeared as its ability to reinstate punished responding grew stronger. After 22 drug trials tolerance to the depressant effect was complete, but the antipunishment effect remained at its strongest. Thus, according to this test tolerance was selective, and within the time limit of the study, did not develop to the anxiolytic effect.

McMillan and Leander (1978) administered rats chlordiazepoxide (average of 50 mg/day) in their drinking water for 7 weeks. This regimen produced tolerance to the effect of injected chlordiazepoxide on the depression of unpunished responding, but not to the antipunishment (anxiolytic) effect. It is worth mentioning that a metabolic/dispositional explanation of tolerance to the depressant action is made improbable by such results, since such a mechanism requires tolerance across all dynamic effects. This phenomenon of selective tolerance has also been observed in monkeys (Sepinwall and Cook, 1980). These authors also noted parallels in the failure to find tolerance to the anxiolytic effect of benzodiazepines and the failure to find changes in binding properties with chronic administration in rat brain.

Lahti and Barsuhn (1974) made some interesting observations on the effect of benzodiazepines on a biochemical index of stress. High doses of diazepam (among other drugs) produce an elevation in corticosteroid levels. Lower doses, however, block the elevation in corticosteroids produced by various environmental stressors (noise, handling) in rats. Whereas tolerance developed rapidly to the former effect, it did not to the latter. However, since the extent of exposure was limited (2 doses per day for 4 days), these results are only suggestive of an unusual lack of tolerance to a potential index of anxiolytic action.

File and Hyde (1978) used a measure of social interaction in rats as an index of anxiolytic activity. Chlordiazepoxide attentuated a decrease in social interaction that occurred in response to several environmental manipulations. Tolerance to the general locomotor effect of the drug occurred within 5 days of daily dosing, but not to the putative measure of anxiolytic activity. Again, this does not comment on the longevity of the sparing of anxiolytic activity from tolerance, but it does confirm that anxiolytic activity is relatively resistant to the development of tolerance (cf. also Sepinwall et al., 1978, for more evidence on this phenomenon with chlordiazepoxide in monkeys; Stein and Berger, 1971, for evidence on lorazepam in monkeys and rats; Wise et al., 1972, for more on the neurochemical basis of this phenomenon in rats; Torrelas et al., 1980, for an apparently contrary view of tolerance as measured by neurochemical events).

There are at least two studies in which claims of tolerance to anxiolytic activity have been made. Vellucci and File (1979) administered chlordiazepoxide (5 mg/kg) to rats daily for 5, 15, or 25 days. Neurochemical and behavioral indices of anxiolytic activity were used to suggest that tolerance developed within 25 days. Unfortunately, the behavioral evidence was methodologically compromised; the neurochemical evidence, although suggestive of the occurrence of tolerance, remains an unproven indicator of anxiolytic potency compared with the antipunishment test.

Cooper et al. (1981) used an eating test as a possible index of anxiolytic activity. Chlordiazepoxide (15 mg) acutely enhanced eating of familiar food in rats and also reduced the neophobic response that rats displayed toward some novel foods. Within 10 days of chronic treatment, tolerance developed to the antineophobic effect but not to the enhanced eating of familiar food. Cooper et al. speculated that the effect of chlordiazepoxide on eating of novel food is indicative of anxiolytic activity and that tolerance to this effect suggests tolerance to anxiolytic activity. This begs the question of what the effect on eating of *familiar* food reflects by way of pharmacologic activity. Enhancement of food intake in general has been suggested to reflect an antipunishment action of benzodiazepines (Margules and Stein, 1966; although see Wise and Dawson, 1974, for a dissenting view), and there was no tolerance to the effect. Cooper and Francis (1979) found that tolerance to the stimulation of water-drinking by chlordiazepoxide in rats did not occur after 9 days of treatment (5–15 mg/kg), a regimen during which tolerance to sedation would be expected to occur. Thus, lack of tolerance to the *enhancement* of various behaviors by benzodiazepines is common.

There have been several demonstrations of cross-tolerance between benzodiazepines and other drugs. Hoogland et al. (1966) showed that rats would become tolerant to the impairment of their ability to cling to an inclined screen, which was used as an indication of muscle relaxant effect of chlordiazepoxide (100 mg). Cross-tolerance was observed in tests with hexobarbital, pentobarbital, and meprobamate, but not with phenobarbital. These findings plus more direct metabolic studies were taken as evidence that this cross-tolerance was mediated dispositionally (cf. also Jori et al., 1969). McMillan and Leander (1978) also reported cross-tolerance between pentobarbital and chlordiazepoxide. Rosenberg et al. (1983) have reported cross-tolerance for ataxia between flurazepam, pentobarbital, and ethanol using procedures described below (Rosenberg and Chiu 1981). Finally, there has been evidence of cross-tolerance between chlordiazepoxide and morphine in tests of the antinociceptive action of the two drugs; however, the relationship was complex and the potential mechanism unclear (Bodnar et al., 1980).

There have been a variety of conflicting reports about the long-term retention of tolerance to benzodiazepines. File (1982) found that rats acquired tolerance to the sedative effect of lorazepam within 3 days, but that this tolerance was no longer evident after 2 days free of drug treatment. Rosenberg and Chiu (1981) administered flurazepam in drinking water to rats daily for 4 weeks. The mean daily dose achieved by this method over the last 3 weeks was approximately 140 mg/kg. A test dose ranging from 20–200 mg/kg was administered 12 or 25 hr after the drug solution was withdrawn. Although there was some evidence of tolerance to the ataxic effect at 12 hr, tolerance reverted to control levels within 24 hr. The time course of behavioral changes had a parallel in studies of receptor binding. McMillan and Leander (1978) found some residual tolerance to the sedative effect of chlordiazepoxide in rats 6 weeks after cessation of drug treatment. Cook and Sepinwall (1975) observed tolerance to a single, very large dose (160 mg/kg orally) 27 days later in rats.

One of the most detailed and informative studies of tolerance to chlordiazepoxide was reported by Ryan and Boisse (1983). They gave rats twice-daily gastric administration of the drug over a period of 5 weeks. The dosing strategy was to produce a constant drug effect on a gross neurologic test battery. As dose was increased with this technique, rats were able to acquire tolerance over a fivefold increase of dosage from approximately 160 to 840 mg/kg. They described tolerance as "without apparent ceiling." Thus tolerance can be quite extensive in rats. The analog of such extensive tolerance is apparent in human polydrug abusers, who are able to function reasonably well at very high daily dosages (Griffiths et al., 1984b,c). Tolerance can be acquired rapidly as well as extensively, as shown by Rosenberg and Chiu (1985) in work with cats. They administered flurazepam (5 mg/kg/day) using a gastric fistula. Loss of muscle tone was the drug effect by which tolerance was monitored. Significant tolerance was observed as early as the second exposure to flurazepam. This appears to be consistent with expectation based on the clinical literature in humans (e.g., Warner, 1965).

Greeley and Cappell (1985) showed that tolerance to the hypothermic effect of chlordiazepoxide, which was rapidly acquired in rats, is subject to behavioral conditioning. Tolerance acquired in one set of environmental conditions was diminished if the test for tolerance was carried out in a distinctive environment. This showed that tolerance to chlordiazepoxide is not simply a matter of pharmacological stimulation. The observation of conditional control of tolerance places chlordiazepoxide in a class with many other drugs of abuse to which tolerance develops, and raises some interesting speculation about treatment of long-term dependence on benzodiazepines (cf. Poulos et al., 1981).

Overview. The foregoing literature shows clearly that tolerance develops relatively quickly to the sedative/depressant effects of a variety of benzodiazepines. Although there have been a few exceptions, most experiments suggest that tolerance to anxiolysis occurs very gradually and to a limited extent. The latter observation has support in both experimental work and clinical observation. Thus it would appear that the profile of tolerance to benzodiazepines is a relatively beneficial one.

## Physical Dependence

It is often assumed that one of the most important properties of a drug related to abuse liability is its capacity to produce physical dependence. A detailed appraisal of this issue from a theoretical point of view has been made by Cappell and LeBlanc (1983). Briefly, they suggested that physical dependence may be important as a problem of medical management should there be a clinically significant withdrawal reaction upon cessation of drug use, or as a substrate for reinforcement of drug-taking (Harrison et al., 1984).

Studies in Animals. The earliest methodology for assessing physical dependence on sedative-hypnotics in animals was introduced in 1931 by Seevers and Tatum and subsequently was employed by others (Deneau and Weiss, 1963, 1968; Fraser and Isbell, 1954). More recently, additional techniques have been added (Ryan and Boisse, 1983, 1984; Yanagita and Takahashi, 1973). The available techniques may be grouped into three categories. This section is intended to be illustrative of the reliable observations that have occurred in the animal literature rather than to be exhaustive.

*Direct Measurement.* One method for assessing the capacity of a drug to produce dependence is to administer different doses chronically (several days to weeks) and then to observe the nature and time course of symptoms that occur after its abrupt discontinuation. Various observational methods and rating scales have been developed to quantitate signs and symptoms of withdrawal.

The central feature of benzodiazepine withdrawal is CNS hyperexcitability, which is reflected in a lowering of electroshock seizure frequency or an increase in seizure activity (Yanagita and Takahashi, 1973). Weight loss and anorexia can be used as nonspecific measures of withdrawal. Most studies in this category have been performed on rats after chronic oral doses of diazepam or chlordiazepoxide. The doses given have usually been very high (e.g., 150 to 300 mg/kg/day; Yoshimura and Yamamoto, 1979). A common occurrence after discontinuation of high doses of benzodiazepines is convulsions. At lower doses, tremor, muscle rigidity, weight loss, and increased locomotor activity have been commonly observed signs of withdrawal (Martin and McNicholas, 1981; Suzuki et al., 1980).

Yanagita and co-workers (e.g., Yanagita, 1983) have performed numerous studies in primates. Their procedure consists of the oral or subcutaneous administration of a benzodiazepine on a once- or twice-daily schedule. Experiments have typically been double-blind and have involved 4–6 naive rhesus monkeys exposed to the drug for up to 4 weeks. Symptoms have been classified as "mild" (mild tremor, anorexia, hyperirritability, and piloerection), "intermediate" (muscle rigidity, retching or vomiting, weight loss), and "severe" (convulsions, hallucinatory behavior, nystagmus, hyperthermia). With this methodology, they have been able to establish the physical dependence liability of several benzodiazepines. For example, severe withdrawal was observed after alprazolam (Yanagita et al., 1981), chlordiazepoxide (Yanagita, 1973), clorazepate (Yanagita et al., 1977*a*), lorazepam (Yanagita et al., 1975*a*), and triazolam (Yanagita et al., 1977*b*). Intermediate withdrawal was observed for halazepam (Yanagita et al., 1975*b*) and nitrazepam (Yanagita et al., 1975*c*). Withdrawal from all the remaining benzodiazepines tested was classified as mild.

An important limitation of the foregoing studies is that they included no systematic attempt to standardize the benzodiazepine doses or to allow for pharmacokinetic differences. A typical example is provided in a recent publication comparing diazepam and lorazepam in dogs (McNicholas et al., 1983). The authors claimed that two qualitatively different withdrawal syndromes were observed, and they speculated that two different mechanisms might be involved. However, when the relative potency of these drugs is considered, the dose of lorazepam administered (100 mg/kg/day) was about eight times larger than the dose of diazepam (60 mg/kg/day).

A methodology has been developed in which drugs of different potency and duration of action may be directly compared. This methodology is based on the "chronically equivalent" dosing principle (Boisse and Okamoto, 1978*b*). Administration of drugs is titrated to produce comparable degrees of CNS effects over equivalent periods of time. The measure of CNS effects consists of scores on a variety of neurological functions (e.g., activity, righting reflex). The application of this methodology has permitted a direct comparison of chlordiazepoxide and phenobarbital. Withdrawal was qualitatively similar for the two drugs, but convulsions occurred only with phenobarbital (Boisse et al., 1981). Other studies have been confined to a description of chlordiazepoxide withdrawal (Ryan and Boisse, 1983, 1984). Legitimate comparisons of the dependence-inducing potential of benzodiazepines will require application of this methodology or one similar to it.

*Substitution Test (Cross-dependence).*   The second general procedure used to determine the physical dependence potential of drugs in animals is the substitution test. Animals are made physically dependent by administration of a high dose of a barbiturate (e.g., phenobarbital or barbital) for several days or weeks. The drug is then withdrawn and replaced by the test drug. The capacity of the test drug to suppress barbiturate withdrawal provides an estimate of its dependence-producing potential.

Most studies of the benzodiazepines in this category involved substitution of chlordiazepoxide or diazepam in rats or mice first made dependent on barbital or phenobarbital. Such studies indicate that several barbiturate withdrawal signs (e.g., tremor, audiogenic convulsions, etc.) can be reversed by benzodiazepines (Deneau and Weiss, 1968; Reigel and Bourn, 1982). Using rhesus monkeys, Yanagita and co-workers have been able to show that all benzodiazepines tested to date suppress barbital withdrawal. In addition, there is a high correlation between the potencies of different benzodiazepines in suppressing barbital withdrawal and their corresponding affinity for the benzodiazepine receptor. These findings suggest high specificity of the benzodiazepines for the suppression of barbital withdrawal (Yanagita et al., 1977*b*).

*Precipitated Withdrawal.*   The most accurate method of diagnosing physical dependence is the use of a specific antagonist to unmask the signs and symptoms of withdrawal. The classical example is naloxone, which is widely used to diagnose physical dependence on opiate drugs.

The first benzodiazepine receptor antagonist identified, Ro 15-1788, has been the most studied and has provided a unique tool for investigating physical dependence. The acute and subacute toxicity of the antagonist have been found to be very low. No intrinsic or other observable effects are detected when the antagonist is administered alone in naive animals (Bonetti et al., 1982; Crawley et al., 1984; Mohler et al., 1981; Rosenberg and Chiu, 1982; although see De Vry and Slangen, 1985). Several studies have shown that Ro 15-1788 elicited withdrawal symptoms when administered to animals treated with a high dose of diazepam, lorazepam, or triazolam (Cumin et al., 1982; Rosenberg et al., 1983). The symptoms typically appeared 10 to 20 min after the administration of the antagonist and differed somewhat among species. The intensity of the antagonist-precipitated withdrawal syndrome appears to be related to both the doses of benzodiazepines and of the antagonist. For example, Cumin et al. observed that the precipitated withdrawal induced by 10 mg of Ro 15-1788 in rats treated for 12 days with diazepam (10 or 100 mg/kg PO daily) was most severe in those animals receiving the highest dose of diazepam (Cumin et al., 1982). A dose–response relationship between Ro 15-1788 and the intensity of observed withdrawal has also been reported (McNicholas and Martin, 1982).

Lukas and Griffiths (1982) gave diazepam to baboons intragastrically (10 mg/kg) in two doses 8.5 hr apart and then administered Ro 15-1788 (5 mg/kg) on days 7 and 35 of the regimen. A mild to intermediate withdrawal syndrome, including lipsmacking, nausea, and rigidity, was observed after 7 days, and a more severe and frequent syndrome including convulsions occurred after 35 days.

All doses of benzodiazepines used in these studies are considerably higher than the effective anxiolytic and sedative doses. Thus, whether dependence is produced at lower doses and shorter durations of use has not yet been explored.

Studies in Humans. The development of physical dependence on drugs in humans is usually identified by the appearance of a withdrawal syndrome after drug discontinuation. Experimental procedures for assessing a drug's dependence potential in humans were initially developed for opiates (Kolb and Himmelsbach,

1938) and comparable strategies have since been adapted to characterize withdrawal from alcohol (Gross et al., 1974) and benzodiazepines (Jasinsky et al., 1984). With all of these compounds, dependence has been induced by repeated administration of the drug of interest to volunteers with a previous history of drug dependence. During a period of forced abstinence, the severity of physical dependence has been rated by a variety of procedures including psychophysiological measurements and observer rating scales. Withdrawal signs of low to moderate intensity with early onset (24 hr) are assigned lower scores in the scale than more severe signs with later onset. Such scales provide reliable and sensitive measures and are used in clinical settings to assess the severity of withdrawal (Shaw et al., 1981).

Shortly after the introduction of chlordiazepoxide, the first clinical report of physical dependence on benzodiazepines appeared (Hollister et al., 1961). Eleven chronic psychiatric patients receiving chlordiazepoxide (300–600 mg/day for 2 to 6 months) were abruptly switched to placebo. These doses of chlordiazepoxide are 8 to 20 times the usual therapeutic dose prescribed for anxiety. Ten of the 11 patients experienced new symptoms and signs consistent with a withdrawal reaction. The signs included agitation, insomnia, depression, nausea, and loss of appetite; typically they appeared between 2–8 days after discontinuation of chlordiazepoxide. Two patients had seizures on days 7 and 8. By the 10th day symptoms had disappeared or were waning. These data are not without problems of interpretation, since at least 5 of the 11 subjects were schizophrenic and were receiving other medications; moreover, the investigators were not blind to the agent that the patients were taking (Hollister et al., 1961). A second study by the same group noted that in 25 schizophrenic patients given 20 to 80 mg/day of diazepam for six weeks, 6 of 13 patients abruptly switched to placebo developed clinical signs of withdrawal and one patient had a major seizure on day 8 (Hollister et al., 1963).

Below we describe studies in several categories. Because of their large number, they have been for the most part summarized in tabular form.

*Case Reports.*    Over the ensuing years a number of nonexperimental reports of physical dependence on *high* doses of benzodiazepines have been published. These are summarized in Table 3, which gives a number of important details of each study. Most are single case reports involving doses 2–4 times the recommended maximum therapeutic durations of use, ranging from 6 months to 8 years. In most of these cases drug withdrawal was abrupt. Some of the symptoms described are indistinguishable from those of severe anxiety. However, auditory and visual hallucinations, confusion, and involuntary movements are seen in withdrawal from most CNS depressants (e.g., alcohol, barbiturates). Epileptic seizures seem to be rare, and this has been attributed to the long half-lives of many benzodiazepines and their metabolites. The importance of kinetic properties

Table 5. Case Studies of Withdrawal Reactions at High Doses of Benzodiazepines

| Benzodiazepine | Number of patients | Dose (mg/day) | Duration of use | Withdrawal symptoms | Source |
|---|---|---|---|---|---|
| Alprazolam | 1 | 5–6 | 4 months | Seizures, delirium | Levy, 1984 |
| Alprazolam | 2 | 8 | 4 months | Seizures | Breier et al., 1984 |
| Chlordiazepoxide | 11 | 300–600 | 1–7 months | Psychosis, agitation | Hollister et al., 1961 |
| Diazepam | 3 | 60–160 | 6–24 months | Anxiety, agitation, psychosis | Preskorn and Denner, 1977 |
| Diazepam | 1 | 60 | 12 months | Agitation, tremor | Gordon, 1967 |
| Diazepam | 1 | 60–80 | 8 years | Anxiety, agitation | Miller and Nielsen, 1979 |
| Diazepam | 1 | 20–200 | Several years | Delusions, hallucinations, confusion, agitation | Minter and Murray, 1978 |
| Diazepam | 1 | 80 | 1 year | Convulsions, confusion | Barten, 1965 |
| Diazepam | 1 | 80 | 4 years | Hallucinations, disorientation, seizures, coma | DeBard, 1979 |
| Diazepam | 1 | 140–240 | 3 months | Extreme restlessness, vomiting, psychosis | Agrawal, 1978 |
| Diazepam | 3 | 150–200 | 2–12 years | Anxiety, anorexia, nausea, agitation, seizures (3) | Robinson and Sellers, 1982 |
| Diazepam | 10 | 60–120 | 3–14 years | Tremor, anorexia, insomnia, "toxic psychosis" alteration in perception | Mellor and Jain, 1982 |
| Lorazepam | 1 | 10 | 4 months | Disorientation, nausea, confusion, hallucinations | Stewart et al., 1980 |
| Lorazepam | 3 | 5–15 | Several months to 2 years | Seizures | Svensson, 1980 |
| Lorazepam | 2 | 7.5 | 4 years | Fear, panic, involuntary movements, seizures (3) | Howe, 1980 |
| Lorazepam | 2 | 8–12 | 5–6 months | Involuntary movements, seizures (2), anxiety, agitation | de la Fuente et al., 1980 |
| Lorazepam | 2 | 12 | 1 to "several" months | Seizures | Einarson, 1980 |
| Oxazepam | 3 | 120–600 | ? | Anxiety, hallucinations, confusion, seizures (1) | Lennane, 1982 |
| Temazepam | 1 | 100 | 6 months | Tremor, depression | Ratna, 1981 |
| Nitrazepam | 1 | 50–70 | 1 year | Paranoid, psychosis | Misra, 1975 |

is suggested by the late appearance of seizures in cases of dependence on diazepam (2 to 12 days) as compared with lorazepam (24 to 68 hr).

The question of whether long-term use of *therapeutic* doses of benzodiazepines can produce physical dependence has been a subject of investigation for some years. Covi and co-workers (1973) compared a group of anxious neurotic outpatients who received 20 weeks of treatment with chlordiazepoxide (15 mg TID) with a group who received the drug for only 10 weeks; the former group of patients showed reliable worsening of anxiety and other symptoms of abstinence. Although withdrawal signs were measured for only 1 week, which makes it difficult to distinguish actual withdrawal from recrudescence of psychophysiological symptoms, this study marked the starting point of a long controversy.

Many additional case reports of physical dependence on therapeutic doses of benzodiazepines have since appeared (Table 4). Most have been letters to medical journals and give very limited information. Drug use history has typically been poorly documented and based primarily on the report of the subject without objective confirmation. Information on concurrent use or abuse of other drugs is also usually lacking. Those cases incorporating actual observations of withdrawal have been collected without benefit of validated procedures or rating scales. However, case reports usually provide the first evidence for a phenomenon and are valuable despite their clear limitations.

*Controlled Studies.*   There have been some controlled but conflicting evaluations of physical dependence after long-term use of therapeutic doses of benzodiazepines (Table 5). It has been difficult to decide whether symptoms observed are withdrawal symptoms and not a simple recrudescence of pre-existing anxiety. The frequency of these reactions and the predisposing factors related to the appearance of withdrawal symptomatology are also unclear. Some studies report no withdrawal signs at all (Laughren et al., 1982*a*) and others report some withdrawal signs in only a small percentage of subjects (Rickels et al., 1983). Still others report withdrawal in a substantial proportion of patients (cf. Table 3). The better controlled the study, the more difficult it is to find strong evidence of withdrawal. For example, Laughren and co-workers (1982*a*) specifically studied withdrawal of therapeutic doses of diazepam in 24 anxious outpatients who had been taking the medication for several years (mean = 5 years). Following a baseline period of 1 week, patients were either abruptly withdrawn, reduced by 5 mg/week, or maintained on the dose they were taking previously. At weekly intervals symptoms of withdrawal were recorded on a symptom checklist, and blood levels of diazepam and its metabolite were taken to assess compliance. No evidence of withdrawal was seen in either group. Petursson and Lader (1981*b*) studied 16 patients who had been receiving benzodiazepines (diazepam, lorazepam, clobazam) in therapeutic doses for several years (range 1–16 years). After baseline assessments they were tapered to a half-dose for 2 weeks and then

Table 4. Case Reports of Withdrawal Reactions at Therapeutic Doses of Benzodiazepines

| Benzodiazepine | Number of patients | Dose (mg/day) | Duration of use | Withdrawal symptoms | Source |
| --- | --- | --- | --- | --- | --- |
| Clobazam | 2 | 20–30 | 2.5–16 years | Blurred vision, tremor, sweating, stiffness | Petursson and Lader, 1981a |
| Diazepam | 1 | 30 | 3 months | Seizures | Rifkin et al., 1977 |
| Diazepam | 1 | 30 | 3 years | Seizures | Vyas and Carney, 1975 |
| Diazepam | 1 | 30–45 | 20 months | Tremor, anxiety, anorexia, nausea, numbness | Pevnick et al., 1978 |
| Diazepam | 1 | 15–30 | 7 years | Confusion, delirium, disorientation | Dysken and Chan, 1977 |
| Diazepam | 1 | 15–25 | 6 years | Anxiety, blurred vision, tinnitus, palpitations, headache | Winokur et al, 1980 |
| Diazepam | 5 | 30–40 | 5 months to several years | Anxiety, insomnia, numbness, hallucinations | Floyd and Murphy, 1976 |
| Diazepam | 2 | 15–30 | 10 years | Gastric distress, anxiety, agitation, dizziness | Khan et al., 1980 |
| Flurazepam | 1 | 30 | 8 years | Anorexia, numbness, dizziness, insomnia | Berlin and Connell, 1983 |
| Lorazepam | 3 | 3–7.5 | 6 months to 5 years | Seizures (1) confusion, anxiety | Bismuth et al., 1980 |
| Lorazepam | 2 | 2–6 | 2–3 months | Seizures (2) | Barton, 1981 |
| Lorazepam | 3 | 3–7.5 | 1 month to 3 years | Hallucinations, disorientation | Khan et al., 1980 |
| Oxazepam | 4 | 60–250 | 6 months to 5 years | Tremor, anxiety | Bliding, 1978 |
| Oxazepam | 1 | 45–90 | 1 year | Agitation, restlessness, depression | Harry, 1972 |
| Oxazepam | 1 | 45–90 | 2 years | Agitation, depression, depersonalization | Hanna, 1972 |
| Oxazepam | 1 | 30 | 3 years | Agitation, depression, depersonalization | Khan et al., 1980 |
| Oxazepam | 1 | 50 | ? | Insomnia, agitation, seizures | Bismuth et al., 1980 |
| Chlordiazepoxide | 1 | 100 | 4 | Sweating, cramps, numbness | Slater, 1966 |

Table 5. Controlled Studies of Withdrawal at Long-Term Therapeutic Doses of Benzodiazepines

| Benzodiazepine | Number of patients | Dose (mg/day) | Duration of use | Methodology | Results | Source |
|---|---|---|---|---|---|---|
| Chlordiazepoxide | 39 | 45 | 10–20 weeks | Abrupt withdrawal, double blind | Patients on 20 treatment experienced more withdrawal symptoms than those on 10 weeks | Covi et al., 1973 |
| Flunitrazepam Nitrazepam Placebo | 117 | 1–5 | 4 weeks | Abrupt withdrawal, double blind, placebo-controlled, randomized | No difference between drug and placebo groups with respect to withdrawal | Hartelius, 1978 |
| Diazepam | 16 | 10–30 | 1–16 years | Gradual withdrawal, double-blind, placebo controlled | All patients experienced withdrawal symptoms | Petursson and Lader, 1981*b* |
| Oxazepam Halazepam | 29 | H–120 O–45 | 3 weeks | Abrupt discontinuation, half continued on placebo had more symptoms | No interdrug differences patients *not* receiving placebo | Pecknold et al., 1982 |

| Diazepam | 24 | Mean, 17 | 1–12 years | Randomly assigned to abrupt withdrawal, gradual withdrawal or maintenance | No differences among groups with respect to withdrawal symptoms | Laughren et al., 1982*b* |
| Diazepam | 180 | 15–40 | 6–22 weeks | Randomly assigned to 6, 14, or 22 weeks of diazepam treatment followed by placebo withdrawal | 8% had "withdrawal reactions," 32% had "return of symptoms," withdrawal was a function of duration treatment | Rickels et al., 1983 |
| Diazepam | 41 | 5–20 | Mean, 18–50 months | Gradual withdrawal, "constrained" randomized procedure, placebo controlled | 16 experienced "withdrawal," 8 "pseudo withdrawal" | Tyrer et al., 1983 |

given placebo for the following 2 weeks. In contrast to the results of Laughren et al., all these patients reported a withdrawal syndrome including agitation and anxiety.

These conflicting observations probably arise because of differences in patient populations and study design and the care with which the critical variable of drug use was characterized. The study designs typically contain an important bias; tapering has been the usual detoxification strategy in studies involving long-term (i.e., years) users, whereas patients receiving drug for only 4–20 weeks have been abruptly withdrawn. It is possible that in these studies, a withdrawal state was not observed because gradual tapering obscured the appearance of a withdrawal syndrome, or conversely, because the duration of exposure was not sufficient to produce physical dependence. Both types of studies are of clinical interest since one suggests a treatment strategy to avoid withdrawal and the other indicates the risk of symptoms after the recommended course of therapy. Neither, however, satisfactorily answers the fundamental question about the intrinsic pharmacological properties that produce physical dependence in humans or about the clinical importance of physical dependence. In many studies, the patterns of use of other drugs, including alcohol, were not reported, and patient self-reports of benzodiazepine use were not validated by objective procedures. Moreover, reliability and validity of measures of benzodiazepine use before, during, and after withdrawal was omitted. Finally, these studies are subject to the criticism that the signs and symptoms that arise after stopping benzodiazepines are simply a recurrence of symptoms present before the drug was prescribed.

A number of these problems were avoided in a double-blind, placebo-controlled study of long-term benzodiazepine users (Busto et al., 1985). Patients were assessed medically and psychologically and then randomly assigned to receive either diazepam ($n = 23$) or placebo ($n = 19$) over a 4–6 week period during which the psychological treatment was combined with gradual tapering from the medication. The groups were comparable with respect to age, sex, daily dose, social adjustment, duration of use, and other characteristics that might have influenced outcome. Patients receiving placebo showed more withdrawal symptoms and rated their symptoms as more severe than those receiving diazepam. A group of symptoms emerged that were qualitatively different from those of anxiety, and the time course of the reaction was different from what would be expected if the syndrome was simply a recrudescence of symptoms that were suppressed by the drug. These results support the view that a mild but detectable withdrawal syndrome occurs after discontinuation of long-term therapeutic use of benzodiazepines.

Overview. In summary, animal studies (direct dependence, substitution, and precipitated withdrawal) provide substantial evidence that all benzodiazepines tested to date can induce physical dependence, especially at high doses. For all tests considered above, they are much less potent than barbiturates, opiates, or other known drugs of abuse (Yanagita and Takahashi, 1973; Griffiths et al., 1979). Whether there are differences among benzodiazepines with respect to their potential to produce physical dependence has not been systematically tested. Most studies have used very high doses to induce dependence and the doses have been much higher than those needed to produce therapeutic effects. It still remains to be determined at what dosage level and duration a benzodiazepine must be administered to induce physical dependence. The procedures developed by Ryan and Boisse have the potential to clarify such questions. In addition, specific benzodiazepine antagonists provide powerful tools for studying the mechanisms involved in the development of dependence.

Human studies show substantial evidence that all benzodiazepines can induce physical dependence when administered in high doses. The precise clinical features of the withdrawal syndrome after discontinuation of chronic therapeutic use of benzodiazepines have been difficult to characterize. However, it has been conclusively shown that there is a mild but clinically detectable withdrawal syndrome after discontinuation of therapeutic doses of benzodiazepines. This syndrome has unique features that appear to be distinguishable from a mere return of pre-existing anxiety.

## 5. ADVERSE EFFECTS OF BENZODIAZEPINES

One of the most important aspects of the abuse potential of a drug is whether there are patterns of use that can cause adverse effects or problems of some kind. For example, Fischman and Schuster (1978) asserted that "implicit in the word abuse is the idea that there are toxic behavioral and physiological consequences associated with drug self-administration." Virtually any therapeutic drug may cause adverse side effects even during responsible therapeutic use, and the inevitable side effects of therapeutic use of benzodiazepines should not automatically be characterized as problems of abuse. However, chronic use or use in excessive doses may increase the severity of various problems, or the likelihood that some effects will become problems. In the following sections we touch briefly upon many of the potentially adverse effects of benzodiazepines not covered elsewhere in this chapter. This coverage is by no means intended to be exhaustive.

### Acute Effects

The division of adverse effects into acute and chronic categories is somewhat arbitrary. A review of abuse implies chronic use, and it could be argued that the distinction could be ignored. However, we have attempted to address separately those effects that can occur with a single dose or episode of use (acute) from those that are unlikely to occur except with long-term use.

Intoxication and Impairment. The most common adverse effects of benzodiazepines are those associated with the primary action of these drugs on the central nervous system. Commonly reported side effects include excessive somnolence, drowsiness, ataxia, dysarthria, apathy, reduced motor coordination, and memory impairment (Garattini et al., 1973; Greenblatt and Shader, 1974). Drowsiness, for example, appears in 4–9% of subjects receiving benzodiazepines, while ataxia and paradoxical excitement occur in less than 1% of recipients (Greenblatt and Shader, 1974). In a large survey of 287 clinical studies involving 17,935 patients receiving chlordiazepoxide, an adverse event occurred at an overall rate of 2 per 1,000. Drowsiness was the most common side effect, reported in 3.9% of patients receiving the drug at therapeutic (75 mg/day) doses (Svenson and Hamilton, 1966). Another prospective epidemiological study of adverse drug reactions to drugs in hospitalized inpatients showed that the most commonly prescribed benzodiazepines were diazepam and chlordiazepoxide (19.8% and 15.6% of patients) and that the most common unwanted effects were drowsiness (diazepam: 5.7% of patients; chlordiazepoxide: 8.7%) followed by paradoxical reactions such as excitement, agitation, hallucinations, and nightmares in 0.7% and 0.3%, respectively, of patients (Miller, 1973). Most of these adverse effects are relatively innocuous; they occur most frequently during the first week of treatment and then wane as tolerance develops (Greenblatt and Shader, 1978).

Definite cognitive and psychomotor impairment follows single-dose and short-term benzodiazepine administration to normal subjects (Wittenborn, 1979). Studies in laboratory settings show that benzodiazepines produce a dose-dependent impairment of psychomotor performance in a wide variety of tasks (e.g., O'Hanlon et al., 1982; Seppala et al., 1979). However, virtually all laboratory studies have been performed with young healthy volunteers in very controlled situations and are therefore difficult to generalize to field conditions.

Epidemiological data on the *actual* consequences (e.g., traffic accidents) of benzodiazepine use are limited. Two studies have reported that the incidence of use of psychotropic drugs (mostly benzodiazepines) among injured drivers was about 2–4% and 1.1–1.8%, respectively (Benjamin, 1977; Kibrick and Smart, 1970). This incidence of use among injured drivers is not significantly different than that described for the general population. However, both studies lacked a control group. Bø et al. (1976) found diazepam alone or in combination with alcohol in the blood of 20% of injured drivers in comparison to only 2% of controls. In a prospective study, Skegg and co-workers (1980) also found that the use of minor tranquilizers increased the risk of a traffic accident, but the confounding factors of other drug use (including alcohol) and the contribution of the underlying disease being treated were not controlled. In addition, subjects in the control group were not matched for driving experience.

In a carefully controlled and conducted study (Honkanen et al., 1980), 201 injured drivers who presented at emergency departments within 6 hr after a traffic accident were interviewed about their state of health and drug use. Plasma concentrations of alcohol and other drugs were also obtained. Drug use of the subjects was compared with that of 325 randomly selected controls. Positive cases of psychotropic drug use were relatively infrequent, and benzodiazepines were identified in 5% of the patients and 2.5% of the controls. This difference was not significant. However, the authors concluded that diazepam may have been a contributory factor to the accidents because it was overrepresented among injured drivers (Honkanen et al., 1980). The experimental literature shows that benzodiazepines impair psychomotor performance that may be related to driving. However, given the present state of knowledge, it is probably reasonable to conclude that the contribution of benzodiazepines to the risk of traffic accidents remains unknown.

Impairment of short-term memory has been shown in laboratory settings on many occasions for all benzodiazepines tested (e.g., McKay and Dundee, 1980; Petersen and Ghoneim, 1980; Ghoneim et al., 1984). Effects at therapeutic doses tend to be modest in size. Although it is clear from the literature in anesthesiology that significant amnesia can be produced (Brandt and Oakes, 1965), the prevalence of substantial amnesic effects of benzodiazepines in ordinary use is not known but is probably not great.

"Paradoxical" Reactions. A "paradoxical" reaction is characterized by the insidious or abrupt appearance of rare behavior in an individual who would not otherwise be thought to be predisposed to such behavior. Paradoxical reactions to benzodiazepines have been described (Hall and Jaffe, 1972; Ingram and Timbury, 1960; Ryan et al., 1968; Salzman et al., 1974). They have typically been rage or hostile and aggressive behavior, but paranoid ideas have also been reported. Other studies have suggested that diazepam therapy was associated with onset of suicidal thoughts and tendencies in depressed patients (Ryan et al., 1968). Most paradoxical reactions have occurred in patients with psychiatric disorders, and many were observed in psychiatric patients having a wide variety of other symptoms. However, increased hostility has also been described in male volunteers receiving chlordiazepoxide when compared with subjects receiving placebo (Salzman et al., 1974). Paradoxical rage reactions have been described for alprazolam (Rosenbaum et al., 1984), chlordiazepoxide (Ingram and Timbury, 1960), diazepam (Hall and Jaffe, 1972), and oxazepam (Zucker, 1973). It is not known whether such reactions are related to misuse or abuse of benzodiazepines or constitute a risk associated with therapeutic use.

The published evidence on paradoxical reactions suggests that occurrences such as angry outburst, assault, psychosis, hallucinations, depression, and nightmares are rare, especially considering the prevalence of benzodiazepine use. Moreover, most of the reported reactions have occurred in patients with underlying psychiatric disorders and in older subjects (Evans and Jarvis, 1972; Taylor, 1973). Notwithstanding the low prevalence of paradoxical reactions, there have been occasional sensationalized reports. For example, van der Kroef (1979) described four cases of psychiatric patients who had suicidal tendencies, nightmares, and hallucinations while receiving triazolam. The contribution of triazolam to the reaction was not clearly established (Lasagna, 1980). Nonetheless, in the face of sensational media coverage, triazolam was suspended from the Dutch market amidst controversy. There have been no further reports of similar reactons to triazolam.

Overdose. A striking relationship between general patterns of psychotropic drug use and the occurrence of cases of overdose has been well documented (Sellers et al., 1981). Because benzodiazepines are the most frequently prescribed psychotropics, it is not surprising they are the drugs most frequently used in instances of overdose (Busto et al., 1980; Proudfoot and Park, 1978; Sellers et al., 1981).

All the available evidence supports the view that benzodiazepines are the least hazardous of psychoactive drugs with respect to overdose (Divoll et al., 1981; Finkle et al., 1979; Greenblatt et al., 1983*b;* Sellers and Busto, 1982*a*). There have been no reported cases of fatal overdoses due to a benzodiazepine taken alone, even after large doses (Greenblatt et al., 1978). However, a substantial proportion of patients with benzodiazepine overdose also mentioned the ingestion of other drugs. In such cases, morbidity increases significantly depending on the coingested drug. For example, an epidemiological study performed in 21 hospitals in Toronto showed that of 2723 patients seen in emergency rooms because of drug overdose, 1071 (39%) had used benzodiazepines and 67% of those patients had also ingested other drugs. The percentage of unconscious patients presenting in emergency rooms increased from 2.6% when the benzodiazepines were taken alone to 15.7% when they were combined with ethanol (Busto et al., 1980).

The type of problems and the drugs alleged by patients with drug abuse or overdose differ according to age and sex (Busto et al., 1981). With benzodiazepines, overdose and abuse are more common in females and in patients over 40 years of age. Since epidemiological studies have shown that benzodiazepine prescription and use are also higher in females and older subjects (Balter et al., 1984), this supports the view that availability plays a major role in the type of drugs used for overdose and abuse purposes. Because availability of drugs plays such a major role on the incidence of overdose and abuse problems, one can hypothesize that not only overdose but dependence and abuse problems with benzodiazepines would be correlated with long-term chronic use of these drugs. Evidence related to this issue is discussed in detail in other sections of this chapter.

Drug Interactions. Since it is likely that other drugs will be consumed concurrently with benzodiazepines, especially during long-term use, the issue of drug interactions deserves some discussion. Comprehensive reviews on the area are available (Linnoila et al., 1979; Sellers and Busto, 1982*b*).

Since benzodiazepines do not cause clinically relevant enzyme induction, they can be administered without having important interactions with other drugs (e.g., anticoagulants) metabolized by liver enzymes (Greenblatt and Shader, 1974; Greenblatt et al., 1983*b*). Of much more importance is the interaction of benzodiazepines and alcohol. Compared with most other psychoactive agents, the likelihood of concurrent ingestion of this pair is relatively great. Numerous studies have shown that measurable interactions can occur. Both dispositional and functional mechanisms have been proposed to explain these interactions (Sellers and Holloway, 1978).

Although there have been negative results in some studies, most have shown significant interactions between chronic exposure to benzodiazepines and acute administration of ethanol. Psychomotor, memory, and learning tests seem to be most affected (Sellers and Busto, 1982b). However, the epidemiological relevance of controlled experimental studies has not been established. The consequences and relative importance of ethanol ingestion combined with chronic benzodiazepine use with respect to motor vehicle accidents, machinery operations, and other behavioral toxicity are difficult to assess separately because ethanol is the predominant risk factor (Hollister, 1974; Sellers and Busto, 1982*b*; Honkanen et al., 1980; Skegg et al., 1980). Proper epidemiological studies controlling for this confounding would be very difficult to do.

In summary, benzodiazepines and ethanol interact in a relatively predictable way. This combination is probably less important pharmacologically and epidemiologically than interactions involving alcohol and many other psychotropic drugs (e.g., cannabis).

## Effects of Long-Term Use

Aside from the phenomena of tolerance and physical dependence, very little is known of the consequences of long-term use of benzodiazepines. Here we summarize the few studies that have addressed this question.

Central Nervous System. The effects of long-term use of benzodiazepines on the CNS have not been extensively studied and are not well understood. Well-controlled epidemiological studies to determine the adverse consequences of long-term benzodiazepine use are difficult to do. Psychomotor or memory deficits, for example, are difficult to demonstrate in anxious patients, since the performance of such patients may improve as a consequence of benzodiazepine use (Bond et al., 1974).

Neuropsychological deficits and morphological cerebral changes have been documented in patients using high doses of benzodiazepines (Bergman et al., in press). Only two studies have reported cerebral abnormalities after long-term therapeutic use of benzodiazepines. In one (Lader et al., 1984), 20 long-term benzodiazepine users had brain scan examinations by computerized tomography

(CT). Definite cerebral abnormalities were reported by the radiologist in three benzodiazepine users as compared with one control subject and three alcoholic patients. The abnormalities were ventricular enlargement, widening of sulci, and Sylvian and interhemispheric fissures. Abnormalities observed on the CT scan of patients using benzodiazepines were more extensive than those of the control subjects and less extensive than those of the alcoholics. There was no significant relationship between CT scan abnormalities and the duration of benzodiazepine therapy. However, a similar study failed to confirm Lader's findings. Poser et al. (1983) compared the CT scan of 28 patients with benzodiazepine dependence, 28 controls, and 28 alcoholic patients matched by age and sex. Indices of brain size were comparable for benzodiazepine-dependent patients and controls and different for alcoholics and patients with combined or sequential alcohol–benzodiazepine dependence. Since the study by Lader et al. did not control for heavy drinking, the data from Poser et al. suggest that the former results are confounded by heavy alcohol use. Moreover, both studies included fewer than 30 patients, and thus the possibility of a Type II error also cannot be ruled out.

Impairment of intellectual, cognitive, and psychomotor functions after long-term benzodiazepine use is also of concern. These effects can be assessed by a battery of specialized tests, although the methodological and interpretative difficulties are considerable. Motivational factors, the underlying condition for which the drug is taken, and practice effects may affect the results and interpretation of neuropsychological tests.

One study on psychological impairment after long-term benzodiazepine use found an improvement in the performance (measured by a digit–symbol substitution test) of patients during withdrawal from long-term benzodiazepine treatment (Petursson and Lader, 1982). These findings were interpreted as possible evidence for psychological impairment due to long-term benzodiazepine treatment. However, Nicolson and Spencer (1982) challenged the interpretation of the results by Petursson and Lader. They found that improved performance could be obtained in a comparable test with normal volunteers simply with practice. Consequently, their interpretation was that long term use of benzodiazepines is unlikely to lead to psychological impairment.

A recent and carefully controlled study of psychomotor performance (Petursson et al., 1983) compared long-term benzodiazepine users and controls. There were substantial and prolonged practice effects on most tests. No performance decrement was found in the long-term benzodiazepine users prior to withdrawal, but a rebound performance increment during withdrawal was observed in one task (tapping rate). Performance was impaired on tasks requiring the combined use of sensory and motor skills.

The data in this area are so limited that no strong conclusions can be drawn. Overall, the few available studies suggest that although chronic benzodiazepine use may result in some very selective but chronic psychological deficits, other causes of the deficits cannot be ruled out. Although there has been a suggestion of behavioral impairment as a result of long-term use of benzodiazepines, more data will be required before any firm conclusions can be drawn.

Adverse Medical Consequences. Cardiovascular and respiratory depression, apnea, and hypotension sometimes leading to fatal outcome have been reported after the acute intravenous administration of diazepam (Buskopp et al., 1967; Greenblatt and Koch-Weser, 1973). Although rare, these reactions appear more frequently in patients having underlying disease or receiving other CNS depressant drugs. Allergic reactions to benzodiazepine use have also been reported. They are very infrequent and the ones described in the literature (Almeyda, 1971; Gaul, 1961; Nackie and Nackie, 1965) take the form of urticaria or macropapular eruptions. Only one anaphylactic reaction has been reported in a patient receiving chlordiazepoxide and other drugs as well (O'Grady and Pokorny, 1964).

Teratogenic, hepatotoxic, and hematologic toxic effects have also been attributed to benzodiazepine use (Cunningham, 1965; Litvak and Kaebling, 1971; Lo et al., 1967; Menon, 1965; Pickering, 1966; Roy-Birne et al., 1983). These reports require cautious interpretation as other causes for the adverse reactions were not ruled out. The evidence supporting a causal role of benzodiazepines in teratogenic, hepatic, and hematologic adverse effects is very weak, and well-controlled epidemiological studies are needed to determine whether long-term use of benzodiazepines may cause such effects.

Anecdotal and experimental data in animals have suggested a link between tumor growth and diazepam. Enhanced growth rate of mammary tumors in rats after the administration of 1 mg/kg of diazepam has been reported (Horrobin et al., 1979; Karmali et al., 1978, 1980). No significant influence on tumor growth rate was seen after the administration of 5 mg/kg of diazepam (Horrobin et al., 1979), although the authors claimed a dose–response relationship based on data from another of their studies (Karmali et al., 1980). Other investigators have failed to reproduce these results with similar doses of diazepam or oxazepam (Guaitani et al., 1979; Jackson and Harris, 1981). In addition, three clinical studies of the incidence of breast cancer among diazepam users and controls have failed to provide any evidence for the claim that regular use of diazepam increases the risk of breast cancer (Friedman and Ury, 1980; Kaufman et al., 1982; Kleinerman et al., 1981). Thus, the clinical relevance of the animal studies on tumor growth and diazepam remains to be proven.

## Overview of Adverse Effects

Compared with many other drugs of abuse (e.g., cocaine, alcohol) for which significant toxic behavioral and organic effects have been reasonably well established, the data on benzodiazepines are relatively poor. Even in the case of those pharmacologic effects that can produce psychomotor impairment, the practical relevance (e.g., for motor vehicle safety) is unknown. There is virtually no lethal dose and no firm evidence of organic pathology. Many of the effects of the benzodiazepine state that might be considered adverse in one context (e.g., sedation, amnesia) are also central to its therapeutic application. Given the extent of their use, the extent of adverse reactions attributable to benzodiazepines alone seems quite limited.

## 6.  BENZODIAZEPINE USE IN MEDICAL VERSUS NONMEDICAL CONTEXTS

Problematic use of benzodiazepines may arise in at least two types of context. The most common is supervised therapeutic use, where the goal is pharmacologic management in response to a *bona fide* indication. Another is frankly illicit use, in which the drug may be obtained illegally and used without medical supervision. Studies of the use of benzodiazepines in clinical practice do not typically focus on issues of chronic use or abuse. Many are concerned with prescribing patterns as they reflect such variables as age, sex, diagnostic category, physician attitudes, and education (see Rosser, 1984*a*, for a review). Other approaches may emphasize data from national rates of prescription (see Marks, 1980, for a review). However, these do not address one of the most prominent features of problematic drug use, namely chronicity in individual cases. In a survey of the published literature from 1960 to mid-1977, Marks (1980) was able to estimate a risk of psychological and physical dependence at approximately one case per 5–10 million patient-months at risk. This estimate included both therapeutic use and use in "the clear drug abuse situation." If only the therapeutic (here described as medical) context was considered, the rate was 5–10 times less. Needless to add, he concluded that the risk of dependence was very low in relation to availability of and exposure to benzodiazepines. Our review will concentrate on studies that illustrate a feature of chronic or heavy use as opposed to patterns of use in general populations.

### Medical Use

A reasonable amount of information describing the clinical use of benzodiazepines is available from studies of family practice clinics. The majority of prescriptions for benzodiazepines originate with primary care physicians (Tessler et al., 1978).

A recent study (Rosser, 1984*b*) reports on a family practice clinic in Ottawa, Ontario. A census of the practice was taken on January 1, 1982, July 1, 1982, and January 1, 1983. Prescriptions for oxazepam. flurazepam, chlordiazepoxide, diazepam, and lorazepam were considered since 95% of all benzodiazepine prescriptions were for these drugs. During the period covered, anxiety was identified as a presenting problem in 1140 of nearly 26,000 visits by approximately 1400 patients. A total of 742 prescriptions were written for 410 patients, who represented 3.2% of the population older than 15 years. The dosages of each benzodiazepine were converted to equivalent "units" (details not specified), and the annual consumption of these units computed for the 410 patients. Of the patients given prescriptions, 21 (5%) received more than 700 units in the year. Seventeen of these heavy users were females, four were males, and all were described as having "few social supports." The heavy users constituted 0.165% of the total population of the clinic. The general conclusion was that the risk of heavy use or abuse in this population was approximately 0.1%, and that this represented a low risk compared with many other abusable substances.

An interesting study of benzodiazepine "abuse" was reported by Ladewig (1983). It is especially valuable because it is based on a survey that covered 72.9% of the entire population of physicians practicing in Switzerland. The data were collected to comment upon "isolated abuse" of benzodiazepines, or abuse of benzodiazepines in the absence of abuse of any other drug. Abuse was defined as use that was not medically necessary or which was excessive in amount. Of 434 documented cases of benzodiazepine abuse yielded by the survey, 180 were identified as the isolated type. Virtually all cases had their origins in an appropriate medical indication, with anxiety and insomnia accounting for nearly 80% of the total. (Interestingly, Ladewig disqualified 142 cases reported as abuse which he deemed also to be appropriate therapy.) In the judgment of those surveyed, roughly 75% of the abuse was to achieve anxiety reduction or sedation (i.e., appropriate medical indications) but at doses considered too high or without physician consent. Roughly 10% were judged to be seeking a form of intoxication, and the rest fell into miscellaneous categories. Interestingly, the avoidance of withdrawal was judged as a reason for abuse in only 2% of the cases. In 50% of the cases for which data were available it was judged that there were no consequences of the abuse, that the consequences were negative in 38.5% and positive in 11% (negative and positive were not defined). Ladewig estimated that there would be 30 new cases of benzodiazepine abuse of this type per year, or about 0.6 cases per 100,000, in a population of 4.5 million. To quote his conclusion: "This is a morbidity in the order of that for breast carcinoma in males and scarcely represents a danger for public health." Whether such a characterization would apply universally is not known, but it appears to confirm Marks' (1980) suggestion, based on the entire literature, that abuse of benzodiazepines is rare in an epidemiological sense.

In their review of opinion on the subject of dependence liability of benzodiazepines, Petursson and Lader (1984) took note of the common view that the risk of dependence is low, but also stated that "it is difficult to reconcile these views with the more recent reports that appear to show that a substantial proportion of patients taking benzodiazepines will develop some form of dependence." In support of this statement, Petursson and Lader pointed out studies by themselves and others (Kemper et al., 1980; Tyrer et al., 1981) focusing on cases of long-term use of benzodiazepines, usually associated with a specific syndrome upon withdrawal. Of course, such findings based on patients referred specifically for long-term or problematic use of benzodiazepines may yield a high incidence of a physical dependence syndrome, but this is very different from saying that a large proportion of patients who are *exposed* to benzodiazepines will be recruited to this dependent population. This population at *risk* for physical dependence is relatively easy to identify, since relatively long-term use, even within a therapeutic dose range, seems to be required. Thus, the total number of cases of physical dependence in a medical/therapeutic context could not exceed the number of patients using the drug for an extended period. Conclusions about the rarity of dependence or abuse clearly depend on whether one begins with the total population *exposed* to benzodiazepines or that fraction of the exposed population that progresses to persistent use. Moreover, an appraisal of the risks of developing dependence in the context of medical use appears to depend upon whether the estimate is based on epidemiological (Ladewig, 1983) or clinical experience (Petursson and Lader, 1984).

An important source of information on the issue of limited use versus protracted use is data from surveys. The most recent of these was reported by Balter et al. (1984). Ten European countries and the United States were sampled in 1981, with sample sizes ranging from 1486 to 2018 respondents. The data were collected by personal interviews conducted in the households of respondents. The survey concentrated on antianxiety agents and sedatives generally rather than benzodiazepines alone, and appropriate cautions should therefore be used in drawing conclusions. The percentage of users, regardless of duration, varied widely from a range of 7.4% to 17.6% across nations. The more pertinent question is how long daily users persisted over time. This statistic also ranged widely when daily use exceeding three months was at issue. In the category of 4–11 months of daily use, the low was 1.5% of daily users (in Spain) and the high was 14.9% of daily users (in Denmark). In the category of 12 months or more, the low was 6.2% (in Sweden) and the high was 33.2% (in Belgium). Overall prevalence of use was a poor predictor of duration of use. This suggests clearly that the risk of dependence, given initial exposure, is not simply a matter of extraordinary biological vulnerability to the reinforcing effects of these drugs, but rather reflects variations in attitudes and values about their use. Some countries emerged as "conservative" in that the majority of use was for less than 3 months, and indeed conservative use predominated. However, according to arguments reviewed earlier, a substantial proportion of the using population was either dependent or at risk of becoming so in some of the countries surveyed. The most striking similarity across countries was that tranquilizers were used by a greater proportion of women than men in 1981, and that the relative rates of use had not changed from 10 years previously.

This recent survey by Balter et al. (1984) may eventually provide some clinically oriented perspectives, but to date only these provocative data on prevalence and duration are available. However, data from an older survey of a sample of American respondents (Mellinger et al., 1978) have permitted hypothesis-testing. Again the survey was not restricted to benzodiazepines, but it did focus on antianxiety drugs. In view of the year of the data collection (1971), however, it is likely that the preponderance of benzodiazepines over other antianxiety agents was less than it is currently. The main conclusion with regard to "regular" use (in this survey defined as daily or near daily use for 2 months or more in the previous year) of tranquilizers was that it was associated with high levels of severe and persistent psychic distress. Thus there may be a psychological/functional basis for persistent use in addition to any role that physical dependence may play in this phenomenon. These investigators also suggested that tranquilizer and alcohol use were alternative ways of coping pharmacologically with psychic distress, one within the medical/therapeutic sphere and the other beyond it. For example, it was hypothesized that distressed men were more likely to use alcohol as a pharmacologic support whereas women were inclined to use tranquilizers. This difference was in turn ascribed to the differential social acceptability of these coping methods according to sex.

One group of investigators (Rickels et al., 1984) has suggested that survey data probably overestimate the incidence of long-term daily use of benzodiazepines, perhaps by as much as 50%. In interviews with patients and prescribing physicians, they found that many of these patients used benzodiazepines as needed rather than daily. For example, they might remain drug-free on vacation or on weekends.

Few surveys have been as extensive as those just described. However, at least one of modest scope (Tessler et al., 1978) did permit some detailed clinical observations not reported elsewhere. This survey was based on questionnaires distributed by five primary care physicians and 55 pharmacists in a part of Massachusetts (probably in 1976 or 1977). The return rate was low, but it was argued that the demographic characteristics of those who responded were representative of a national random sample of psychoactive drug users. The most remarkable feature of this survey was that more than 65% of respondents ($n = 236$) reported using diazepam for 1 year or more, with nearly 25% reporting use for 3 years or more. Even though it was not stated whether use was daily, these reports are considerably discrepant from those in the survey of the US sample by Balter et al. (1984). Only six persons who responded reported taking 40 mg or more of diazepam per day. Nearly all reported that their usage had remained stable or decreased since the initial prescription; therefore if any of these patients were dependent, their dependence did not lead to an escalation in dosage.

A slightly different view of use in a medical context comes from a study by Maletzky and Klotter (1976). They collected data on referrals to the neuro-psychiatry clinic on a military installation in the latter half of 1972. The subjects were the first 50 patients who used diazepam and were referred to the study by local physicians. There was the usual preponderance of females (31 : 19) and a variety of indications for prescribing diazepam. The mean duration of use in this group was approximately 26 months. Maletzky and Klotter's objective was to refute an existing literature that claimed chlordiazepoxide and diazepam not to be addictive. Within this population of medical users, they did find signs of addiction in the sense of persistent use, withdrawal symptoms during spontaneous attempts to reduce or discontinue drug use, and inability to stop using diazepam. Some patients reduced dosage on their own, but some increased it. However, this was done within the context of a mean daily dose of 16 mg, with 36 of 50 subjects using not more than 20 mg per day. (It was stated that half the patients used other minor tranquilizers, mainly chlordiazepoxide and meprobamate, but no quantitative information about other drug use was supplied.) One of the more interesting results was obtained when a panel of physicians rated the degree of addiction of each patient before and after being informed that the substance in question was diazepam. The mere information that the drug was diazepam significantly lowered their ratings of the severity of the addiction. It would be interesting to know whether a similar study would have the same result today.

## Nonmedical and Multiple Drug Use

A recent report of the World Health Organization Review Group (1983) asserted that benzodiazepines "are commonly used by multidrug users in many regions of the world, but are seldom used as the primary drug of abuse." The two major bases for nonmedical use were said to be self-medication of withdrawal symptoms produced by another drug (such as heroin) when it was not available and pursuit of the experience of euphoria.

Recently (Busto et al., 1983) there has been a study that addressed the anecdotal reports that benzodiazepine abuse is a particular risk in alcoholics. This prospective study covered 261 consecutive cases of patients referred to an outpatient clinic with a diagnosis of alcoholism. Self-report of the use of various drugs was obtained, and urine was screened for benzodiazepines. Roughly one-third of these patients had a positive urine screen. Although 77% of the patients were males, the rate of positives among females was almost twice that for males. Thus, the relative rate of use was similar to that for the general population of benzodiazepine users. The overwhelming majority (48 of 61) patients obtained their benzodiazepine from a medical source. Busto et al. developed an operational definition of benzodiazepine abuse. Abuse was considered to exist if the urine screen was positive and the benzodiazepine had come from a nonmedical source, or if the patient reported a lifetime cumulative intake of 2700 mg of diazepam (or its equivalent). By these criteria, half the patients using benzodiazepines and 17% of the total sample were abusers. From these data it is impossible to determine whether alcoholics use benzodiazepines any differently than nonalcoholics, and whether indeed most of the use detected in this study should be described as "nonmedical." In fact, the greatest current need in this area is for more detailed data on patterns and consequences of combined use of alcohol and benzodiazepines. For example, Schuster and Humphries (1981) reported three case descriptions of what they called benzodiazepine dependency in alcoholics. Although these patients clearly had numerous problems requiring clinical management, the relative importance of the benzodiazepine component in the total spectrum of the problems of substance abuse was difficult to assess.

Kryspin-Exner and Demel (1975) reported data on the use of diazepam and chlordiazepoxide by alcoholics in treatment in Vienna. Over a period from 1961 to 1966, 323 patients were prescribed these compounds as part of their treatment. Each patient was followed for approximately 1 year, during which time 11 of the 323 were described as "dependent." No operational definition of dependence was given, with the exception that there was a "tendency toward increased doses and corresponding psycho-organic deficit symptoms." Of these 11, 7 were reported to have a history of abusing drugs other than alcohol. In a study of inpatient alcoholics in treatment, patients were allowed to self-administer diazepam ($n = 24$, up to 50 mg per day) or placebo ($n = 23$, 10 tablets per day) over a 32-day trial. There was no difference between the two groups in number of tablets self-administered, which led Kryspin-Exner and Demel to conclude that diazepam had a low abuse potential. Of the 240 doses available, 16 of 24 subjects who had access to active diazepam consumed less than 100. The authors hypothesized that the relatively weak attraction of diazepam was due to the fact that these were "pure alcoholics" with little or no prior experience with tranquilizers.

The use of benzodiazepines by opiate addicts in methadone maintenance programs appears to be common. In one study of patients in two urban clinics (Stitzer et al., 1981), between 65 and 70% of patients had urine positive for benzodiazepines during a 1-month period. Diazepam was favored by far and instances of daily dosage in excess of 100 mg were frequent. Daily dosages as high as 300 mg were reported. The primary reason for use was to "boost" the effect of methadone. Stitzer et al. concluded that this benzodiazepine usage was abusive rather than therapeutic. Other investigators (Budd et al., 1979; Kleber and Gold, 1978) have reported essentially confirmatory results.

A recent study of patients at the Addiction Research Foundation (Busto et al., 1986) suggests that there are essentially two distinct populations of benzo-diazepine "abusers." In a sample of 163, 92 patients reported using only ben-zodiazepines and 71 used them among various other substances. In general, the multiple-drug abusers were younger and used much higher doses than those who used benzodiazepines exclusively. In the latter group the daily median dose (diazepam-equivalent units) was 15 mg, whereas the daily median was 50 mg for multiple-drug abusers. Moreover, despite a shorter duration of lifetime use, the multiple abusers had a cumulative dosage exposure more than twice that of those who used only benzodiazepines.

## Overview

With virtually every abusable drug there are different patterns and conse-quences of use that can reasonably be described as abusive. When benzodiazepine use is persistent for a medical indication, the mean daily dose is likely to remain low (approximately 15 mg per day of diazepam or equivalent). Although there is a distinct risk of physical dependence, our own clinical experience with this population is that acute intoxication almost never occurs. The reinforcement sought by this group of users would normally be considered therapeutically legitimate, notwithstanding that some nonpharmacological means of reinforce-ment (e.g., to achieve anxiety reduction) would be preferred. This pattern of dependence often occurs (cf. Ladewig, 1983) in the absence of problems with other drugs.

A form of abuse totally apart from persistent use for medical indications is likely to involve multiple drug use and higher doses of benzodiazepine. The achievement of intoxication is also more likely to be a function of benzodiazepine use in this population.

## 7.   THE TREATMENT OF BENZODIAZEPINE ABUSE AND DEPENDENCE

The treatment of benzodiazepine abuse and dependence involves two basic issues. First, there is the question of what is the most effective strategy for achieving reduction in dosage or abstinence in cases where outpatient manage-ment is acceptable because the maintenance dose is low. Secondly, there is the question of medical management of withdrawal (detoxification), which is typi-cally relevant when the dose level has been relatively high over a protracted period.

## Detoxification from a Low Dose

Studies on the management of withdrawal from long-duration, low-dose benzodiazepine therapy have only recently begun to appear. Most involve descriptions of the occurrence and management of the symptomatic consequences of becoming abstinent from benzodiazepine use. Occasionally there has been information about the proportion of patients who complete a withdrawal program, including status at follow-up. However, there is as yet relatively little by way of information comparing various treatment modalities as there is, for example, in the case of alcoholism treatment (cf. Tyrer et al., 1985).

Typically, the initial goal of treatment is to achieve complete abstinence. Because the vast majority of cases occur in the context of therapeutic medical use, it is often physicians who are confronted with the following question, as expressed in a commentary by Tyrer (1980): "If patients are dependent but otherwise well is it right to deprive them of their drugs?" His own answer was that it should then be up to the patient to decide whether to attempt to stop taking medication if dependence is suspected. When cessation is indicated, a gradual withdrawal, perhaps over months, was suggested. The goal of total abstinence is approached by steps, with a view to avoiding an unnecessarily harsh withdrawal reaction. Once withdrawal is completed, it is important according to Tyrer to be alert to the possibility that the patients' anxiety symptoms may return: "Once this is demonstrated it is reasonable to prescribe the benzodiazepine, and in doing so I reassure myself that the alternatives of handicapping anxiety or recourse to alcohol or similar drugs are a great deal worse." Another attempt, however, may be indicated at a later date.

Tyrer and his associates have reported attempts at outpatient withdrawal in considerable detail. In one study (Tyrer et al., 1981), an attempt was made to withdraw patients from either diazepam or lorazepam. It is interesting first to note that of 86 patients who were eligible in principle, only 40 agreed to participate. The mean duration of use was 3.6 years and the mean daily doses for diazepam and lorazepam were 10 and 4 mg, respectively. The benzodiazepine was stopped abruptly and replaced with either propranolol or placebo. During an observation period of 14 days, 18 of the 40 patients dropped out of the study and resumed use of their benzodiazepine. Drop-out was significantly more likely to occur among patients using lorazepam as compared with diazepam. Although propranolol appeared to suppress withdrawal symptoms somewhat, it had no effect on drop-out rate. Of the 22 patients who completed the withdrawal successfully, 17 were reported to be taking neither benzodiazepines nor other drugs at a 3-month follow-up. By comparison, only 6 of 46 patients who had refused to participate, and 2 of the 18 who resumed benzodiazepine use within the study, were reported to be taking no benzodiazepine at 3-month follow-up. Although

Tyrer et al. placed a great deal of emphasis on the management of withdrawal symptoms, they did not report whether successful and unsuccessful quitters differed in symptom quality, duration, or intensity.

In a later study (Tyrer et al., 1983), patients who had been taking diazepam (5–20 mg daily) for 6 months or more were studied. The major independent variable was whether dosage reduction was attempted relatively early (during weeks 2–6) or later (during weeks 8–12) in a 14-week course of treatment. In either case, dosage reduction was gradual over the 4-week period of withdrawal. It was reduced by 50% in the first 2-week interval, by 75% in the second 2-week interval, and to zero after 4 weeks had elapsed. Although there was much of interest in this study, one of the main points was a comparison of the completion rates for withdrawal with the gradual schedule in this study and the "sudden" method used in the earlier one (Tyrer et al., 1981). This comparison was made in a review by Tyrer (1982) in which it was noted that 18 of 40 patients dropped out when the sudden method was used, whereas 5 of 36 did when withdrawal was gradual over a 4-week period. What is perplexing about this is the role of withdrawal reactions in drop-out. Using the same criteria, it was estimated that 45% of patients experienced withdrawal reactions when the withdrawal was sudden and 44% had withdrawal reactions when it was gradual. Yet despite this similarity, the drop-out rate was significantly less when withdrawal was gradual. This may suggest that it is not the *mere occurrence* of withdrawal reactions, but *how they are managed* that is important. No information was given on this, but it seems clear that patients receiving gradual withdrawal received at least more contact overall with the investigators. Whether the quality of this contact differed cannot be determined from the report.

A direct comparison of different withdrawal regimens was made by Laughren et al. (1982b). The subjects were 37 male patients diagnosed by *DSM-III* criteria as having generalized anxiety disorder. All were taking diazepam. The mean daily dose was 17 mg and the mean duration of use was 5 years. Subjects were assigned to groups that continued on diazepam maintenance (M), were withdrawn gradually by 5 mg of the daily dose each week (GW), or abruptly withdrawn (AW). Thirteen patients were excluded for various reasons, leaving eight in each condition. The only significant difference to emerge was a higher self-report of anxiety in the AW group compared with M and GW, which did not differ from each other. A specific assessment of withdrawal revealed no difference among groups, and even abrupt withdrawal resulted in no notable withdrawal reaction. In discussing their results, Laughren et al. acknowledged reports of more prominent withdrawal reactions from other studies, but noted also that "given the perhaps hundreds of thousands of anxious patients who periodically stop their benzodiazepines abruptly for a variety of reasons without experiencing any major symptoms, the risk of significant physical dependence at therapeutic doses of these drugs must be minimal." This view of course differs from reports mentioned earlier, and radically from one to be discussed below.

Cappell et al. (1985) compared two groups of subjects attempting to achieve abstinence. Both groups were placed on a withdrawal program in which their daily intake of tablets was tapered very gradually over 4–6 weeks from an initial level of approximately 15 mg per day. However, one group was tapered on active diazepam and the other on placebo in a double-blind design. There were two important differences in behavior that emerged during the course of treatment. First, 84% (16 of 19) of subjects in the placebo condition supplemented their benzodiazepine intake from their own supplies at least once, whereas only 33% (7 of 21) of subjects did so when active drug was given ($p < .01$). In addition, six subjects in the placebo condition spontaneously requested a switch from the medication provided in the study back to their own supplies, whereas only one receiving active drug did so. These data show that there is a genuine drug-deprivation state that requires some form of management, which patients will take initiatives to address on their own. At the same time, it is evident that use of a gradual withdrawal regimen is effective in reducing these management problems compared with abrupt withdrawal. These results are in contrast to those reported by Laughren et al. (1982b). However, given the proclivity of patients to supplement their drug intake when withdrawal is too abrupt, it is possible that the manipulation of abrupt versus gradual withdrawal by Laughren et al. was compromised.

Hopkins et al. (1982) described a benzodiazepine withdrawal program initiated in a general practice setting. All patients who had been taking a benzodiazepine for at least 3 months were invited to attempt withdrawal. Those with a history of psychosis, concurrent psychiatric treatment, or acute physical illness were excluded. A total of 78 patients entered the study, during which they were withdrawn gradually over a period of up to 6 weeks. Forty-six patients (59%) were able to achieve abstinence and 26 (33%) achieved a significant reduction in dosage within a median duration of approximately 3 weeks. Those who were able to quit could be identified early in treatment; those who elected to reduce dosage most gradually were least likely to achieve abstinence. Upon follow-up at 3 to 5 months after the withdrawal phase, 49 (63%) of patients reported that they remained abstinent and a further 15 (19%) were taking a reduced dosage.

Petursson and Lader (1984) reported outcomes of 33 patients withdrawn from long-term, low-dose treatment with benzodiazepines. Withdrawal reactions were common, but according to Petursson and Lader, in most cases not more severe than withdrawal from treatment at higher doses. Although detailed data were not given, there were three categories of outcome at 1-year follow-up. Approximately one-third of patients were "doing very well" and presumably remained abstinent. Another one-third still had "occasional episodes of anxiety and tension," but were "pleased to discover they can manage without chronic sedative medication." The remaining one-third were having more severe problems with anxiety and found it "very difficult to stay off their drugs."

Rickels et al. (1984) reported outcomes of 31 patients withdrawn from long-term benzodiazepine therapy. The mean duration of use was approximately 8 years, and the mean daily dose was from 15–25 mg of diazepam or its equivalent. In 28 of the 31 patients, withdrawal was abrupt. The supportive treatment was not well described, but the observation was made that most patients needed further psychiatric management and treatment. Of these, only 20% were being managed without medication of some kind. Thirty percent were switched to antidepressant medication and another 30% reverted to prescribed benzodiazepines. Since this was a medically based program, pharmacotherapy was emphasized; whether satisfactory results could have been achieved in more patients without drug therapy is an open question.

All of the previous studies give little information on the details of the nonpharmacological supportive care given to patients during withdrawal, although it is likely that some form of reassurance was the most common. Recently, results of attempts at specific behavioral treatments have been reported, with an interesting commonality. In one study (Cormack and Sinnott, 1983), 50 patients in general practice were contacted with invitations to attempt to reduce their use of benzodiazepines. The letter of invitation was sent by a physician, but the treatment offered was a behaviorally oriented program offered by a psychologist, with abstinence as the goal. Only 13 patients ever entered treatment. The results of these patients are less important than the fact that so many of the patients did not wish to get involved. A recent study (Nathan et al., 1986) reported a similar phenomenon. In 1981, letters were sent to 115 general practice physicians. Within 3 months no referrals had been generated. A newspaper article on the project generated 30 inquiries and 20 appropriate subjects, of whom 12 entered the study. Four were excluded and referred elsewhere for treatment and another dropped out, leaving seven in the study. Because of the small sample, detailed discussion is gratuitous. The point is that self-identification as being in need of treatment appears rare.

There is one study (Ashton, 1984) that emerges as totally distinct from all others that have appeared in the recent literature. In this highly detailed description, 12 patients were referred by physicians concerned with dependence on therapeutic doses. Most had attempted to reduce dosage or abstain on their own, but without success. Nine were treated in hospital and three as outpatients. All were switched to diazepam if not already using it (many were using lorazepam to help sleep) and then withdrawn gradually over a 2-week period during which they were seen frequently by Ashton. Many patients were treated with other drugs (nonbenzodiazepine hypnotics, including butabarbitone, major tranquilizers, propranolol, tricyclic antidepressants) during withdrawal and therefore the description of the clinical course of withdrawal was compromised. A surprisingly high incidence of severe psychological symptoms was described in these patients. Although it was asserted that only two patients had a history of psychiatric symptoms *prior* to using benzodiazepines, *11 of 12* developed agoraphobia *while taking benzodiazepines*, but this was generally alleviated after withdrawal. Seven of 12 experienced depression during benzodiazepine use. Nearly all complained of depersonalization and a "fear of going mad" which was improved, but not eliminated, by withdrawal. In patients who were dependent on lorazepam, there was "craving" for the drug while they were taking it, but this abated after switching to diazepam or after withdrawal. A variety of severe somatic symptoms was described well. The incidence of severe symptoms appears relatively high in this study compared with others (e.g., Tyrer et al., 1981, 1983). However, what is most interesting is the suggestion that *significant psychological problems were acquired while taking benzodiazepines that did not exist beforehand.* Ashton concluded because of the relative uniformity of the symptoms that they "resulted from benzodiazepine use and not from an underlying anxiety neurosis." Of course, it is possible that these patients first began on benzodiazepines when symptoms were mild and deteriorated thereafter. However, if this is so it is difficult to understand why withdrawal of benzodiazepines led to symptomatic improvement. Ashton attributed the development of symptoms while still taking benzodiazepines to tolerance, but this is no real explanation for the emergence of novel complications. Ashton's general characterization is clearly at odds with other experts in this field, but if valid, these findings have clear relevance for case management.

## Detoxification from a High Dose

Very little information on detoxification of patients using high doses of benzodiazepines is available. However, one systematic study was published recently (Harrison et al., 1984). Twenty-three subjects were admitted for in-patient detoxification. Benzodiazepine use was confirmed by blood or urine assays in all but one patient, and multiple drug abuse was diagnosed in eight patients. The median daily dosage was 150 mg (range 40–500 mg) of diazepam or its equivalent. In all cases, however, the assessment of the attending physician was that a benzodiazepine was the primary drug of abuse. Sixteen subjects completed the procedure in hospital. Detoxification was begun with a diazepam loading equivalent to 40% of the reported daily dose at admission. This was reduced by an average of 12% daily until a level of zero was attained. There were no complications except in one case in which the dosage given in hospital was much lower than it should have been according to the protocol. This occurred because the patient's claim of using 450 mg of diazepam per day was doubted, and the drop in plasma level was more precipitous than expected. The authors concluded that their tapering procedure was "simple and effective."

It appears that the medical consequences of dependence on high doses can be managed with relative ease and little risk. The contrast between this study, in which high doses were documented, and Ashton's (1984), in which dependence was on therapeutic levels, is striking. None of the extreme responses to withdrawal reported by Ashton were noted in the study by Harrison et al., nor was there any suggestion of the induction of psychological problems as a result of taking benzodiazepines.

## Overview

At present there is an extremely limited body of evidence on the treatment of benzodiazepine dependence and little basis for generalizations about the relative effectiveness of different approaches. In cases of dependence on either very high or therapeutic daily doses, it appears that gradual tapering is safe and effective for the patient in the short term. There are no data about the characteristics of patients in relation to long-term outcome, about the relationship between duration of use and treatment outcome, and so forth. The sample sizes needed to address such questions are unlikely to be readily achieved, especially given the apparent reluctance of therapeutic users to enter formal treatment. On the other hand, there are indications that treatments developed for other substances can be adapted to this population. As for inpatient detoxification in the case of high dosages, a simple and effective procedure is available.

## 8.   GENERAL CONCLUSIONS

As drugs of abuse, benzodiazepines have received less scientific attention than many other commonly abused substances. Unlike these other compounds, benzodiazepines have a clinical legitimacy that is endorsed by many practitioners, albeit questioned by many as well. Much of the controversy surrounding benzodiazepines has abated, and prescribing practices appear to have become more conservative. Indeed, there has been some concern expressed (Shader and Greenblatt, 1984) that these practices have become *over*conservative, with the result that patients who would benefit from pharmacotherapy do not receive it because of inappropriate resistance from them or from their physicians. In the end, all drug use occurs within the context of a value system that is more than a reflection of objective evidence. The huge variations in prescribing practices across nations (Balter et al., 1984) amply support this view. However, there are some generalizations that are supportable on scientific grounds:

1. Benzodiazepines possess the cardinal characteristics of abusable compounds. Although there appear to be variations in potency among compounds and important variations in sensitivity among individuals, benzodiazepines are reinforcing according to accepted criteria of evaluation. With the proliferation of compounds with varying kinetic and (possibly) dynamic profiles, important differences in reinforcing efficacy can be created. The study of such variations is still in its infancy, but it promises to be an important source of predictive data on the relative abuse liability of different benzodiazepine compounds.

2. Tolerance is also associated with abuse liability. Tolerance to the depressant effects of benzodiazepines is acquired rapidly, but tolerance to the anxiolytic effects develops very slowly and to an apparently limited extent. Although some users achieve remarkably high daily doses, many long-term users persist at daily doses that are relatively low (15–20 mg diazepam or its equivalent daily) for years without escalation.

3. Physical dependence is clearly a risk associated with protracted use even at therapeutic doses. The withdrawal syndrome is comparatively mild at low doses and only likely to require medical management at very high doses. Relatively simple management techniques can be applied in either case so that abstinence can be achieved quickly with little discomfort. The role of physical dependence in maintaining use is not clear; however, it is clear that the continued self-administration of even low therapeutic doses is based on redressing a drug-deprivation state.

4. Like any CNS depressant, benzodiazepines produce acute impairments of various types. Some may be quite subtle. There is little evidence on the practical significance of such impairments in relation to such phenomena as traffic accidents. Adverse reactions are not common. Therapeutic users appear rarely if ever to seek intoxication deliberately. Whether there are any social, behavioral, or organic deficits associated with benzodiazepines and attributable to a direct toxic effect of exposure is simply unknown. If there are such effects, they are not gross.

5. There is more than one pattern of benzodiazepine abuse. Many therapeutic users persist under medical supervision for long periods at low doses. Although dosage may not escalate, both physical and psychological dependence can occur. Aside from the dependence as such, it is often difficult to identify other complications of this behavior. Another distinct group uses much higher doses, seeks intoxication, and often uses a benzodiazepine in conjunction with other drugs.

6. Except in relation to detoxification, treatment for those who are dependent and wish to become abstinent is a relatively unexplored area. It seems probable that behavioral management techniques developed for other substances can be applied to benzodiazepines, but no appropriate outcome study exists.

The foregoing conclusions are based principally on experience with diazepam. How completely they will apply to newer formulations remains to be seen. Indeed, it will be interesting to see whether benzodiazepines will remain in such common use. The increased understanding of the biological basis of anxiety that is now so exciting to the research community may well result in the development of new anxiolytic compounds that are significantly different from benzodiazepines in their pharmacological action. Notwithstanding that the hazards of benzodiazepines have on occasion been exaggerated, it is to be hoped that agents with greater anxiolytic specificity and less or no depressant action will give rise to even fewer problems than their progenitors.

## REFERENCES


Agrawal, P., 1978, Diazepam addiction: A case Report, *Can. Psychiatr. Assoc. J.* **23**:35.

Almeyda, J., 1971, Subcutaneous reactions to imipramine and chlordiazepoxide, *Br. J. Dermatol.* **84**:298.

American Psychiatric Association, 1980, *Diagnostic and Statistical Manual of Mental Disorders,* Third Edition, APA, Washington, D.C.

Amit, Z., Corcoran, M. E., Charness, M. E., and Shizgal, P., 1973, Intake of diazepam and hashish by alcohol preferring rats deprived of alcohol, *Physiol. Behav.* **10**:523.

Aranko, K., Mattila, M. J., and Seppala, T., 1983, Development of tolerance and cross-tolerance to the psychomotor actions of lorazepam and diazepam in man, *Br. J. Clin. Pharmacol.* **15**:545.

Arendt, R. M., Greenblatt, D. J., De Jong, R. H., Bonin, J. D., Abernethy, D. R., Ehrenberg, B. L., Giles, H. G., Sellers, E. M., and Shader, R. I., 1983, In vitro correlates of benzodiazepine cerebrospinal fluid uptake, pharmacodynamic action and peripheral distribution, *J. Pharmacol. Exp. Ther.* **227**:98.

Ashton, H., 1984, Benzodiazepine withdrawal: An unfinished story, *Br. Med. J.* **288**:1135.

Balmer, R., Battegay, R., and von Marschall, R., 1981, Long-term treatment with diazepam— Investigation of consumption habits and the interaction between psychotherapy and psychopharmacotherapy: A prospective study, *Int. Pharmacopsychiatry* **16**:221.

Balter, M. B., Manheimer, D. I., Mellinger, G. D., and Uhlenhuth, E. H., 1984, A cross-national comparison of anti-anxiety/sedative drug use, *Curr. Med. Res. Opin.* **8**:5.

Barten, H. H., 1965, Toxic psychosis with transient dysmnestic syndrome following from valium, *Am. J. Psychiatry* **121**:1210.

Barton, D. F., 1981, More on lorazepam withdrawal, *Drug Intell. Clin. Pharm.* **15**:134.

Benjamin, F. B., 1977, A review of safety hazards due to poor health, drugs and their interactions, *Hum. Factors* **2**:127.

Bergman, H., Borg, S., Engelbrekton, K., and Vikander, B., Neuropsychological deficits, field dependence and morphological cerebral changes in a follow-up study of patients with sedative hypnotic abuse, *Acta Psychiatr. Scand., in press.*

Bergman, J., and Johanson, C. E., 1985, The reinforcing properties of diazepam under several conditions in the rhesus monkey, *Psychopharmacology* **86**:108.

Berlin, R. M., and Connell, J., 1983, Withdrawal symptoms after long-term treatment with therapeutic doses of flurazepam: A case report, *Am. J. Psychiatry* **140**:488.

Bigelow, G., Griffiths, R. R., and Liebson, I., 1976, Experimental human drug self-administration: Methodology and application to the study of sedative abuse, *Pharmacol. Rev.* **27**:523.

Bismuth, C., Le Bellec, M., Dalby, S., and Lagier, G., 1980, Dependence physique aux benzodiazepines, *Nouvelle Press Med.* **9**:1941.

Blackwell, B., and Cooperstock, R., 1983, Benzodiazepine use and the biopsychosocial model, *J. Fam. Prac.* **17**:451.

Bliding, A., 1978, The abuse potential of benzodiazepines with special reference to oxazepam, *Acta Physiol. Scand.* **274**:111.

Bø, O., Haffner, J. F. W., Langard, O., Trumpy, J. H., Bredesen, J. E., and Lunde, P. K. M., 1976, Ethanol and diazepam as causative agents in road traffice accidents, in: *Alcohol, Drugs and Traffic Safety* (S. Israelstam and S. Lambert, eds.), pp. 439–448, Addiction Research Foundation, Toronto.

Bodnar, R. J., Kelly, D. D., Thomas, L. W., Mansour, A., Brutus, M., and Glusman, M., 1980, Chlordiazepoxide antinociception: Cross-tolerance with opiates and with stress, *Psychopharmacology* **69**:107.

Boisse, C., and Okamoto, M., 1978a, Physical dependence to barbital compared to pentobarbital. IV. Influence of elimination kinetics, *J. Pharmcol. Exp. Ther.* **204**:526.

Boisse, N. R., and Okamoto, M., 1978b, Physical dependence to barbital compared to pentobarbital. I. "Chronically equivalent" dosing method, *J. Pharmacol. Exp. Ther.* **204**:497.

Boisse, N. R., Ryan, G. P., Guarino, J. J., and Gay, M. H., 1981, Comparison of benzodiazepine and barbiturate tolerance and physical dependence in the rat, *Pharmacologist* **23**:192.

Bond, A., James, D. C., and Lader, M., 1974, Sedative effects on physiological and psychological measures in anxious patients, *Psychol. Med.* **4**:374.

Bond, A., Lader, M., and Shrotriya, R., 1983, Comparative effects of a repeated dose regime of diazepam and buspirone on subjective ratings, psychological tests and the EEG, *Eur. J. Clin. Pharmacol.* **24**:463.

Bonetti, E. P., Pieri, L., Cumin, R., Schaffner, R., Peri, M., Gamzu, E. R., Muller, R. K., and Haefely, W., 1982, Benzodiazepine antagonist RO 15-1788. Neurological and behavioral effects, *Psychopharmacology* **78**:8.

Bowden, C. L., and Fisher, J. G., 1980, Safety and efficacy of long-term diazepam therapy, *Southern Med. J.* **73**:1581.

Braestrup, C., and Squires, R. F., 1977, Specific benzodiazepine receptors in rat brain characterized by high-affinity 3H-diazepam binding, *Proc. Nat. Acad. Sci.* **74**:3805.

Braestrup, C., Albrechtsen, R., and Squires, R. F., 1977, High densities of benzodiazepine receptors in human cortical areas, *Nature* **269**:702.

Brandt, A. L., and Oakes, F. D., 1965, Preanesthesia medication: Double-blind study of a new drug, diazepam, *Anesth. Analg. (Cleveland)* **44**:125.

Breier, A., Charney, D. S., and Nelson, J. C., 1984, Seizures induced by abrupt discontinuation of alprazolam, *Am. J. Psychiatry* **141**:1606.

Brodie, B. B., Kurz, H., and Schanker, L. S., 1960, The importance of dissociation constant and lipid-solubility in influencing the passage of drugs in the cerebrospinal fluid, *J. Pharmacol. Exp. Ther.* **130**:20.

Budd, R. D., Walkin, E., Jain, N. C., and Sneath, T. C., 1979, Frequency of use of diazepam in individuals on probation and in methadone maintenance programs, *Am. J. Drug Alcohol Abuse* **6**:511.

Buskopp, J. J., Price, I., and Molnar, I., 1967, Untoward effect of diazepam, *N. Engl. J. Med.* **277**:316.

Busto, U., Kaplan, H. L., and Sellers, E. M., 1980, Benzodiazepine-associated emergencies in Toronto, *Am. J. Psychiatry* **137**:224.

Busto, U., Kaplan, K., and Sellers, E. M., 1981, Age- and sex-related differences in patterns of drug overdose and abuse, *Soc. Sci. Med.* **15**:275.

Busto, U., Simpkins, J., Sellers, E. M., Sisson, B., and Segal, R., 1983, Objective determination of benzodiazepine use and abuse in alcoholics, *Br. J. Addict.* **78**:429.

Busto, U., Sellers, E. M., Naranjo, C. A., Cappell, H., Sanchez-Craig, M., and Kaplan, K., 1985, Withdrawal symptoms after long-term therapeutic use of benzodiazepines: A randomized, double-blind placebo controlled trial, *Clin. Pharmacol. Ther.* **37**:185.

Busto, U., Sellers, E. M., Naranjo, C. A., Cappell, H. D., Sanchez-Craig, M., and Simpkins, J., 1986, Patterns of benzodiazepine abuse and dependence, *Br. J. Addict.* **81**:87–94.

Cappell, H. D., and LeBlanc, A. E., 1981, Tolerance and physical dependence: Do they play a role in alcohol and drug self-administration? in: *Research Advances in Alcohol and Drug Problems,* Vol. 6 (Y. Israel, F. B. Glaser, H. Kalant, R. Popham, W. Schmidt, and R. G. Smart, eds.), pp. 159–196, Plenum Press, New York.

Cappell, H. D., and LeBlanc, A. E., 1983, The relationship of tolerance and physical dependence to alcohol abuse and alcohol problems, in: *The Biology of Alcoholism,* Vol. 7 (B. Kissin and H. Begleiter, eds.), pp. 359–414, Plenum Press, New York.

Cappell, H. D., Busto, U., Kay, G., Naranjo, C., Sellers, E. M., and Sanchez-Craig, M., 1985, A comparative study of abrupt and gradual cessation of long-term therapeutic self-administration of benzodiazepines. Poster presented at the Committee on Problems of Drug Dependence in Baltimore, Maryland, June, 1985.

Catalan, J., Gath, D., Edmonds, G., and Ennis, J., 1984, The effects of non-prescribing of anxiolytics in general practice I. Controlled evaluation of psychiatric and social outcome, *Br. J. of Psychiatry* **144**:593.

Clare, A. W., 1983, Benzodiazepines, alcohol or nicotine? in: *Benzodiazepines Divided* (M. R. Trimble, ed.), pp. 1–16, Wiley, New York.

Cohen, S., 1983, Current attitudes about the benzodiazapines: Trial by media, *J. Psychoact. Drugs* **15**:109.

Cook, L., and Sepinwall, J., 1975, Behavioral analysis of the effects and mechanisms of action of benzodiazepines, in: *Mechanism of Action of Benzodiazepines* (E. Costa and P. Greengard, eds.), pp. 1–28, Raven Press, New York.

Cook, P. J., Flanagan, R., and James, I. M., 1984, Diazepam tolerance: Effect of age, regular sedation, and alcohol, *Br. Med. J.* **289**:351.

Cooper, S. J., and Francis, R. L., 1979, Water intake and time course of drinking after single or repeated chlordiazepoxide injections, *Psychopharmacology* **65**:191.

Cooper, S. J., Burnett, G., and Brown, K., 1981, Food preference following acute or chronic chlordiazepoxide administration: Tolerance to an antineophobic action, *Psychopharmacology* **73**:70.

Cooperstock, R., and Hill, J., 1982, *The Effects of Tranquillization: Benzodiazepine Use in Canada,* Health and Welfare Canada, Ottawa.

Cooperstock, R., and Lennard, H. L., 1979, Some social meanings of tranquilizer use, *Soc. Health Ill.* **1**:331.

Cormack, M. A., and Sinnott, A., 1983, Psychological alternatives to long-term benzodiazepine use, *J. R. Coll. Gen. Pract.* **33**:279.

Covi, L., Lipman, R. S., Pattison, J. A., Derogatis, L. R., and Uhlenhuth, E. H., 1973, Length of treatment with anxiolytic sedatives and response to their sudden withdrawal, *Acta Psychiatr. Scand.* **49**:51.

Crawley, J. N., Skolnick, P., and Paul, S. M., 1984, Absence of intrinsic antagonist actions of benzodiazepine antagonists on an exploratory model of anxiety in the mouse, *Neuropharmacology* **23**:531.

Cumin, R., Bonetti, E. P., Scherschlicht, R., and Haefely, W. E., 1982, Use of the specific benzodiazepine antagonist, Ro 15-1788, in studies of physiological dependence on benzodiazepines, *Experientia* **38**:833.

Cunningham, M. L., 1965, Acute hepatic necrosis following treatment with amitriptyline and diazepam, *Br. J. Psychiatry* **111**:1107.

DeBard, M. L., 1979, Diazepam withdrawal syndrome: A case of psychosis, seizure and coma, *Am. J. Psychiatry* **136**:105.

Deneau, G. A., and Weiss, S., 1963, A substitution technique for determining barbiturate-like physical dependence capacity in the dog, *Pharmacologist* **5**:239.

Deneau, G. A., and Weiss, S., 1968, A substitution technique for determining barbiturate-like physiological dependence capacity in the dog, *Pharmacopsychiat. Neuro-Psychopharmacol.* **1**:270.

De Vry, J., and Slangen, J. L., 1985, Stimulus control induced by benzodiazepine antagonist Ro 15-1788 in the rat, *Psychopharmacology* **85**:483.

deWit, H., Johanson, C. E., Uhlenhuth, E. H., and McCracken, S., 1982, The effects of two non-pharmacological variables on drug preference in humans, in: *National Institute on Drug Abuse Research Monograph No. 43* (L. S. Harris, ed.), pp. 251–257, U.S. Government Printing Office, Washington D.C.

de Wit, H., Johanson, C. E., and Uhlenhuth, E. H., 1984a, Reinforcing properties of lorazepam in normal volunteers, *Drug Alcohol Depend.* **13**:31.

de Wit, H., Johanson, C. E., and Uhlenhuth, E. H., 1984b, The dependence potential of benzodiazepines, *Curr. Med. Res. Opin.* **8**:48.

de Wit, H., Uhlenhuth, E. H., and Johanson, C. E., 1984c, Lack of preference for flurazepam in normal volunteers, *Pharmacol. Biochem. Behav.* **21**:865.

de la Fuente, J. R., Rosenbaum, A. H., Martin, H. R., and Niven, R. G., 1980, Lorazepam related withdrawal seizures, *Mayo Clin Proc.* **55**:190.

Divoll, M., Greenblatt, D. J., Lacasse, Y., and Shader, R. J., 1981, Benzodiazepine overdosage: Plasma concentrations and clinical outcome, *Psychopharmacology* **73**:381.

Dysken, M. W., and Chan, C. H., 1977, Diazepam withdrawal psychosis: A case report, *Am. J. Psychiatry* **134**:573.

Einarson, T. R., 1980, Lorazepam withdrawal seizures, *Lancet* **1**:151.

Ellinwood, E. H., Jr., Linnoila, M., Easler, M. E., and Molter, D. W., 1981, Onset of peak impairment after diazepam and after alcohol, *Clin. Pharmacol. Ther.* **30**:534.

Ellinwood, E. H., Jr., Linnoila, M., Easler, M. E., and Molter, D. W., 1983, Profile of acute tolerance to three sedative anxiolytics, *Psychopharmacology* **79**:137.

Evans, J. G., and Jarvis, E. H., 1972, Nitrazepam and the elderly, *Br. Med. J.* **4**:487.

Ferrero, P., Guidotti, A., Conti-Tronconi, B., and Costa, E., 1984, A brain octadecaneuropeptide generated by tryptic digestion of DBI (diazepam binding inhibitor) functions as a proconflict ligand of benzodiazepine recognition sites, *Neuropharmacology* **23**:1359.

File, S. E., 1982, Recovery from lorazepam tolerance and the effects of a benzodiazepine antagonist (RO 15-1788) on the development of tolerance, *Psychopharmacology* **77**:284.

File, S. E., and Hyde, J. R. G., 1978, Can social interaction be used to measure anxiety? *Br. J. Pharmacol.* **62**:19.

File, S. E., and Lister, R. G., 1983, Does tolerance to lorazepam develop with once weekly dosing? *Br. J. Clin. Pharmacol.* **16**:645.

Finkle, B. S., McCloskey, K. L., and Goodman, L. S., 1979, Diazepam and drug associated deaths: A survey in the United States and Canada, *J. Am. Med. Assoc.* **242**:429.

Fischman, M. W., and Schuster, C. R., 1978, Drug seeking: A behavioral analysis in animals and humans, in: *Self-administration of Abused Substances: Methods for Study* (N. Krasnegor, ed.), pp. 4–23, National Institute of Drug Abuse Research Monograph No. 20, Washington, D.C.

Floyd, J. B., and Murphy, C. M., 1976, Hallucinations following withdrawal from valium, *J. Kentucky Med. Assoc.* **74:**549.

Fontaine, R., Chouinard, G., and Annable, L., 1984, Rebound anxiety in anxious patients after abrupt withdrawal of benzodiazepine treatment, *Am. J. Psychiatry* **141:**848.

Fraser, H. F., and Isbell, H., 1954, Abstinence syndrome in dogs after chronic barbiturate medication, *J. Pharmacol. Exp. Ther.* **112:**261.

Friedman, G. A., and Ury, H. K., 1980, Initial screening for carcinogenicity of commonly used drugs, *J. Natl. Cancer Inst.* **65:**723.

Fuchs, V., Burbes, E., and Coper, H., 1984, The influence of haloperidol and aminooxyacetic acid on etonitazene, alcohol, diazepam and barbital consumption, *Drug Alcohol Depend.* **14:**179.

Garattini, S., Mussini, E., and Randall, L. O. (eds.), 1973, *The Benzodiazepines*, Raven Press, New York.

Gaul, L. E., 1961, Fixed drug eruption from chlordiazepoxide, *Arch. Dermatol.* **83:**1010.

Ghoneim, M. M., Mewaldt, S. P., Berie, J. L., and Hinrichs, J. V., 1981, Memory and performance effects of single and 3-week administration of diazepam, *Psychopharmacology* **73:**147.

Ghoneim, M. M., Mewaldt, S. P., and Hinrichs, J. V., 1984, Dose–response analysis of the behavioral effects of diazepam: II. Psychomotor performance, cognition and mood, *Psychopharmacology* **82:**296.

Giles, H. G., MacLeod, S. M., Wright, J. R., and Sellers, E. M., 1978, Influence of age and previous use on diazepam dosage required for endoscopy, *Can. Med. Assoc. J.* **118:**513.

Gordon, F. B., 1967, Addiction to diazepam (Valium), *Br. Med. J.* **1:**112.

Greeley, J., and Cappell, H., 1985, Associative control of tolerance to the sedative and hypothermic effects of chlordiazepoxide, *Psychopharmacology* **86:**487.

Greenblatt, D. J., and Koch-Weser, J., 1973, Adverse reactions to intravenous diazepam. A report from the Boston Collaborative Drug Surveillance Program, *Am. J. Med. Sci.* **266:**261.

Greenblatt, D. J., and Shader, R. I., 1974, *Benzodiazepines in Clinical Practice*, Raven Press, New York.

Greenblatt, D. J., and Shader, R. I., 1978, Dependence, tolerance and addiction to benzodiazepines: Clinical and pharmacokinetic considerations, *Drug Metab. Rev.* **8:**12.

Greenblatt, D. J., Woo, E., Divoll, M., Orsulak, P. J., and Shader, R. I., 1978, Rapid recovery from massive diazepam overdose, *J. Am. Med. Assoc.* **240:**1872.

Greenblatt, D. J., Shader, R. I., Harmatz, J. S., and Georgotas, A., 1979, Self-rated sedation and plasma concentrations of desmethyldiazepam following single doses of clorazepate, *Psychopharmacology* **66:**289.

Greenblatt, D. J., Sellers, E. M., and Shader, R. I., 1982, Drug disposition in old age, *N. Engl. J. Med.* **306:**1081.

Greenblatt, D. J., Shader, R. I., and Abernethy, D. R., 1983a, Current status of benzodiazepines (Part one), *N. Engl. J. Med.* **309:**354.

Greenblatt, D. J., Shader, R. I., and Abernethy, D. R., 1983b, Current status of benzodiazepines (Part two), *N. Engl. J. Med.* **309:**410.

Griffiths, A. N., Marshall, R. W., and Richens, A., 1984a, Tolerance to a sedative effect of diazepam after 6 nights nitrazepam pretreatment in man, *Br. J. Clin. Pharmacol.* **18:**305.

Griffiths, R. R., and Ator, N. A., 1980, Benzodiazepine self-administration in animals and humans: A comprehensive literature review, in: *Benzodiazepines: A Review of Research Results* (S. I. Szara and J. Ludford, eds.), pp. 22–36, National Institute of Drug Abuse Research Monograph Series No 33, Washington, D.C.

Griffiths, R. R., Bigelow, G. E., and Liebson, I., 1976, Human sedative self-administration: Effects of interingestion interval and dose, *J. Pharmacol. Exp. Ther.* **197:**488.

Griffiths, R. R., Bigelow, G., and Liebson, I., 1979, Human drug self-administration: Double-blind comparison of pentobarbital, diazepam, chlorpromazine and placebo, *J. Pharmacol. Exp. Ther.* **210**:301.

Griffiths, R. R., Bigelow, G. E., Liebson, I., and Kaliszak, J. E., 1980, Drug preference in humans: Double-blind choice comparison of pentobarbital, diazepam, and placebo, *J. Pharmacol. Exp. Ther.* **215**:649.

Griffiths, R. R., Lukas, S. E., Bradford, L. D., Brady, J. V., and Snell, J. D., 1981, Self-injection of barbiturates and benzodiazepines in baboons, *Psychopharmacology* **75**:101.

Griffiths, R. R., Bigelow, G. E., and Liebson, I., 1983, Differential effects of diazepam and pentobarbital on mood and behavior, *Arch. Gen. Psychiatry* **40**:865.

Griffiths, R. R., McLeod, D. R., Bigelow, G. E., Liebson, I. A., Roache, J. D., and Nowowieski, P., 1984c, Comparison of diazepam and oxazepam: Preference, liking and extent of abuse, *J. Pharmacol. Exp. Ther.* **229**:501.

Griffiths, R. R., McLeod, D. R., Bigelow, G. E., Liebson, I. A., and Roache, J. D., 1984b, Relative abuse liability of diazepam and oxazepam: Behavioral and subjective dose effects, *Psychopharmacology* **84**:147.

Gross, M. M., Lewis, E., and Hastey, J., 1974, Acute alcohol withdrawal syndrome, in: *The Biology of Alcoholism* (B. Kissin and H. Begleiter, eds.), pp. 191–263, Plenum Press, New York.

Guaitani, A., Carli, M., Rocchetti, M., and Garattini, S., 1979, Diazepam and experimental tumor growth, *Lancet* **1**:1147.

Hall, R. C. W., and Jaffe, J. R., 1972, Aberrant response to diazepam: A new syndrome, *Am. J. Psychiatry* **129**:114.

Hanna, S. M., 1972, A case of oxazepam (Serenid) dependence, *Br. J. Psychiatry* **120**:443.

Harrison, M., Busto, U., Naranjo, C. A., Kaplan, H. L., and Sellers, E. M., 1984, Identification and detoxification of high dose benzodiazepine abusers, *Clin. Pharmacol. Ther.* **36**:527.

Harry, T. V. A., 1972, Oxazepam (Serenid D) dependence, *Br. J. Psychiatry* **121**:235.

Hartelius, H., Larsson, A-K., Lepp, M., Malm, U., Arvidsson, A., and, Dahlstrom, H., 1978, A controlled long-term study of flunitrazepam, nitrazepam and placebo with special regard to withdrawal effects, *Acta Psychiatr. Scand.* **58**:1.

Hillestad, L., Hansen, T., and Melsom, H., 1974, Diazepam metabolism in normal man II. Serum concentration and clinical effect after oral administration and cumulation, *Clin. Pharmacol. Ther.* **16**:485.

Hollister, L. E., 1974, Psychotherapeutic drugs and driving, *Ann. Intern. Med.* **80**:413.

Hollister, L. E., 1980, Benzodiazepines 1980: A look at the issues, *Psychosomatics* **21**:4.

Hollister, L. E., Motzenbecker, F. P., and Degan, R. O., 1961, Withdrawal reactions from chlordiazepoxide ("Librium"), *Psychopharmacologia* **2**:63.

Hollister, L. E., Benneth, J. L., Kimbell, C., Savage, C., and Overall, J. E., 1963, Diazepam in newly admitted schizophrenics, *Dis. Nerv. Sys.* **24**:746.

Honkanen, R., Ertama, L., and Linnoila, M., 1980, Role of drugs in traffic accidents, *Br. Med. J.* **281**:1309.

Hoogland, D. R., Miya, T. S., and Bousquet, W. F., 1966, Metabolism and tolerance studies with chlordiazepoxide-2-14 C in the rat, *Toxicol. Appl. Pharmacol.* **9**:116.

Hopkins, D. R., Sethi, K. B. S., and Mucklow, J. C., 1982, Benzodiazepine withdrawal in general practice, *J. R. Coll. Gen. Pract.* **32**:758.

Horrobin, D. F., Ghayur, T., and Karmali, R. A., 1979, Mind and cancer, *Lancet* **1**:978.

Howe, J. G., 1980, Lorazepam withdrawal seizures, *Br. Med. J.* **280**:1163.

Hughes, R., and Brewin, R., 1979, *The Tranquillizing of America*, Harcourt Brace Jovanovich, New York.

Ingram, I. M., and Timbury, G. C., 1960, Side effects of librium, *Lancet* **2**:766.

Jackson, M. R., and Harris, P. A., 1981, Absence of effect of diazepam on tumors, *Lancet* **1**:104.

Jaffe, J. H., Ciraulo, D. A., Nies, A., Dixon, R. B., and Monroe, L. L., 1983, Abuse potential of halazepam and of diazepam in patients recently treated for acute alcohol withdrawal, *Clin. Pharmacol. Ther.* **34**:623.

Jasinsky, D. R., Johnson, R. E., and Henningfield, J. E., 1984, Abuse liability assessment in human subjects, *Trends Pharm. Sci.* **5**:196.

Johanson, C. E., and Uhlenhuth, E. H., 1980, Drug preference and mood in humans: Diazepam, *Psychopharmacology* **71**:269.

Jori, A., Prestini, P. E., and Pugliatti, C., 1969, Effect of diazepam and chlordiazepoxide on the metabolism of other drugs, *J. Pharm. Pharmacol.* **21**:387.

Kales, A., and Scharf, M. B., 1973, Sleep laboratory and clinical studies of the effects of benzodiazepines on sleep: Flurazepam, diazepam, chlordiazepoxide, and RO 5-4200, in: *The Benzodiazepines* (S. Garattini, E. Mussini, and L. O. Randall, eds.), pp. 577–598, Raven Press, New York.

Karmali, R. A., Horrobin, D. F., Ghayur, T., Manku, M. S., Cunnane, S. C., Morgan, R. O., Ally, A. I., Karmazyn, M., and Oka, M., 1978, Influence of agents which modulate thromboxane A-2 synthesis or action on R-3230 AC mammary carcinoma, *Cancer Lett.* **5**:205.

Karmali, R. A., Volkman, A., Muse, P., and Louis, T. M., 1980, Effects of diazepam on tumor growth and prostaglandin E2 in rats, *Adv. Prostaglandin Res.* **6**:523.

Kaufman, D. W., Shapiro, M., Slone, D., Rosenberg, L., Helmrich, S. P., Miettinen, O. S., Stolley, P. D., Levy, M., and Schottenfeld, D., 1982, Diazepam and the risk of breast cancer, *Lancet* **1**:537.

Kemper, N., Poser, W., and Poser, S., 1980, Benzodiazepin-Abhangigkeit, *Dtsch. Med. Wochenschr.* **105**:1707.

Khan, A., Joyce, P., and Jones, A. J., 1980, Benzodiazepine withdrawal syndromes, *N. Z. Med. J.* **92**:94.

Khan, A., Hornblow, A. R., and Walshe, J. W. B., 1981, Benzodiazepine dependence: A general practice survey, *N. Z. Med. J.* **93**:19.

Kibrick, E., and Smart, R., 1970, Psychotropic drug use and driving risk: A review and analysis, *J. Safety Res.* **2**:73.

Kleber, H. D., and Gold, M. S., 1978, Use of psychotropic drugs in treatment of methadone maintained narcotic addicts, *Ann. N.Y. Acad. Sci.* **311**:81.

Kleinerman, R. A., Brinton, L. A., Hoover, R., and Fraumeni, J. F., 1981, Diazepam and breast cancer, *Lancet* **1**:1153.

Klepner, C. A., Lippa, A. S., Benson, D. I., Sano, M. C., and Beer, B., 1979, Resolution of two biochemically and pharmacologically distinct benzodiazepine receptors, *Pharmacol. Biochem. Behav.* **11**:457.

Kolb, L., and Himmelsbach, C. K., 1938, Clinical studies of drug addiction, III. A critical review of the withdrawal treatment methodology of evaluating abstinence syndrome, *Am. J. Psychiatry* **94**:759.

Kryspin-Exner, K., and Demel, I., 1975, The use of tranquilizers in the treatment of mixed drug abuse, *Int. J. Clin. Pharmacol.* **12**:13.

Lader, M., 1978, Benzodiazepines—the opium of the masses? *Neuroscience* **3**:159.

Lader, M., and Petursson, H., 1983, Long-term effects of benzodiazepines, *Neuropharmacology* **22**:527.

Lader, M. H., Curry, S., and Baker, W. J., 1980, Physiological and psychological effects of clorazepate in man, *Br. J. Clin. Pharmacol.* **9**:83.

Lader, M. H., Ron, M., and Petursson, H., 1984, Computed axial brain tomography in long-term benzodiazepine users, *Psychol. Med.* **14**:203.

Ladewig, D., 1983, Abuse of benzodiazepines in Western European society—incidence and prevalence, motives, drug acquisition, *Pharmacopsychiatry* **16**:103.

Lahti, R. A., and Barsuhn, C., 1975, The effect of various doses of minor tranquilizers on plasma corticosteroids in stressed rats, *Res. Commun. Chem. Pathol. Pharmacol.* **11:**595.

Lasagna, L., 1980, The halcion study: Trial by media, *Lancet* **1:**815.

Laughren, T. P., Battey, Y. W., and Greenblatt, D. J., 1982a, Chronic diazepam treatment in psychiatric outpatients, *J. Clin. Psychiatry* **43:**461.

Laughren, T. P., Battey, Y., Greenblatt, D. J., and Harrop, D. S., 1982b, A controlled trial of diazepam withdrawal in chronically anxious outpatients, *Acta Psychiatr. Scand.* **65:**171.

Lennane, K. J., 1982, Oxazepam withdrawal syndrome, *Med. J. Australia* **1:**287.

Levy, A. B., 1984, Delirium and seizures due to abrupt alprazolam withdrawal, *J. Clin. Psychiatry* **45:**38.

Linnoila, M., Mattila, M. J., and Kitchell, B. S., 1979, Drug interactions with alcohol, *Drugs* **18:**299.

Litvak, R., and Kaebling, R., 1971, Agranulocylosis, leukopenia and psychotropic drugs, *Arch. Gen. Psychiatry* **24:**265.

Lo, K.-J., Eastwood, I. R., and Eidelman, S., 1967, Cholestatic jaundice associated with chlordiazepoxide. Report of a case and review of the literature, *Am. J. Dig. Dis.* **12:**845.

Lucki, I., Rickels, K., and Geller, A. M., 1985, Psychomotor performance following the long-term use of benzodiazepines, *Psychopharmacol. Bull.* **21:**93.

Lukas, S. E., and Griffiths, R. R., 1982, Precipitated withdrawal by a benzodiazepine receptor antagonist (RO 15-1788) after 7 days of diazepam, *Science* **217:**1161.

Maletzky, B. M., and Klotter, J., 1976, Addiction to diazepam, *Int. J. Addict.* **11:**95.

Margules, D. L., and Stein, L., 1966, Neuroleptics vs. tranquilizers: Evidence from animal behavior studies of mode and site of action, in: *Neuro-Psycho-Pharmacology,* Vol. 5 (H. Brill, ed.), pp. 108–120, Elsevier, Amsterdam.

Margules, D. L., and Stein, L., 1968, Increase of "antianxiety" activity and tolerance of behavioral depression during chronic administration of oxazepam, *Psychopharmacologia* **13:**74.

Marks, J., 1980, The benzodiazepines—use and abuse, *Arzneim.-Forsch.* **30:**898.

Marks, J., 1981, The benzodiazepines—use and abuse: Current status, *Pharm. Int.* **2:**84.

Martin, W. R., and McNicholas, L. F., 1981, Benzodiazepine dependence studies in rodents, in: *Benzodiazepines: A Review of Research Results* (S. I. Szara and J. P. Ludford, eds.), pp. 37–42, National Institute of Drug Abuse Research Monograph No. 33, Washington, D.C.

McClearn, G. E., and Erwin, V. G., 1980, The effect of diazepam on alcohol consumption in C57BL mice (abstr.), *Drug Alcohol Depend.* **6:**117.

McKay, A. C., and Dundee, J. W., 1980, Effects of oral benzodiazepines on memory, *Br. J. Anaesth.* **52:**1247.

McMillan, D. E., and Leander, J. D., 1978, Chronic chlordiazepoxide and pentobarbital interactions on punished and unpunished behavior, *J. Pharmacol. Exp. Ther.* **207:**515.

McNicholas, L. F., and Martin, W. R., 1982, The effect of a benzodiazepine antagonist, RO 15-1788 in diazepam dependent rats, *Life Sci.* **31:**731.

McNicholas, L. F., Martin, W. R., and Cheriau, S., 1983, Physical dependence on diazepam and lorazepam in the dog, *J. Pharmacol. Exp. Ther.* **226:**783.

Mellinger, G. D., Balter, M. B., Manheimer, D. I., Cisin, I. H., and Parry, H. J., 1978, Psychic distress, life crisis, and use of psychotherapeutic medications, *Arch. Gen. Psychiatry* **35:**1045.

Mellor, C. S., and Jain, V. K., 1982, Diazepam withdrawal syndrome: Its prolonged and changing nature, *Can. Med. Assoc. J.* **127:**1093.

Menon, G. N., 1965, Hypoplastic anaemia—an unusual complication of chlordiazepoxide therapy, *Postgrad. Med. J.* **41:**282.

Miller, F., and Nielsen, J., 1979, Diazepam (Valium) detoxification, *J. Nerv. Ment. Dis.* **167:**637.

Miller, R. R., 1973, Drug surveillance utilizing epidemiological methods: A report from the Boston Collaborative Drug Surveillance Program, *Am. J. Hosp. Pharm.* **30:**584.

Minter, R., and Murray, G. B., 1978, Diazepam withdrawal: A current problem in recognition, *J. Fam. Practice* **7**:1233.

Misra, P. C., 1975, Nitrazepam (Mogadon) dependence, *Br. J. Psychiatry* **126**:81.

Mohler, H., and Okada, T., 1977, Benzodiazepine receptor: Demonstration in the central nervous system, *Science* **198**:849.

Mohler, H., Burkard, W. P., Keller, H. H., Richard, J. G., and Haefely, W., 1981, Benzodiazepine antagonist RO 15-1788. Binding characteristics and interaction with drug-induced changes in dopamine turnover and cerebellar cGMP levels, *J. Neurochem.* **37**:714.

Murray, J., 1981, Long-term psychotropic drug-taking and the process of withdrawal, *Psych. Med.* **11**:853.

Nackie, B. S., and Nackie, L. E., 1965, Antihistamines in the treatment of drug eruptions, *Med. J. Australia* **2**:1034.

Nathan, R. G., Robinson, D., Cherek, D. R., Sebastian, C. S., Hack, M., and Davison, S., 1986, Alternative treatments for withdrawing the long-term benzodiazepine user: A pilot study, *Int. J. Addict.* **21**:195.

Nicolson, A. N., and Spencer, M. B., 1982, Psychological impairment and low dose benzodiazepine treatment, *Br. Med. J.* **285**:99.

O'Grady, J. A., and Pokorny, C., 1964, Drug allergy: Severe allergic reaction to librium, *J. Kansas Med. Soc.* **65**:65.

O'Hanlon, J. F., Haak, T. W., Blaauw, G. J., and Riemersma, J. B. J., 1982, Diazepam impairs lateral position control in highway driving, *Science* **217**:79.

Orzack, M. H., Cole, J. O., Ionescu-Pioggia, M., Beake, B. J., Bird, M. P., and Lobel, M., 1982, A comparison of some subjective effects of prazepam, diazepam, and placebo, in: *Problems of Drug Dependence, 1981* (L. S. Harris, ed.), pp. 309–317, National Institute of Drug Abuse Research Monograph No. 41, Washington, D.C.

Oswald, I., French, C., Adam, K., and Gilham, J., 1982, Benzodiazepine hypnotics remain effective for 24 weeks, *Br. Med. J.* **284**:860.

Pecknold, J. C., McClure, D. J., Fleuri, D., and Chang, H., 1982, Benzodiazepine withdrawal effects, *Prog. Neuropsychopharmacol.* **6**:517.

Petersen, R. C., and Ghoneim, M. M., 1980, Diazepam and human memory: Influence on acquisition, retrieval and state dependent learning, *Prog. Neuropsychopharmacol.* **4**:87.

Petursson, H., and Lader, M., 1981*a*, Withdrawal reaction from clobazam, *Br. Med. J.* **282**:1931.

Petursson, H., and Lader, M., 1981*b*, Withdrawal from long-term benzodiazepine treatment, *Br. Med. J.* **283**:643.

Petursson, H., and Lader, M. H., 1982, Psychological impairment and low-dose benzodiazepine treatment, *Br. Med. J.* **285**:815.

Petursson, H., and Lader, M., 1984, Benzodiazepine dependence, tolerance and withdrawal syndrome, *Adv. Hum. Psychopharmacol.* **3**:89.

Petursson, H., Gudjonsson, G. H., and Lader, M., 1983, Psychometric performance during withdrawal from long-term benzodiazepine treatment, *Psychopharmacology* **81**:345.

Pevnick, J. S., Jasinski, D. R., and Haertzen, C. A., 1978, Abrupt withdrawal from therapeutically administered diazepam, *Arch. Gen. Psychiatry* **35**:955.

Pickering, D., 1966, Hepatic necrosis after chlordiazepoxide therapy, *N. Engl. J. Med.* **274**:1449.

Pilotto, R., Singer, G., and Overstreet, D., 1984, Self-injection of diazepam in naive rats: Effects of dose, schedule and blockade of different receptors, *Psychopharmacology* **84**:174.

Poser, W., Poser, S., Roscher, D., and Argyrakis, A., 1983, Do benzodiazepines cause cerebral atrophy? *Lancet* **1**:715.

Poulos, C. X., Hinson, R. E., and Siegel, S., 1981, The role of Pavlovian processes in drug tolerance and dependence: Implications for treatment, *Addict. Behav.* **6**:205.

Preskorn, S. H., and Denner, L. J., 1977, Benzodiazepines and withdrawal psychosis, *J. Am. Med. Assoc.* **237**:36.

Preston, K. L., Bigelow, G. E., and Liebson, I. A., 1984, Self-administration of clonidine and oxazepam by methadone detoxification patients, in: *Problems of Drug Dependence, 1983* (L. S. Harris, ed.), pp. 192–198, National Institute of Drug Abuse Research Monograph No. 49, Washington, D.C.

Proudfoot, A. T., and Park, J., 1978, Changing patterns of drugs used for self-poisoining, *Br. J. Med.* **1**:90.

Ratna, L., 1981, Addiction to temazepam, *Br. Med. J.* **282**:1837.

Riegel, C. E., and Bourn, W. M., 1982, Low incidence of audiogenic convulsions upon withdrawal of diazepam cross-substituted for sodium barbital in dependent rats, *Fed. Proc.* **41**:1542.

Rickels, K., 1981, Benzodiazepines: Use and misuse, in: *Anxiety: New Research and Changing Concepts* (D. F. Klein and J. Rabkin, eds.), pp. 1–26, Raven Press, New York.

Rickels, K., Case, W. G., Downing, R. W., and Winokur, A., 1983, Long-term diazepam therapy and clinical outcome, *J. Am. Med. Assoc.* **250**:767.

Rickels, K., Case, G. W., Winokur, A., and Swenson, C., 1984, Long-term benzodiazepine therapy: Benefits and risks, *Psychopharmacol. Bull.* **20**:608.

Rifkin, A., Klein, D. F., and Quintin, F., 1977, Withdrawal from diazepam, *J. Am. Med. Assoc.* **238**:306.

Roache, J. D., and Griffiths, R. R., 1985, Comparison of triazolam and pentobarbital: Performance impairment, subjective effects and abuse liability, *J. Pharmacol. Exp. Ther.* **234**:120.

Robinson, G. M., and Sellers, E. M., 1982, Diazepam withdrawal seizures, *Can. Med. Assoc. J.* **126**:944.

Roehrs, T., Yang, O., and Samson, H., 1984, Chlordiazepoxide's interaction with ethanol intake in the rat: Relation to ethanol exposure paradigms, *Pharmacol. Biochem. Behav.* **20**:849.

Rosenbaum, J. F., Woods, S. W., Groves, J. E., and Klerman, G. L., 1984, Emergence of hostility during alprazolam treatment, *Am. J. Psychiatry* **141**:792.

Rosenberg, H. C., and Chiu, T. H., 1981, Tolerance during chronic benzodiazepine treatment associated with decreased receptor binding, *Eur. J. Pharmacol.* **70**:453.

Rosenberg, H. C., and Chiu, T. H., 1982, An antagonist-induced benzodiazepine abstinence syndrome, *Eur. J. Pharmacol.* **81**:153.

Rosenberg, H. C., and Chiu, T. H., 1985, Time course for development of benzodiazepine tolerance and physical dependence, *Neurosci. Biobehav. Rev.* **9**:123.

Rosenberg, H. C., Smith, S., and Chiu, T. H., 1983, Benzodiazepine-specific and nonspecific tolerance following chronic flurazepam treatment, *Life Sci.* **32**:279.

Rosser, W. W., 1982, Benzodiazepine prescription to middle-aged women—Is it done indiscriminately by family physicians? *Postgrad. Med.* **71**:115.

Rosser, W. W., 1984*a*, Physician education in tranquilizer prescribing, *Adv. Hum. Psychopharmacol.* **3**:165.

Rosser, W. W., 1984*b*, Benzodiazepines: Part of lifestyle in the 1980's, *Can. Fam. Physician* **30**:193.

Roy-Birne, P., Vittone, B. J., and Uhde, T. W., 1983, Alprazolam-related hepatotoxicity, *Lancet* **2**:786.

Ryan, G. P., and Boisse, N. R., 1983, Experimental induction of benzodiazepine tolerance and physical dependence, *J. Pharmacol. Exp. Ther.* **226**:100.

Ryan, G. P., and Boisse, N. R., 1984, Benzodiazepine tolerance, physical dependence and withdrawal: Electro-physiological study of spinal reflex function, *J. Pharmacol. Exp. Ther.* **231**:464.

Ryan, H. F., Merrill, F. B., Scot, G. E., Krebs, R., and Thompson, B. L., 1968, Increase in suicidal thoughts associated with diazepam therapy, *J. Am. Med. Assoc.* **203**:1135.

Salzman, C., Kochansky, G. E., Shader, R. I., Porrino, L. J., Harmatz, J. S., and Swett, C. P., 1974, Chlordiazepoxide-induced hostility in small group settings, *Arch. Gen. Psychiatry* **10**:401.

Samson, H. H., and Grant, K. A., 1985, Chlordiazepoxide effects on ethanol self-administration: Dependence on concurrent conditions, *J. Exp. Anal. Beh.* **43**:353.

Scharf, M. B., and Jacoby, J. A., 1982, Lorazepam—Efficacy, side effects and rebound phenomena, *Clin. Pharmacol. Ther.* **31**:175.

Schuster, C. L., and Humphries, R. H., 1981, Benzodiazepine dependency in alcoholics, *Conn. Med.* **45**:11.

Seevers, M. H., and Tatum, A. L., 1931, Chronic experimental barbital poisoning, *J. Pharmacol. Exp. Ther.* **42**:217.

Sellers, E. M., and Busto, U., 1982a, Recent advances with benzodiazepines. (One step forward or two steps back?), in: *Clinical Pharmacology and Therapeutics* (M. Velasco, ed.), pp. 121–127, Excerpta Medica, Amsterdam.

Sellers, E. M., and Busto, U., 1982b, Benzodiazepines and ethanol. Assessment of the effects and consequences of psychotropic drug interactions, *J. Clin. Psychopharmacol.* **2**:249.

Sellers, E. M., and Holloway, M. R., 1978, Drug kinetics and alcohol ingestion, *Clin. Pharmacokin.* **3**:440.

Sellers, E. M., Marshman, J. A., Kaplan, H. L., Giles, H. G., MacLeod, S. M., Kapur, B. M., Stapleton, C., Sealy, F., and Busto, U., 1981, Acute and chronic drug abuse emergencies in Toronto, *Int. J. Addict.* **16**:283.

Sepinwall, J., and Cook, L., 1980, Mechanism of action of the benzodiazepines: Behavioral aspect, *Fed. Proc.* **39**:3024.

Sepinwall, J., Grodsky, F. S., and Cook, L., 1978, Conflict behavior in the squirrel monkey: Effects of chlordiazepoxide, diazepam and N-desmethyldiazepam, *J. Pharmacol. Exp. Ther.* **204**:88.

Seppala, T., Linnoila, M., and Mattila, M. J., 1979, Drugs, alcohol and driving, *Drugs* **17**:389.

Seppala, T., Palva, E., Mattila, M. J., Korttila, K., and Shrotriya, R. C., 1980, Tofisopam, a novel 3,4,-Benzodiazepine: Multiple-dose effects on psychomotor skills and memory. Comparison with diazepam and interactions with ethanol, *Psychopharmacology* **69**:209.

Shader, R. J., and Greenblatt, D. M., 1984, Benzodiazepine overuse-misuse (Editorial), *J. Clin. Psychopharmacol.* **4**:123.

Shapiro, A. K., Struening, E. L., Shapiro, E., and Milcarek, B. I., 1983, Diazepam: How much better than Placebo? *J. Psychiatr. Res.* **17**:51.

Shaw, J. M., Kolesar, G. S., Sellers, E. M., Kaplan, H. L., and Sandor, P., 1981, Development of optimal treatment tactics for alcohol withdrawal. I Assessment and effectiveness of supportive care, *J. Clin. Psychopharmacol.* **1**:382.

Skegg, D. C. G., Richard, S. M., and Doll, R., 1980, Minor tranquilizers and road accidents, *Br. Med. J.* **1**:917.

Slater, J., 1966, Suspected dependence on chlordiazepoxide hydrochloride (Librium), *Can. Med. Assoc. J.* **95**:416.

Stein, L., and Berger, B. D., 1971, Psychopharmacology of 7-chloro-5-(*o*-chlorophenyl)-1,3-di-hydro-3-hydroxy-2*H*-1,4-benzodiazepine-2-one (lorazepam) in squirrel monkey and rat, *Arzneim.-Forsch.* **21**:1073.

Stewart, R. B., Salem, R. B., and Springer, P. K., 1980, A case report of lorazepam withdrawal, *Am. J. Psychiatry* **137**:1113.

Stitzer, M. L., Griffiths, R. R., McLellan, A. T., Grabowski, J., and Hawthorne, J. W., 1981, Diazepam use among methadone maintenance patients: Patterns and dosages, *Drug Alcohol Depend.* **8**:189.

Suzuki, T., Fukumori, R., and Yoshii, T., Yanaura, S., Satoh, T., and Kitagawa, H., 1980, Effects of *p*-chlorophenylalanine on diazepam withdrawal signs in rats, *Psychopharmacology* **71**:91.

Svenson, S. E., and Hamilton, R. G., 1966, A critique of overemphasis on side effects with the psychotropic drugs: An analysis of 18,000 chlordiazepoxide-treated cases, *Curr. Ther. Res.* **8:**455.

Svensson, C. R., 1980, Lorazepam withdrawal, *Drug Intell. Clin. Pharm.* **14:**628.

Swift, C. G., Swift, M. R., Hamley, J., Stevenson, I. H., and Crooks, J., 1983, CNS effects of chronic benzodiazepine hypnotic ingestion in the elderly, *Br. J. Clin. Pharmacol.* **16:**P217.

Taylor, F., 1973, Nitrazepam in the elderly, *Br. Med. J.* **1:**113.

Tessler, R., Stokes, R., and Pietras, M., 1978, Consumer response to valium, *Drug. Ther.* **8:**178.

Torrellas, A., Guaza, C., and Borrell, J., 1980, Effects of acute and prolonged administration of chlordiazepoxide upon the pituitary-adrenal activity and brain catecholamines in sound stressed and unstressed rats, *Neurosci.,* **5:**2289.

Tyrer, P., 1980, Dependence on benzodiazepines, *Br. J. Psychiatry* **137:**576.

Tyrer, P., 1982, Clinical and pharmacological evidence of dependence on benzodiazepine hypnotics, in: *Hypnotics in Clinical Practice,* Vol. 7 (A. Nicholson, ed.), pp. 59–66, Pembroke, Oxford.

Tyrer, P., Rutherford, D., and Huggett, T., 1981, Benzodiazepine withdrawal symptoms and propranolol, *Lancet* **1:**520.

Tyrer, P., Owen, R., and Dawling, S., 1983, Gradual withdrawal of diazepam after long-term therapy, *Lancet* **1:**1402.

Tyrer, P., Murphy, S., Oates, G., and Kingdon, D., 1985, Psychological treatment for benzodiazepine dependence, *Lancet* **1:**1042.

Uhlenhuth, E. H., Balter, M. B., and Lipman, R. S., 1978, Minor tranquilizers: Clinical correlates of use in an urban population, *Arch. Gen Psychiatry* **35:**650.

van der Kroef, C., 1979, Reactions to triazolam, *Lancet* **2:**526.

Vellucci, S. V., and File, S. E., 1979, Chlordiazepoxide loses it anxiolytic action with long-term treatment, *Psychopharmacology* **62:**61.

Vyas, I., and Carney, M. W. P., 1975, Diazepam withdrawal fits, *Br. Med. J.* **4:**44.

Warner, R. S., 1965, Management of the office patient with anxiety and depression, *Psychosomatics* **6:**347.

Winokur, A., Rickels, K., Greenblatt, D. J., Snyder, P. J., and Schatz, N. J., 1980, Withdrawal reaction from long-term low-dosage administration of diazepam, *Arch. Gen. Psychiatry* **37:**101.

Winstead, D. K., Anderson, A., Eilers, M. K., Blackwell, B., and Zaremba, A. L., 1974, Diazepam on demand: Drug-seeking behavior in psychiatric inpatients, *Arch. Gen. Psychiatry* **30:**349.

Wise, C. D., Berger, B. D., and Stein, L., 1972, Benzodiazepines: Anxiety-reducing activity by reduction of serotonin turnover in the brain, *Science* **177:**180.

Wise, R. A., and Dawson, V., 1974, Diazepam-induced eating and lever pressing for food in sated rats, *J. Comp. Physiol. Psychol.* **86:**930.

Wittenborn, J. R., 1979, Effects of benzodiazepines on psychomotor performance, *Br. J. Clin. Pharmacol.* **1:**61S.

World Health Organization, 1982, Final Report of the 6th Review of Psychoactive Substances for International Control.

World Health Organization Review Group, 1983, Use and abuse of benzodiazepines, *Bull. W. H. O.* **61:**551.

Yanagita, T., 1981, Dependence-producing effects of anxiolytics, in: *Psychotropic Agents,* Part II: *Anxiolytics, Gerontopsychopharmacological Agents, and Psychomotor Stimulants* (F. Hoffmeister and G. Stille, eds.), pp. 395–406, Springer-Verlag, New York.

Yanagita, T., 1983, Dependence-related pharmacological characteristics of benzodiazepines, 8th Review of Psychoactive Substances, Geneva, Sept. 12–16.

Yanagita, T., and Takahashi, S., 1973, Dependence liability of several sedative-hypnotic agents evaluated in monkeys, *J. Pharmacol. Exp. Ther.* **185:**307.

Yanagita, T., Takahashi, S., and Oinuma, N., 1975a, Drug dependence potential of lorazepam evaluated in the rhesus monkey, *CIEA Preclin. Rep.* **1**:1.

Yanagita, T., Oinuma, N., and Takahashi, S., 1975b, Drug dependence potential of halazepam evaluated in the rhesus monkey, *CIEA Preclin. Rep.* **1**:231.

Yanagita, T., Takahashi, S., and Oinuma, N., 1975c, Drug dependence liability of 5-1530 and nitrazepam evaluated in the rhesus monkey, *CIEA Preclin. Rep.* **1**:151.

Yanagita, T., Nijasato, K., Takahashi, S., and Kijohara, H., 1977a, Dependence potential of depotassium chlorazepate tested in rhesus monkeys, *CIEA Preclin. Rep.* **3**:67.

Yanagita, T., Miyasato, K., and Kijohara, H., 1977b, Drug dependence potential of triazolam evaluated in rhesus monkeys, *CIEA Preclin. Rep.* **3**:1.

Yanagita, T., Wakasa, Y., and Kato, S., 1981, Dependence potential of alprazolam tested in rhesus monkeys, *CIEA Preclin. Rep.* **7**:91.

Yoshimura, K., and Yamamoto, K. L., 1979, Neuropharmacological studies on drug dependence I. Effects due to differences in strain, sex and drug administration time on physical dependence development and characteristics of withdrawal, *Folia Pharmacol. Jpn.* **75**:805.

Zucker, H. S., 1973, Strange behaviour with oxazepam, *N.Y. State J. Med.* **72**:974.

3

# Some Implications of Alcohol-Induced Lipid Changes

JOSEPH J. BARBORIAK and LAWRENCE A. MENAHAN

## 1. INTRODUCTION

The increasing number of reports indicating an inverse association between cardiovascular morbidity or mortality and moderate alcohol intake (Gordon and Kannell, 1983; Kagan et al., 1981; Klatsky et al., 1974; Kozarevic et al., 1982; Salonen et al., 1983) represent a perplexing dilemma for the providers of health care (Ashley, 1984; Lieber, 1984; Popham et al., 1983).

On the one hand, cardiovascular diseases (CD) continue to be the leading cause of death in the industralized nations (Fuchs and Scheidt, 1983), and any avenue offering a reduction or prevention of CD will be and should be extensively investigated. On the other hand, alcohol consumption, especially if excessive and chronic, has been associated with so many serious biological, social, and economic ills (Eckart et al.., 1981; Lieber, 1982; Mishara and Kastenbaum, 1980) that one should be rightly concerned about recommending a toxic agent with considerable addicting potential as a means for general treatment or prevention of any disease.

Little definitive information is available on the possible mechanism(s) involved in the CD-attenuating properties of alcohol. Most research activity at this time is focused on the effect of alcohol on blood lipid levels, especially on the

JOSEPH J. BARBORIAK ● Department of Pharmacology and Toxicology, The Medical College of Wisconsin, and Biochemistry Section, Research Service, Veterans Administration Medical Center, Milwaukee, Wisconsin 53193. LAWRENCE A. MENAHAN ● School of Pharmacy, University of Wisconsin, Madison, Wisconsin 53208.

high-density lipoprotein cholesterol, since the changes in this lipid class may offer a plausible explanation of the clinical and experimental findings. However, alcohol has been known to affect metabolism of practically all lipid classes in several body organs (Baraona and Lieber, 1979). These changes need to be considered to better understand the overall implication of alcohol effects on circulating and tissue lipid levels. The present review describes some properties of the individual components of the circulating lipid classes and their modification by alcohol, both in the tissues and in blood.

## 2. LIPID AND LIPOPROTEIN STRUCTURE AND METABOLISM

### Lipid and Lipoprotein Constituents

To better understand the effects of alcohol consumption on individual lipid classes, a short discussion of their properties and functions seems appropriate.

Fatty Acids. Fatty acids (Fig. 1) are considered to be the basic units of most lipid classes. They consist of a chain of carbon atoms with hydrogen atoms attached and a carboxylic acid group at one end. They differ in length of the carbon chain and the presence of one or several double bonds. Those with double bonds are called unsaturated fatty acids. Most naturally occurring fatty acids have an even number of carbon atoms. Under normal biological conditions, they are either noncovalently bound to protein (albumin) or form an integral part of other more complex lipids such as triacylglycerols, phospholipids, or cholesterol esters. Fatty acids are an excellent source of energy, producing 9 kcal/gm as compared with 7 kcal/gm from ethanol and 4 kcal/gm for carbohydrate or protein. Certain unsaturated fatty acids, especially arachidonic acid, serve as precursors of prostaglandins, a group of biologically active and important agents.

Triacylglycerols. Triacylglycerols (triglycerides, neutral fat) (Fig. 2) are esters of fatty acids with glycerol, a three-carbon polyalcohol. They are the primary or only components of the natural fats such as cooking oils (corn oil, olive oil) or lard. In the natural state, an individual triacylglycerol molecule contains several different fatty acids. Occasionally, only one or two of the three hydroxyl groups of glycerol are esterified; these are called monoacylglycerols

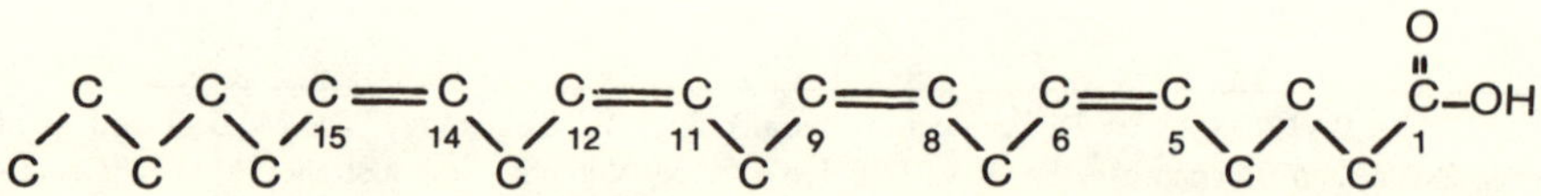

Figure 1.   Chemical structure of polyunsaturated fatty acid (arachidonic acid).

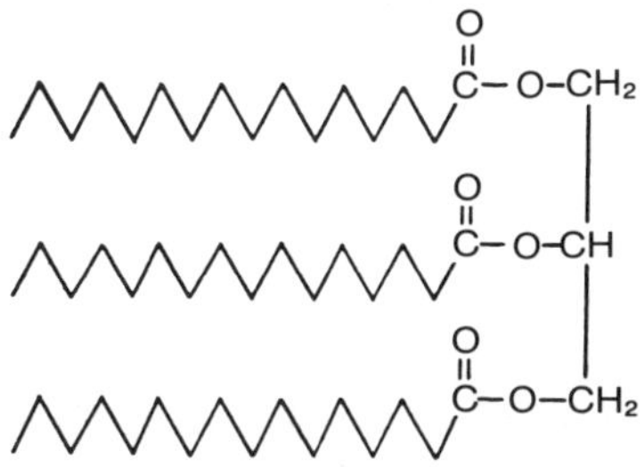

Figure 2.   Chemical structure of a triacylglycerol (tripalmitin).

or diacylglycerols. Triacylglycerols are nearly insoluble in water, but are quite soluble in organic solvents such as ether or chloroform.

Phospholipids. Phospholipids are a group of lipids characterized by a phosphate ester and the presence of one or more fatty acid residues. The most common phospholipids in human tissues are lecithins and cephalins. The structure of lecithin, a glycerophospholipid (Fig. 3), is similar to that of a triacylglycerol, but one of the outer fatty acids on the glycerol backbone is replaced by a phosphate group esterified with choline. Phospholipids are present universally in all cells as components of their membranes.

Cholesterol. Chemically cholesterol is a sterol alcohol (Fig. 4), yet is included as a lipid because of its frequent metabolic association with true lipids and similar solubility, i.e., cholesterol is readily soluble in organic solvents but sparingly in water. It may exist as free sterol, especially in the nervous system and cellular membranes, or esterified with a fatty acid, mostly in the liver and blood. Cholesterol serves as a precursor of bile acids and steroid hormones such as androgens, estrogens, and adrenocortical steroids.

Apolipoproteins. Most of the body lipids are not water-soluble. Yet, in order to be utilized or stored, they need to be transported by the blood, which is mostly water. The solubility is achieved by combining the lipids with phospholipids and specific proteins, i.e., apolipoproteins. Complexes of lipids with

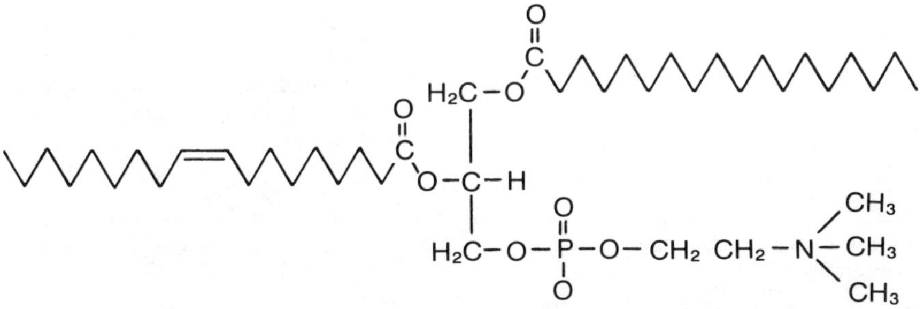

Figure 3.   Chemical structure of a phospholipid (lecithin).

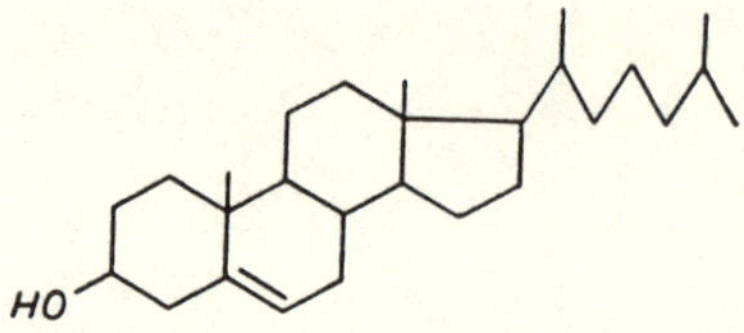

Figure 4.   Chemical structure of cholesterol.

these proteins are called lipoproteins. Thus, the apolipoproteins need to be considered as an integral part of the overall lipid metabolism. They also serve as markers directing the lipoproteins to the appropriate receptors on the cell membrane surface and as activators of some enzymes involved in lipid breakdown. Some of the properties and functions of the more common apolipoproteins are given in Table 1.

## Lipoprotein Structure and Metabolism

The various plasma lipoproteins can be separated by preparative ultracentrifugation, electrophoretic methods, column chromatography, or precipitation by polyanions. The conditions for separation of the lipoproteins by ultracentrifugation from human plasma and their physical properties are given in Table 2. The chemical composition of the various lipoproteins of normal human plasma is summarized in Table 3.

Chylomicrons. These particles are large and, on electrophoresis, remain at our very near the origin (Table 2). Chylomicrons contain mostly triacylglycerols and a small but significant amount of protein (Table 3). When secreted

Table 1. The Major Apolipoproteins[a]

| Designation | Serum levels (mg/deciliter) | Contained mainly in | Function |
| --- | --- | --- | --- |
| A-I | 110–150 | HDL | LCAT[b]/activation |
| A-II | 30–60 | HDL | |
| A-IV | 10–20 | Chy[c] | |
| B-48 | | Chy | |
| B-100 | 80–100 | IDL[d], LDL | Lipid transport |
| C-I | 5–8 | Chy, IDL, HDL | LCAT/activation |
| C-II | 3–6 | Chy, VLDL, LDL | LPL[e]/activation |
| C-III | 8–18 | Chy,VLDL, LDL | |
| D | 6–12 | HDL | Lipid transport |
| E | 3–12 | VLDL, IDL, LDL | |

[a] Adapted from Schonfeld (1983).
[b] LCAT, lecithin cholesterol acyltransferase.
[c] Chy, Chylomicrons.
[d] IDL, Intermediate-density lipoproteins.
[e] LPL, lipoprotein lipase.

Table 2. Characterization of Human Plasma Lipoproteins

|  | Chylomicrons | VLDL | LDL | HDL |
|---|---|---|---|---|
| Source |  |  |  |  |
| Major | Intestine | Liver | Plasma | Intestine |
| Minor |  | Intestine |  | Liver |
| Density (g/ml) | < 0.95 | 0.95–1.006 | 1.019–1.063 | 1.063–1.210 |
| Diameter (Å) | 1000–10,000 | 250–800 | 200–250 | 30–150 |
| Electrophoretic |  |  |  |  |
| Mobility | Origin | Pre-β | β | $\alpha_1$ |
| Apolipoproteins |  |  |  |  |
| (Major) | B, C-I, C-II, C-III | B, C-III, E | B | A-I, A-II |

[a] Adapted from Eder and Bergman (1983) and Sabesin (1981).

by the intestine, chylomicrons are deficient in apolipoprotein C-II but acquire apolipoprotein C from high-density lipoproteins (HDL) after entering the plasma (Havel et al., 1973). During their metabolism by peripheral tissue, chylomicrons become depleted of triacylglycerols and enriched in cholesterol esters. These chylomicron remnants are removed from the plasma and catabolized by the liver (Floren and Nilsson, 1977; Sherrill and Dietschy, 1978).

Very-Low Density Lipoprotein. Very-low density lipoproteins (VLDL) consist of a range of particles of various sizes whose origins are both hepatic and intestinal in nature (Wu and Windmueller, 1979). In electrophoresis, VLDL have pre-β mobility (Table 2). VLDL have a high content of triacylglycerols but somewhat more protein and cholesterol than chylomicrons (Table 3). During fasting, VLDL carry the major portion of triacylglycerols in plasma of healthy humans. Although the lipid composition of nascent VLDL secreted by the liver and of the VLDL lipoprotein found in plasma is similar, their apolipoprotein content is strikingly different (Fig. 5). The nascent VLDL, as secreted by the liver, consist almost entirely of Apo B (Ragland et al., 1978). The apoprotein content of circulating VLDL includes Apo C (approximately 50% of total apo-

Table 3. Chemical Composition of Human Plasma Lipoproteins[a]

| Constituent (% dry weight) | Chylomicrons | VLDL | LDL | HDL |
|---|---|---|---|---|
| Protein | 2 | 10 | 25 | 48 |
| Triacylglycerols | 85 | 60 | 7 | 4 |
| Cholesterol |  |  |  |  |
| Free | 2 | 8 | 10 | 4 |
| Esterified | 4 | 13 | 40 | 14 |
| Phospholipid | 8 | 18 | 20 | 30 |

[a] Adapted from Eder and Bergman (1983)

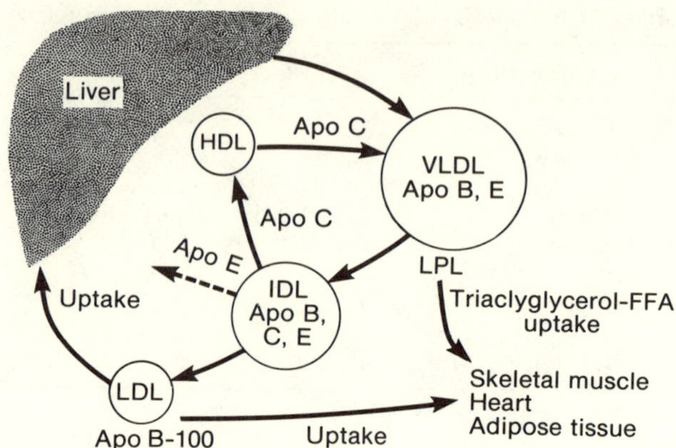



Figure 5.   VLDL and LDL metabolism. VLDL secreted by the liver contain apolipoproteins B and E and acquire Apo C from HDL in the plasm compartment. This permits the hydrolysis of the triacylglycerols in VLDL by lipoprotein lipase and uptake of triacylglycerol-fatty acids by peripheral tissues (skeletal muscle, heart, adipose tissue). Removal of surface components, including phospholipids and Apo C, results in smaller particles (IDL). During the conversion of IDL to LDL, Apo E is lost by a mechanism that still remains uncertain. The uptake of LDL by peripheral cells and liver is indicated. (Adapted from Eder and Bergman, 1983.)

lipoprotein), Apo B (about 37%), and Apo E (about 13%) (Sabesin, 1981). Thus, nascent VLDL apparently acquire Apo C within the plasma compartment.

Low-Density Lipoprotein. The major lipoprotein class in the human, the low-density lipoproteins (LDL), is formed in the plasma as a result of VLDL catabolism (Fig. 5) (Schaefer et al., 1978). LDL apolipoproteins consist almost entirely of Apo B (Table 2) and LDL's principal lipid component is esterified cholesterol. Both peripheral tissues and liver can remove LDL from plasma, and the latter can account for somewhat more than 50% of the metabolized (Steinberg et al., 1980). The LDL is a primary source of cholesterol for extrahepatic cells.

High-Density Lipoprotein. High-density lipoprotein (HDL) represents a heterogeneous fraction. The HDL have been subfractionated into three components: HDL$_{2a}$, HDL$_{2b}$, and HDL$_3$, which vary in density and size (Anderson et al., 1978). HDL contain appreciably more protein than the other lipoproteins (Table 3). HDL are secreted both by liver and intestine (Fig. 6). Nascent HDL is an ideal substrate for lecithin-cholesterol acyltransferase (LCAT), a key enzyme in the esterification of cholesterol in plasma (Fig. 6). The enzyme is activated by Apo A-I. When cholesterol is esterified, Apo E is lost and the HDL is formed. HDL may transport cholesterol from peripheral tissues, thereby preventing its excessive accumulation and providing a source of cholesterol for subsequent esterification by LCAT (Sabesin, 1981).

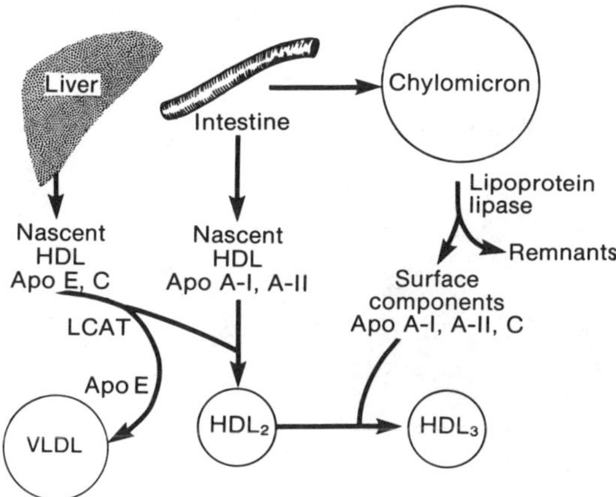


Figure 6. HDL metabolism. There are three differenct sources of HDL apolipoproteins. The nascent HDL produced by the liver contain largely apoproteins E and C, whereas those originating from the intestine have Apo A-I and A-II. Particles in the HDL fraction are also produced during the metabolism of chylomicrons and contain apoproteins A-I, A-III, and C. The free cholesterol in nascent HDL from liver and intestine undergoes esterification by LCAT. The Apo E in the nascent HDL secreted by liver is probably transferred in the plasma to VLDL. The apoproteins derived from the surface components of chylomicrons are involved in the conversion of $HDL_3$ to $HDL_2$. (Adapted from Eder and Bergman, 1983.)

## 3. ALCOHOL-INDUCED CHANGES IN LIVER LIPIDS

Liver represents the main body organ for the formation and transformation of lipoproteins. Therefore, hepatic lipid metabolism is reviewed in detail.

### Hepatic Fatty Acid Metabolism

Hepatic fatty acids, free and esterified, can originate both from plasma free fatty acids (FFA) released from adipose tissue and *de novo* synthesis within the liver itself. As triacylglycerol fatty acids in the form of chylomicrons are not metabolized to any appreciable extent by the isolated perfused liver (Felts and Mayes, 1965) or morphologically intact liver cells (Felts and Berry, 1971), hepatic lipids are not influenced directly by dietary fat. Yet, liver lipids can accumulate from changes in one or more of the following metabolic alterations within the hepatocyte itself: reduced oxidation of fatty acids, increased glyceride formation, enhanced *de novo* fatty acid synthesis, and decreased release of VLDL (Fig. 7).

The accumulation of fat in the liver is the most common disturbance of

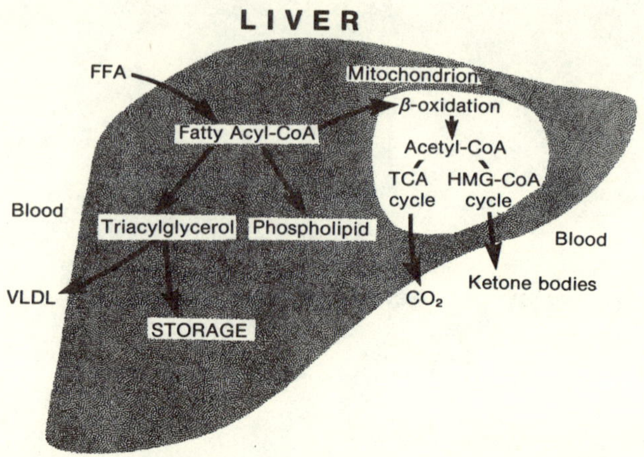


Figure 7.    Pathways of hepatic FFA metabolism.

lipid metabolism produced by alcohol (Baraona and Lieber, 1979). In understanding the effects of ethanol on hepatic lipid metabolism, it is important to
know the factors that control partitioning of fatty acids between β-oxidation and
glycerolipid synthesis (Fig. 7) (Brindley et al., 1979a; Mayes and Felts, 1967;
McGarry and Foster, 1971, 1976; Ontko, 1972). Conversion of $^{14}$C-labeled long-
chain fatty acid to $^{14}CO_2$ is depressed by the presence of ethanol in liver slices
(Blomstrand et al., 1973; Lieber and Schmid, 1961), isolated hepatocytes (Grunnet and Kondrup, 1983; Ontko, 1973; Wiebe and Belfrage, 1980), and the
perfused liver (Fellenius and Kiessling, 1973; Kondrup et al., 1979; Lieber et
al., 1967; Lindros, 1970; McGarry and Foster, 1971; Schapiro et al., 1964;
Williamson et al., 1969). The inhibitory effect of ethanol on the β-oxidation of
fatty acids may range from 0 to 50%, depending on the experimental conditions;
yet, its true importance in ethanol-induced accumulation of hepatic lipids still
remains uncertain (Grunnet and Kondrup, 1983).

The presence of ethanol consistently decreases hepatic conversion of fatty
acids to $CO_2$, most probably due to inhibition of the tricarboxylic acid (TCA)
cycle (Fellenius and Kiessling, 1973; Forsander et al., 1965a,b; Ontko, 1973;
Williamson et al., 1969). The activity of the TCA cycle in liver (Fig. 7) is
arrested during ethanol metabolism (Veech et al., 1981), most likely by the
elevation in the mitochondrial [(NADH)/(NAD$^+$)] ratio. It has been suggested
that the NAD$^+$ requiring dehydrogenases, involving 2-ketoglutarate (Ontko,
1973) and pyruvate (Wieland, 1983), would thereby be inhibited by ethanol
oxidation. The redox change associated with ethanol metabolism could shift the
mitochondrial malate/oxaloacetate couple (Williamson et al., 1969) to the point

where the concentration of oxaloacetate would fall below the $K_m$ of citrate synthase (Srere, 1975). Until more accurate data as to mitochondrial matrix effector concentrations are available, precise mechanism(s) for the ethanol inhibition of the TCA cycle in liver cannot be given (Veech et al., 1981). During chronic alcohol administration, acetaldehyde generated from ethanol may also contribute to the altered hepatic mitochondrial fatty acid oxidation (Baraona and Lieber, 1979).

Thus, ethanol effectively competes for fatty acids as the major hepatic fuel (Baraona and Lieber, 1979; Lieber and Schmid, 1961; Ontko, 1973; Reboucas and Isselbacher, 1961). This extends to the acetate produced from ethanol metabolism, of which only a small part is oxidized in the liver. Most of the acetate is released into the hepatic vein for utilization by peripheral tissues (Juhlin-Dannfelt, 1977; Lundquist, 1975; Lundquist et al., 1962). In the absence of ethanol, acetate is metabolized readily by the perfused liver (Lindros, 1970; Menahan et al., 1968) and with isolated hepatocytes (Veech et al., 1981). However, the metabolism of acetate is severely inhibited by ethanol in the nonrecirculating liver perfusion (Lindros, 1970) and incubated liver slices (Majchrowicz and Quastel, 1961). Acetate oxidation by isolated liver mitochondria was found to be restricted by the rate of recycling of AMP produced during its activation (Lumeng and Davis, 1973). This limitation would be augmented when the TCA cycle is arrested during ethanol metabolism (Veech et al., 1981). The suppression of acetate oxidation by ethanol is much weaker in livers from hyperthyroid rats and stronger in those in the hypothyroid state (Lindros, 1970). Furthermore, the reduction of intramitochondrial redox pairs associated with ethanol metabolism is enhanced in hypothyroid rats but diminished in their hyperthyroid counterpart (Lindros, 1970; Rawat and Lundquist, 1968). These findings support the conclusion that the increased mitochondrial $[(NADH)/(NAD^+)]$ during acute ethanol administration is responsible not only for decreased TCA cycle activity but inhibition of hepatic fatty acid oxidation including that of acetate arising from ethanol itself.

## Liver Triacylglycerols and Phospholipids

The ingestion or administration of alcohol induces the accumulation of triacylglycerols in the liver (Baraona and Lieber, 1979; Barboriak and Menahan, 1979). An ethanol-induced decrease in fatty acid oxidation has been suggested as a possible basis for development of fatty liver (Blomstrand et al., 1973; Fellenius and Kiessling, 1973; Lieber and Schmid, 1961; Lieber et al., 1967; McGarry and Foster, 1971; Ontko, 1973; Reboucas and Isselbacker, 1961). This diminution in hepatic fatty acid oxidation would result in an increased availability of long-chain fatty acyl-CoA for esterification into triacylglycerols (Fig. 7). An increased supply of FFA to isolated liver cells does, indeed, lead to a more marked increase in triacylglycerol synthesis than in phospholipid formation (Ontko, 1972). This change can be considered a major cause of the development of alcoholic fatty liver, the first state of hepatic injury (Baraona and Lieber, 1979; Lieber, 1984).

In addition to an inhibition of hepatic fatty acid oxidation, an elevation of *sn*-glycerol-3-phosphate concentration (Fig. 8) in the liver has been consistently found when ethanol is present (Fellenius et al., 1973; McGarry and Foster, 1971; Nikkila and Ojala, 1963; Rawat, 1970; Thieden and Lundquist, 1967; Williamson et al., 1969; Ylikahri, 1970; Zakim, 1965). It has been generally considered that the primary factor influencing hepatic fatty acid esterification in the fed state is the concentration of free *sn*-glycerol-3-phosphate (Mayes, 1976; Mayes and Felts, 1967). However, rates of hepatic triacylglycerol formation may not be a simple function of hepatic *sn*-glycerol-3-phosphate concentration. Changes in hepatic triacylglycerols during perfusion of rat liver *in situ* have been found not to be a function of liver concentration of *sn*-glycerol-3-phosphate in the absence of added long-chain fatty acid (Fellenius et al., 1973). In cultured hepatocytes from female rats, the accumulation of triacylglycerols continued linearly in the presence of ethanol, while the concentration of *sn*-glycerol-3-phosphate declined almost to control values (Dich et al., 1983). This suggests that the hepatocyte concentration of *sn*-glycerol-3-phosphate was not critical for the increase of triacylglycerols when ethanol is added to the culture medium.

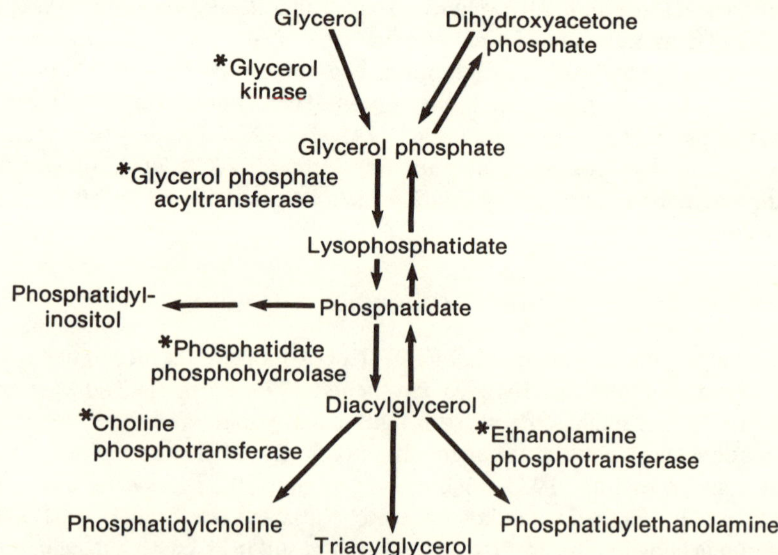


Figure 8.  Pathways of glycerolipid and phosphatidate metabolism. Enzyme activities increased by acute and/or chronic ethanol administration are indicated (*).

The increase of hepatic triacylglycerols that follows ethanol administration might result from an inhibition of lipolysis as well as stimulation of the esterification process. Even though ethanol enhances the rate of recycling of the fatty acid moiety of triacylglycerols (Kondrup et al., 1979), its presence has no or a slight ($\sim$10%) inhibitory effect on lipolysis in hepatocytes (Grunnet and Kondrup, 1983).

Ethanol ingestion also increased the enzymatic capacity of the liver to synthesize triacylglycerols (Fig. 8). In adult rat hepatocytes in primary monolayer culture for 6–24 hr, the ethanol-induced increase in glycerol kinase activity correlated well with monolayer glycerolipid formation (Wood and Lamb, 1979). The first enzyme of hepatic glycerolipid biosynthesis, $sn$-glycerol-3-phosphate acyltransferase (Fig. 8), is increased only after 24 days of alcohol feeding (Joly et al., 1973) and not by short-term exposure of hepatocytes to ethanol in primary monolayer culture (Wood and Lamb, 1979).

Phosphatidate phosphohydrolase (Fig. 8) exhibits many of the properties of an important regulatory enzyme for the synthesis of triacylglycerols in the liver (Brindley and Sturton, 1982). Chronic ethanol intake has been found to stimulate the capacity of neutral glycerolipid production by the liver; this correlated well with an increase in hepatic phosphatidate phosphohydrolase activity (Lamb et al., 1979; Savolainen et al., 1984). Hepatic phosphatidate phosphohydrolase activity is also increased by a single feeding of ethanol (Brindley et al., 1979$a$; Lamb et al., 1979; Pritchard et al., 1977; Savolainen, 1977). The ethanol-induced increase in phosphatidate phosphohydrolase can be prevented to a large degree by adrenalectomy (Brindley et al., 1979$b$). Furthermore, the duration of an ethanol-stimulated increase in plasma corticosterone levels has been found to be decreased in rats pretreated with benfluorex [1-(3-trifluoromethylphenyl)-2-[$N$-(2-benzoyloxyethyl)-amino]propane, or S-780], a hypolipidemic agent (Brindley and Sturton, 1982). Chronic benfluorex administration prior to acute ethanol feeding diminished the increase in phosphatidate phosphohydrolase activity (Pritchard et al., 1977) and the accumulation of triacylglycerols in the liver (Brindley and Sturton, 1982). Acute ethanol feeding may also cause activation of existing phosphatidate phosphohydrolase in rat liver; this could explain some of the stimulation of the enzyme by ethanol remaining following adrenalectomy (Sturton et al., 1981). The acute stimulation of hepatic phosphatidate phosphohydrolase may be mediated by hepatic metabolites, most probably $sn$-glycerol-3-phosphate (Savolainen and Hassinen, 1978). It has also been suggested that the soluble- and membrane-bound phosphatidate phosphohydrolase activities in the liver may have different functions in hepatic glycerolipid biosynthesis (Savolainen and Hassinen, 1980). More recently, it has been found that substrate availability can override enzyme modulations in the control of triacylglycerol synthesis and the role of phosphatidate phosphohydrolase as the primary regulator in the biosynthesis of neutral glycerols has been questioned

(Pikkukangas et al., 1982). In primary monolayer cultures of adult rat hepato-cytes, phosphatidate phosphohydrolase activity did not correlate well with the incorporation of [1,3-$^{14}$C]glycerol into monolayer glycerolipids (Wood and Lamb, 1979).

Diacylglycerol acyltransferase is the last enzyme in the biosynthesis of triacylglycerols and only one unique to their formation (Fig. 8). However, phos-phatidate phosphohydrolase is considered to be the key regulatory enzyme in hepatic triacylglycerol synthesis as diacylglycerol acyltransferase has a relatively high specific activity in the liver (Brindley and Sturton, 1982) and does not change after an acute dose of ethanol (Pritchard et al., 1977). At the early stages of fatty liver development in baboons receiving alcohol for a long time, dia-cylglycerol acyltransferase activity is increased (Savolainen et al., 1984). Re-cently, a cytosolic factor has been found to potentiate the ethanol stimulation of a microsomal diacylglycerol acyltransferase (Väänänen et al., 1981) and may be linked to regulation of the enzyme by phosphorylation-dephosphorylation (Haagsman et al., 1982).

The diacylglycerol formed by phosphatidate phosphohydrolase is prefer-entially incorporated into phosphatidylcholine and phosphatidylethanolamine when the supply of fatty acids to the liver is low (Brindley and Sturton, 1982). Yet, chronic ethanol administration increases hepatic phospholipid content along with producing a fatty liver (Lieber et al., 1965). More recently, ethanol ingestion was found to increase the microsomal activity of choline phosphotransferase and phosphatidylethanolamine transferase, key enzymes in the synthesis of phos-phatidylcholine (Uthus et al., 1976). However, exposure to ethanol for 6 hr had no effect on choline phosphotransferase activity in homogenates prepared from primary monolayer culture of adult rat hepatocytes (Wood and Lamb, 1979).

Although the activities of several enzymes involved with triacylglycerol and phospholipid synthesis do change in response to either acute or chronic ethanol administration, the mechanisms producing these alterations are largely unknown.

## Hepatic Membrane Lipids

Marked alteration in both the function and morphology of hepatic mito-chondria is a consequence of chronic alcohol consumption (Baraona and Lieber, 1979). These structural changes in the mitochondrion are associated with a decreased capacity of the organelle to oxidize substrates and form high-energy phosphates (Bernstein and Penniall, 1978; Cederbaum et al., 1974; Gordon, 1973; Rubin et al., 1970). It has been proposed that these functional alterations in mitochondria from animals receiving alcohol chronically may be directly related to the increased rigidity of mitochondrial membranes which is manifested by shifts in the temperature dependence of their physical properties (Rottenberg et al., 1980, 1981, 1984; Waring et al., 1981, 1982). However, a lack of correlation between hepatic mitochondrial membrane structure and functions in ethanol-fed rats has also been reported (Gordon, 1984; Gordon et al., 1982).

Although a direct link between functional, lipid, and fluidity changes of mitrochondrial membranes during chronic consumption of alcohol remains debatable (Gordon, 1984; Rottenberg et al., 1984), decreased phospholipid content (French et al., 1970) and altered fatty acid composition (French et al., 1971) have been found in hepatic mitochondria from ethanol-fed animals. Ethanol inhalation for a 2-day period decreases the proportion of arachidonic acid in total liver mitochondrial phospholipids, while an increase in polyunsaturated fatty acids were found in the phosphatidylcholine and phosphatidylethanolamine after a 4-day inhalation period (Rouach et al., 1984). In baboons fed ethanol chronically, alterations in the phospholipid composition of the mitochondrial membranes may be responsible, at least in part, for the decrease in activity of cytochrome oxidase found in hepatic mitochondria (Arai et al., 1984). Furthermore, the importance of dietary levels of fat, as well as the sex of the animal, in the alterations of liver mitochondria structure and function following chronic ethanol ingestion has been emphasized (Thompson and Reitz, 1978). Therefore, the relationship between the physical and functional effects of ethanol on hepatic mitochondria is both unclear and seemingly complex.

Also, changes in plasma membrane lipid composition of erythrocytes in ethanol-tolerant mice can partially account for an adaptation to ethanol-induced membrane disordering (Wing et al., 1982). However, contrasting types of plasma membrane preparation from liver, myelin and synaptosomes differed in the patterns of change in their phospholipid-fatty acids from erythrocyte membranes during chronic ethanol administration. More recently, chronic ethanol administration to mice has been shown to cause changes in the total fatty acyl composition of erythrocyte membrane phospholipids (Wing et al., 1984). Furthermore, the particular sensitivity of monounsaturated fatty acyl groups of phosphatidylethanolamine to ethanol administered have led Wing et al. (1984) to suggest that these fatty acids may have a pivotal role in the control of membrane function, at least in the erythrocyte.

Chronic ethanol ingestion by male rats has no effect on the total content of phosphatidylinositol-phosphatidylserine, phosphatidylethanolamine and phosphatidylcholine in isolated hepatocytes (Smith et al., 1983). However, the vasopressin and $\alpha_1$-adrenergic stimulation of [$^{32}$P]-incorporation into phosphatiodylinositol of isolated hepatocytes from rats ingesting ethanol was increased. Also, maternal ethanol ingestion has been found to have a detrimental effect on the postnatal development of rat liver plasma membrane $\alpha_1$-adrenergic receptors (Rovinski and Hosein, 1983). While chronic maternal ethanol administration did not appear to affect the postnatal development of the liver plasma membrane lipids, specific adaptive changes were induced in the membrane lipids of the progeny. Thus, a functional link of changes in hepatic plasma membrane lipids following chronic alcohol ingestion has yet to be elucidated.

## 4. EFFECTS OF ALCOHOL ON PLASMA LIPIDS, LIPOPROTEINS, AND APOLIPOPROTEINS

The alcohol-induced changes in plasma lipid components have been a favorite subject of studies since the initial observations of these effects (Albrink and Klatskin, 1957; Feigl, 1918; Kaffarnik and Schneider, 1970; Lieber et al., 1963; Losowsky et al., 1963; Schapiro et al., 1965). While initially limited to studies of plasma triacylglycerols, cholesterol, and FFA, the investigators have subsequently made good use of the increased sophistication in the lipid analyses and applied the new methodology to examination of the alcohol-induced changes of lipoproteins and apolipoproteins.

### Hyperlipemias

The first reports on the effect of alcohol on plasma lipids described increased plasma lactescence (Albrink and Klatskin, 1957; Feigl, 1918; Zieve, 1958), which has been shown to be due to increased levels of plasma triacylglycerols (Kaffarnik and Schneider, 1970; Losowsky et al., 1963; Schapiro et al., 1965). Originally, the hyperlipemias were observed almost exclusively in chronic alcoholics. Subsequently, individuals were identified who responded to relatively moderate alcohol intake with an excessive rise of plasma triacylglycerols (Fry et al., 1973; Kudzma and Schonfeld, 1971). Other studies have indicated that preprandial ingestion of alcohol before a substantial meal is also associated with a considerable but temporary increase in plasma triacylglycerols in most individuals (Barboriak and Meade, 1968; Wilson et al., 1970). The exact mechanism of the alcohol-induced increase in plasma triacylglycerols has not been elucidated, and both an increased hepatic production and reduced metabolic processing have been implicated (Baraona and Lieber, 1979). It should be mentioned, however, that in most of the above studies alcohol was added to the usual diet. When inclusion of alcohol was compensated isocalorically by reduction of other dietary components, the higher plasma triacylglycerol levels were not maintained (Erkelens and Brunzell, 1980). More recent studies suggest that obese individuals may be more prone to the hyperlipemic effects of alcohol (Crouse and Grundy, 1984).

### Lipoproteins

Since the individual blood lipid classes are constituents of lipoproteins, it may be expected that these complex molecules will reflect the changes in their components.

Very-Low Density Lipoproteins. Plasma triacylglycerols represent one of the main components of VLDL, and it could be expected that the alcohol-induced VLDL changes would parallel the changes seen in triacylglycerols. Experimental data confirm this suggestion (Avogaro and Cazzolato, 1975; Crouse and Grundy, 1984; Wilson et al., 1970). However, some differences in the composition of the normal and alcohol-induced VLDL particles have been observed (Fellin et al., 1974).

Low-Density Lipoproteins. Unlike the marked changes in VLDL, the alcohol-induced changes in LDL appear to be less marked and less consistent (Crouse and Grundy, 1984). Recent reports indicate that the extent of hepatic injury may modify the LDL composition. Alcoholic patients with cirrhosis have been shown to have higher LDL levels and a lower proportion of esterified cholesterol than alcoholic patients with normal liver function or control subjects (Duhamel et al., 1984). Moderate alcohol intake failed to affect the LDL-cholesterol levels of normolipidemic volunteers (Thornton et al., 1983) or of premenopausal women (Stamford et al., 1984). However, in large epidemiological studies, LDL cholesterol was negatively associated with alcohol consumption (Castelli et al., 1977). It appears that additional studies will be needed to clarify the association between alcohol intake and LDL and its components.

High-Density Lipoproteins. It is generally agreed that the most consistent effect of alcohol intake on plasma lipoprotein levels is primarily seen in the HDL. Due to the complexity of the HDL structure and assays, many investigators have measured only the HDL cholesterol (HDL-C) and assumed that this part represents the total HDL. However, cholesterol makes up only about 20% of the total HDL moiety and some of the discrepancies in reports on changes in the HDL levels may be due to the lack of consideration of other HDL components. For instance, recent findings have indicated that 6 hr after alcohol administration there was a considerable increase in HDL phospholipids (Goldberg et al., 1984).

The initial reports on modification of HDL levels by alcohol were published by Swedish investigators (Johansson and Laurell, 1969) and were obtained with chronic alcoholics. Subsequently, epidemiological studies (Barrett-Connor and Suarez, 1982; Castelli et al., 1977; Heiss et al., 1980) and controlled administration of alcohol to volunteers (Belfrage et al., 1977; Crouse and Grundy, 1984; Fraser et al., 1983) have indicated that alcohol intake may increase plasma HDL-C even in nonalcoholic subjects. The increase in HDL-C levels is, to some extent, dose-dependent and ranges from 2.0 to 3.5 mg/dl per drink per day (Barrett-Connor and Suarez, 1982; Fraser et al., 1983; Jacqueson et al., 1983; Stamford et al., 1984).

A consistent effect on HDL-C is usually seen after 2 or more weeks of regular alcohol intake (Belfrage et al., 1977; Fraser et al., 1983), although some temporary changes may be seen within a few hours of alcohol consumption (Goldberg et al., 1984). Chronic alcoholics without extensive liver impairment may show unusually high plasma HDL-C levels well in excess of 100 mg/dl

(Devenyi et al., 1984; Masarei et al., 1982). The average HDL-C level for normal adult men is about 45 mg/dl and for normal adult women 60 mg/dl (Rifkind and Segal, 1983). The alcohol-induced increase in plasma HDL-C is additive to other factors affecting this variable, such as aerobic exercise (Willet et al., 1980). It may also be more pronounced in older subjects (Barrett-Connor and Suarez, 1982; Jacqueson et al., 1983). Abstinence from alcohol for approximately 1–2 weeks duration is usually associated with a reduction in HDL-C levels, both in chronic alcoholics (Barboriak et al., 1980; Danielson et al., 1978; Devenyi et al., 1984) and in social drinkers (Belfrage et al., 1977).

In addition to the amounts imbibed, the *pattern of alcohol consumption* (such as binge drinking versus regular intake of smaller amounts of alcohol) is also an important factor, both in the effect on HDL-C levels and less extensive coronary artery occlusion (Gruchow et al., 1982). It has been found that many imbibers consume smaller amounts of alcohol in a regular pattern and, in addition, tend to "binge," i.e., consume much higher amounts of alcohol at longer intervals. Furthermore, the previously mentioned almost linear relationship between the amounts of alcohol consumed and HDL-C levels fails to occur in the "binge" drinkers. It would seem therefore that the pattern of alcohol consumption is a variable which needs more attention.

The interpretation of alcohol–HDL-C interrelationship has been recently complicated by a study suggesting differences in the effect of moderate amounts of alcohol on the two subclasses of HDL: $HDL_2$ and $HDL_3$ (Haskell et al., 1984). As previously mentioned, the HDL are subdivided on the basis of their densities into $HDL_2$, floating density range (d = 1.063–1.125), and $HDL_3$, (d = 1.125–1.20). It is generally believed that the $HDL_2$ is the "active" component of this lipid class and represents the "protective" effects of HDL against the development of CD (Ballantyne et al., 1982; Miller et al., 1981). Reported data dealing with the effect of alcohol on the two subclasses are rather inconsistent. While some studies indicate a more pronounced effect of alcohol on $HDL_3$ (Buring et al., 1983; Haskell et al., 1984), other investigators have seen changes in both subclasses (Ekman et al., 1981). The use of the $HDL_2/HDL_3$ ratios has been recently proposed for estimating the extent of alcohol-induced damage (Duhamel et al., 1984; Sabesin and Weidman, 1984).

In the previously mentioned study (Haskell et al., 1984) with 12 clinically healthy men, a 6-week abstention from drinking was associated in a reduction of $HDL_3$; no marked change in $HDL_2$ was observed. Resumption of drinking was again followed by a rise in $HDL_3$. No such changes in $HDL_3$ were observed in 12 control subjects who continued their drinking throughout the entire 12-week period. If one assumes that the $HDL_3$ is not involved in the antiatherogenic association of HDL-C with alcohol-induced changes, other factors need to be invoked to explain the cardioprotective effects of alcohol. However, the recent findings on an inverse interrelationship between $HDL_3$ and coronary artery occlusion (Levy et al., 1984) indicate a need for reevaluation of the association of CD and the HDL subclasses.

Changes in Apolipoprotein Levels. Several studies have indicated that the receptor-mediated metabolic processing of plasma lipoproteins is governed by the protein moieties of the lipoproteins, i.e., the apolipoproteins (Havel, 1982). In general, investigators dealing with the effect of alcohol on apolipoproteins have primarily considered the changes in two principal lipoproteins of the HDL, Apo AI and Apo AII. Statistically significant increases in Apo I and Apo II levels in consumers of alcoholic beverages have been reported for both the general population groups (Dedonder-Decoopman et al., 1980; Fex et al., 1982; Fraser et al., 1983) and in alcoholics (Barboriak et al., 1981*a;* Marth et al., 1982). It is of interest that in the study suggesting that alcohol consumption is associated with an increase in $HDL_3$ rather than $HDL_2$ (Haskell et al., 1984) both Apo AI and AII were increased with alcohol intake and showed a reduction on abstinence Camargo et al., 1984). The levels of Apo B, the main apolipoprotein of the LDL, believed to be involved in the atherogenesis, are usually lower in actively imbibing alcoholics and show an increase on abstinence (Barboriak et al., 1981*a*). It would seem therefore that the observed changes in apolipoprotein levels correlate well with changes in the levels of lipoprotein classes and subclasses.


## 5.  ALCOHOL AND CARDIOVASCULAR DISEASE

The relatively recent increased interest in the alcohol-induced blood lipid changes is mainly due to the suggestion that the reported lower morbidity and mortality of imbibers of alcoholic beverages may be due to a favorable blood lipid modification in such individuals.


### Cardiovascular Disease

The notion of a "beneficial" effect of moderate drinking on the development of CD is mostly based on the results of a large number of epidemiological studies indicating less extensive coronary artery disease (Barboriak et al., 1977) and lower cardiovascular mortality and morbidity in individuals consuming alcoholic beverages (Chiba et al., 1983; Gordon and Kannell, 1983; Hennekens et al., 1979; Klatsky et al., 1974; Kono et al., 1983; Kozarevic et al., 1982; Marmot et al., 1981; Poikolainen and Simpura, 1983; Salonen et al., 1983; Stason et al., 1976). The complexities and pitfalls in the interpretation of some of these studies have been discussed in a previous volume of this series (Ashley, 1984).

The reported negative association between alcohol intake and CD is rather unexpected, since most of the previous reports dealing with the alcohol–heart interaction found impairment of heart function (Parker, 1974; Regan, 1971) and abnormal macroscopic and microscopic myocardial morphology (Alexander, 1975; Burch and Giles, 1971; Burch and Walsh, 1960; Pintar et al., 1965). In fact, several studies have reported *increased* cardiovascular morbidity and mortality in individuals with *alcohol abuse* (Dyer et al., 1977; Pell and D'Alonzo, 1973; Wilhelmsen et al., 1978). The main difference between these two groups of investigations appears to be the amount of alcohol consumed, since the studies describing an attenuating effect of alcohol on CD used alcohol amounts usually associated with social drinking.

The possible mechanism(s) that could explain the observed cardioprotective effects of alcohol have not been definitively elucidated. Animal studies, using rabbbits and subhuman primates as experimental species, have shown that inclusion of alcohol into their atherogenic diet has reduced development of atherosclerosis in both species (Goto et al.,1974; Klurfeld and Kritchevsky, 1981; Rudel et al., 1981). It would seem therefore that alcohol does exert a real reducing effect on arterial occlusion. While most investigators believe that this effect is mediated by the alcohol-induced changes in plasma lipids, especially the HDL-C, other factors have also been implicated. They include, among others, modification of the coagulation and fibrinolytic processes (Barboriak, 1984), which have also been implicated in atherogenesis.


## 6.   ALCOHOL-INDUCED CHANGES IN MYOCARDIAL LIPIDS

### Myocardial Lipid Metabolism

In a small proportion of chronic alcoholics, perhaps 1–3%, irreversible degeneration of the myocardium, i.e., cardiomyopathy, develops (Asokan et al., 1972; Rubin, 1982). Alcoholic cardiomyopathy in man is associated with an accumulation of neutral lipids in the myocardial fibers (Ferrans et al., 1965) and the presence of numerous lipid droplets throughout the myocardium, even in areas in which no other alterations are demonstrable with electron microscopy (Hibbs et al., 1965). The administration of alcohol to mice (Alexander et al., 1977*a,b*), rats (Lieber et al., 1966; Williams and Li, 1977), rabbits (Kikuchi and Kako, 1970), or dogs (Marciniak et al., 1968; Regan et al., 1974) is also associated with an increase in the content of triacylglycerols in the heart. Diminished myocardial fatty acid oxidation is also found in the rat (Lochner et al., 1969), rabbit (Kikuchi and Kako, 1970), and dog (Regan et al., 1974) given ethanol. Although mitochondrial damage (Bing, 1982) might contribute to a lessened fatty acid oxidation by the heart, the ability to transport and oxidize fatty acid by the rat myocardium is found to be unaffected by chronic ingestion of alcohol (Williams and Li, 1977).

Since there is neither appreciable activity of alcohol dehydrogenase in the myocardium (Forsyth et al., 1976) nor significant oxidation of ethanol by the heart (Lochner et al., 1969), the factor(s) responsible for the accumulation of triacylglycerols in the heart during alcohol ingestion are not the result of myocardial ethanol oxidation. Recently, the incorporation of [$^{14}$C]ethanol into the neutral lipid fraction of rabbit myocardium was investigated with both the isolated, perfused organ and cell-free homogenate and the formation of fatty acid ethyl esters was identified (Lange et al., 1981). The esterification of fatty acid with ethanol has been shown to be enzymatically mediated by cholesterol esterase (Lange, 1982). Fatty acid ethyl esters were also found in the myocardium from human subjects who had been exposed acutely or chronically to ethanol while they were consistently absent from heart tissue of abstainers (Lange, and Sobel, 1983). With a homogenate of rabbit heart tissue, Mogelson and Lange (1982) reported a concentration-dependent inhibition of triacylglycerol lipolysis by fatty acid ethyl esters. Such an inhibition of lipolysis could be a contributory factor in the accumulation of myocardial triacylglycerols found during exposure to alcohol.

## Secondary Effects of Hepatic Alcohol Metabolism

Acetaldehyde. Acetaldehyde, produced by the liver during alcohol ingestion, might also influence cardiac lipid metabolism. Indeed, acetaldehyde metabolism in isolated myocytes from adult rats decreased the oxidation of palmitate (Farmer and Williams, 1980). As blood acetaldehyde concentrations are below 50 $\mu$M during alcohol ingestion (Ericksson and Peachey, 1980), this inhibition of palmitate oxidation by 500 $\mu$M acetaldehyde in isolated heart cells is probably not significant during ethanol metabolism *in vivo*.

Acetate. A major part (50–100%) of the ethanol metabolized in the splanchnic area in human subjects is found as acetate in the hepatic venous blood (Lundquist et al., 1962). Shortly after a constant infusion of ethanol that yields an arterial alcohol concentration of 2–12 mM, arterial blood acetate concentration reaches a nearly constant level of 0.8–1.0 mM (Juhlin-Dannfelt, 1977; Lundquist et al., 1962). Furthermore, the ingestion of ethanol in amounts sufficient to produce a maximal output of acetate from the liver in human subjects increased myocardial utilization of acetate and lactate in proportion to their arterial concentration (Lindeneg et al., 1964). A corresponding decrease in myocardial uptake of FFA was also found following alcohol ingestion. Acetate seems to replace alternative substrates in the heart. The oxidation of 0.2 mM oleate by the isolated perfused rat heart is inhibited by 35% in the presence of 2 mM acetate (Peuhkurinen et al., 1983). Esterification of [1-$^{14}$C]oleate into myocardial triacylglycerols was also augmented by the addition of acetate to the perfusate.

In addition to decreasing myocardial oxidation of long-chain fatty acids and promoting their esterification, acetate inhibits isoproterenol-stimulated lipolysis (glycerol output) in isolated perfused rat heart (Hron et al., 1978). Thus, acetate, a product of hepatic alcohol metabolism (Lundquist et al., 1962), is not only utilized by the heart but can also promote accumulation of triacylglycerols in the myocardium as is found following the ingestion of alcohol.

Very-Low Density Lipoproteins. The possibility also exists that increased production of triacylglycerols as VLDL following the ingestion of alcohol would result not only in an increased extraction of plasma triacylglycerols by the myocardium but also in an augmentation of neutral lipid deposition in the heart. Indeed, acute administration of ethanol increased extraction of plasma triacylglycerols by the dog myocardium (Regan et al., 1966). While cardiac FAA uptake was also not altered during chronic administration of ethanol, there was a significantly increased incorporation of $[^{14}C]$oleic acid into triacylglycerols, a reduced production of $^{14}CO_2$ and a diminished incorporation of label into phospholipids (Regan et al., 1974). Enhanced esterification during chronic ethanol ingestion has been implicated as a major factor in the accumulation of triacylglycerols by the myocardium, but a reduced activity of the hormone-sensitive (intracellular) lipase cannot be excluded (Regan et al., 1975). As discussed previously, the effects of acetate on lipid metabolism in the isolated perfused rat heart would also be compatible with the data obtained on myocardial triacylglycerols in the dog following chronic administration of alcohol (Regan et al., 1975).

## 7.  CONCLUSIONS

The present knowledge of the effects of alcohol on blood and tissue lipids is undergoing a rapid expansion. This review has attempted to include some of the more recent experimental findings and provide new insights and interpretations.

Considerable interest is being focused on the results of large epidemiological studies, suggesting that consumption of alcohol may be associated with blood lipid changes favoring reduction of cardiovascular morbidity and mortality. However, the presently available information does not allow recommendations or suggestions that alcohol is considered as a possible corrective or preventive treatment for coronary artery disease.

1. The current knowledge of the metabolic processes involved in the alcohol–coronary artery disease interrelationship is too meager to support any definitive recommendations. New observations, such as the reported changes in $HDL_2$ and $HDL_3$ following alcohol administration (Haskell et al., 1984) indicate considerable "fluidity" of the field and suggest the need for more definitive clarification of the mechanism(s) involved.

2. The majority of known heart patients are already consumers of alcoholic beverages (Barboriak et al., 1981*b*). Increasing their usual intake of alcohol may enhance the risk of producing some of the undesirable psychological and biological effects of excessive drinking.
3. "Prescribing" alcohol to a previously abstinent adult carries a risk of possible addiction, with considerable adverse legal implications for the health professional and socioeconomic and medical dangers for the patient.
4. Increasingly, chronic consumption of not necessarily excessive amounts of alcohol, especially in smokers, is believed to be associated with higher prevalence of malignancies (Popham et al., 1983).

Therefore, encouraging chronic alcohol consumption without sufficient evidence and knowledge of the "beneficial" and harmful effects of such a habit should not be considered.

## ACKNOWLEDGMENTS

Work cited as being done by the authors was supported by NIH grants HL-14378 and HL-29011. The typing and editorial assistance of Catherine A. Walther, Marianne Menahan, and Gloria Sullivan is gratefully acknowledged.

## REFERENCES

Albrink, M. J., and Klatskin, G., 1957, Lactescence of serum following episodes of acute alcoholism and its probable relationship to acute pancreatitis, *Am. J. Med.* **23**:26.

Alexander, C. S., 1975, Alcoholic cardiomyopathy, *Postgrad. Med. J.* **58**:127.

Alexander, C. S., Forsyth, G. W., Nagasawa, H. T., and Kohlhoff, J. G., 1977*a*, Alcoholic cardiomyopathy in mice. Myocardial glycogen, lipids and certain enzymes, *J. Mol. Cell. Cardiol.* **9**:235.

Alexander, C. S., Sekhri, K. K., and Nagasawa, H. T., 1977*b*, Alcoholic cardiomyopathy in mice. Electron microscopic observations, *J. Mol. Cell. Cardiol.* **9**:247.

Anderson, D. W., Nichols, A. V., Pan, S. S., and Lindgren, F. T., 1978, High density lipoprotein distribution. Resolution and determination of three major components in a normal population sample, *Atherosclerosis* **29**:161.

Arai, M., Gordon, E. R., and Lieber, C. S., 1984, Decreased cytochrome oxidase activity in hepatic mitochondria after chronic ethanol consumption and the possible role of decreased cytochrome $aa_3$ content and changes in phospholipid, *Biochim. Biophys. Acta* **797**:320.

Ashley, M. J., 1984, Alcohol consumption and ischemic heart disease, *Res. Adv. Alcohol Drug Probl.* **8**:99.

Asokan, S. K., Frank, M. J., and Whittam, A. C., 1972, Cardiomyopathy without cardiomegaly in alcoholics, *Am. Heart J.* **81**:13.

Avogaro, P., and Cazzolato, G., 1975, Changes in the composition and physicochemical characteristics of serum lipoproteins during ethanol-induced lipaemia in alcoholic subjects, *Metabolism* **24**:1231.

Ballantyne, F. C., Clark, R. S., Simpson, H. S., and Ballantyne, D., 1982, High density and low density lipoprotein subfraction in survivors of myocardial infarction and in control subjects, *Metabolism* **31**:433.

Baraona, E., and Lieber, C. S., 1979, Effects of ethanol on lipid metabolism, *J. Lipid Res.* **20**:289.

Barboriak, J. J., 1984, Alcohol, lipids and heart disease, *Alcohol* **1**:341.

Barboriak, J. J., and Meade, R. C., 1968, Enhancement of alimentary lipemia by preprandial alcohol, *Am. J. Med. Sci.* **255**:245.

Barboriak, J. J., and Menahan, L. A., 1979, Hyperlipemic effects of alcohol, in: *Biochemistry and Pharmacology of Ethanol*, Vol. 1 (E. Majchrowicz and E. P. Noble, eds.), pp. 587–601, Plenum Press, New York.

Barboriak, J. J., Rimm, A. A., Anderson, A. J., Schmidhofer, M., and Tristani, F. E., 1977, Coronary artery occlusion and alcohol intake, *Br. Heart J.* **39**:289.

Barboriak, J. J., Jacobson, G. R., Cushman, P., Herrington, R. E., Lipo, R. F., Daley, M. E., and Anderson, A. J., 1980, Chronic alcohol abuse and high density lipoprotein cholesterol, *Alcohol. Clin. Exp. Res.* **4**:346.

Barboriak, J. J., Alaupovic, P., and Cushman, P., 1981a, Abstinence-induced changes in plasma apolipoprotein levels of alcoholics, *Drug Alcohol Depend.* **8**:337.

Barboriak, J. J., Barboriak, D. P., Anderson, A. J., and Hoffman, R. G., 1981b, Drinking patterns and preferences among heart patients, *Curr. Alcohol* **8**:293.

Barrett-Connor, E., and Suarez, L., 1982, A community study of alcohol and other factors associated with the distribution of high density lipoprotein cholesterol in older vs. younger men, *Am. J. Epidemiol.* **115**:888.

Belfrage, P., Berg, B., Hagerstrand, I., Nilsson-Ehle, P., Torqvist, H., and Wieber, T., 1977, Alterations of lipid metabolism in healthy volunteers during long-term ethanol intake, *Eur. J. Clin. Invest.* **7**:127.

Bernstein, J. D., and Penniall, R., 1978, Effects of chronic ethanol treatment upon rat liver mitochondria, *Biochem. Pharmacol.* **27**:2337.

Bing, R. J., 1982, Effect of alcohol on the heart and cardiac metabolism, *Fed. Proc.* **41**:2443.

Blomstrand, R., Kager, L., and Lantto, O., 1973, Studies on the ethanol-induced decrease of fatty acid oxidation in rat and human liver slices, *Life Sci.* **13**:1131.

Brindley, D. N., and Sturton, R. G., 1982, Phosphatidate metabolism and its relation in triacylglycerol biosynthesis, in: *Phospholipids* (J. N. Hawthorne and G. B. Ansell, eds.), pp. 179–213, Elsevier, Amsterdam.

Brindley, D. N., Cooling, J., and Burditt, S. L., 1979a, Dietary and hormonal control of fatty acid esterification in the liver, in: *Obesity—Cellular and Molecular Aspects*, Vol. 87 (G. Alhoud, ed.), pp. 251–262, INSERM, Paris.

Brindley, D. N., Cooling, J., Burditt, S. L., Pritchard, P. H., Pawson, S., and Sturton, R. G., 1979b, The involvement of glucocorticoids in regulating the activity of phosphatidate phosphohydrolase and the synthesis of triacylglycerols in the liver. Effects of feeding rats with glucose, sorbitol, fructose, glycerol and ethanol, *Biochem. J.* **180**:195.

Burch, G. E., and Giles, T. D., 1971, Alcoholic cardiomyopathy. Concept of the disease and its treatment, *Am. J. Med.* **50**:141.

Burch, G. E., and Walsh, J. J., 1960, Cardiac insufficiency in chronic alcoholic, *Am. J. Cardiol.* **6**:864.

Buring, J. E., Willett, W., Goldhaver, S. Z., Mayrent, S. L., Breslow, J., and Hennekens, C. H., 1983, Alcohol HDL and nonfatal MI: Preliminary results from a case control study, *Circulation* **68**(Suppl. III):227.

Camargo, C. A., Jr., Albers, J. J., Vranizan, K., and Wood, P. D., 1984, Effects of moderate alcohol intake on serum apolipoproteins in man—a controlled study, *Fed. Proc.* **43**:1533.

Castelli, W. P., Gordon, T., Hjortland, M. C., Kagan, A., Koyle, J. T., Hames, C. G., Hulley, S. B., and Zukel, W. J., 1977, Alcohol and blood lipids: The Cooperative Lipoprotein Phenotyping Study, *Lancet* **1**:153.

Cederbaum, A. I., Lieber, C. S., and Rubin, E., 1974, Effects of chronic ethanol treatment on mitochondrial functions, *Arch. Biochem. Biophys.* **165**:560.

Chiba, K., Koizumin, A., Kumai, M., Watanabe, T., and Ikeda, M., 1983, Nationwide survey of high density lipoprotein cholesterol among farmers in Japan, *Prev. Med.* **12**:508.

Crouse, J. R., and Grundy, S. M., 1984, Effects of alcohol on plasma lipoproteins and cholesterol and triglyceride metabolism in man, *J. Lipid Res.* **25**:486.

Danielsson, B., Ekman, R., Fex, G., Johansson, B. G., Kristensson, H., Nilsson-Ehle, P. and Wadstein, J., 1978, Changes in plasma high-density lipoproteins in chronic male alcoholics during and after abuse, *Scand. J. Clin. Lab. Invest.* **38**:113.

Dedonder-Decoopman, E., Fiebet-Desreumaux, C., Compos, E., Moulin, S., Dewailly, P., Sezille, G., and Jaillard, J., 1980, Plasma levels of VLDL + LDL-cholesterol, HDL-cholesterol, triglycerides and apoproteins B and A-I in a healthy population, *Atherosclerosis* **37**:559.

Devenyi, P., Kapur, B. M., and Roy, J. H. J., 1984, High density lipoprotein response to alcohol consumption and abstinence as an indicator of liver function in alcoholic patients, *Can. Med. Assoc. J.* **130**:1445.

Dich, J., Bro, B., Grunnet, N., Jensen, F., and Kondrup, J., 1983, Accumulation of triacylglycerol in cultured rat hepatocytes is increased by ethanol and by insulin and dexamethasone, *Biochem. J.* **212**:617.

Duhamel, G., Nalpas, B., Goldstein, S., Lapland, P. M., Berthelot, P., and Chapman, J. M., 1984, Plasma lipoprotein and apolipoprotein profile in alcoholic patients with and without liver disease: On the relative roles of alcohol and liver injury, *Hepatology* **4**:577.

Dyer, A. E., Stamler, J., Berkson, P. O., Lepper, M. H., McKean, H., Shekelle, R. B., Lindberg, H. A., and Garside D., 1977, Alcohol consumption, cardiovascular risk factors, and mortality in two Chicago epidemiologic studies, *Circulation* **56**:1067.

Eckardt, M. S., Harford, T. C., Kaelber, C. T., Parker, E. S., Rosenthal, L. S., Ryback, R. S., Salmoiraghi, G. C., Vanderveen, E., and Rarren, K. R., 1981, Health hazards associated with alcohol consumption, *J. Am. Med. Assoc.* **246**:648.

Eder, H. A., and Bergman, M., 1983, Metabolism of plasma lipids and lipoproteins, in: *Diabetes Mellitus, Theory and Practice*, 3rd ed. (M. Ellenburg and H. Rifkin, eds.) pp. 61–76, Excerpta Medica, Amsterdam.

Ekman, R., Fex, G., Johansson, B. G., Nilsson-Ehle, P., and Wadstein, J., 1981, Changes in plasma high-density lipoproteins and lipolytic enzymes after long-term heavy ethanol consumption, *Scand. J. Clin. Lab. Invest.* **41**:701.

Eriksson, C. J. P., and Peachey, J. E., 1980, Absence of blood acetaldehyde differences between alcoholics and controls after ethanol ingestion, *Drug Alcohol Depend.* **6**:35.

Erkelens, W. D., and Brunzell, J. D., 1980, Effect of controlled alcohol feeding on triglycerides in patients with outpatient "alcohol hypertriglyceridemia," *J. Hum. Nutr.* **34**:370.

Farmer, R. B., and Williams, E. S., 1980, The oxidation of acetaldehyde by isolated adult rat heart cells, *Fed. Proc.* **39**:1753.

Feigl, J., 1918, Neue Untersuchungen zur Chemie des Blutes bei akuter Alkohol Intoxikation und bei chronischem Alkoholismus mit besonderer Berücksichtigung der Fette und Lipoide, *Biochem. Z.* **92**:282.

Fellenius, E., and Kiessling, K. H., 1973, Effect of ethanol on fatty acid oxidation in the perfused livers of starved, fed and fat-fed rats, *Acta Chem. Scand.* **27**:2781.

Fellenius, E., Bengtsson, G., and Kiessling, K.H., 1973, The influence of ethanol-induced changes of the $\alpha$-glycerophosphate level on hepatic triglyceride synthesis, *Acta Chem. Scand.* **27**:2893.

Fellin, R., Agostini, B., Rost, W., and Seidel, D., 1974, Isolation and analysis of human plasma lipoproteins accumulating post-prandial in an intermediate density fraction (d.1.006–1.019 g/ml) *Clin. Chim. Acta* **54**:325

Felts, J. M., and Berry, M. N., 1971, The metabolism of free fatty acids and chylomicron triglyceride fatty acids by isolated rat liver cells, *Biochim. Biophys. Acta* **231**:1.

Felts, J. M., and Mayes, P. A., 1965, Lack of uptake and oxidation of chylomicron triglyceride to carbon dioxide and ketone bodies by the perfused rat liver, *Nature* **206**:195.

Ferrans, V. J., Hibbs, R. G., Weilbaecher, D. G., Black, W. C., Walsh, J. J., and Burch, G. E., 1965, Alcoholic cardiomyopathy. A histochemical study, *Am. Heart J.* **69**:748.

Fex, G., Kristenson, H., and Trell, E., 1982, Correlation of serum lipids and lipoproteins with gamma-glutamyltransferase and attitude to alcohol consumption, *Ann. Clin. Biochem.* **19**:345.

Floren, C. H., and Nilsson, A., 1977, Binding, interiorization and degradation of cholesterol ester-labeled chylomicron-remnant particles by rat hepatocyte monolayers, *Biochem. J.* **168**:483.

Forsander, O. A., Mäenpää, P. H., and Salaspuro, M. P., 1965a, Influence of ethanol on the lactate/pyruvate and beta-hydroxybutyrate/acetoacetate ratios in rat liver experiments, *Acta Chem. Scand.* **19**:1770.

Forsander, O. A., Räihä, H., Salaspuro, M., and Mäenpää, P. H., 1965b, Influence of ethanol on the liver metabolism of fed and starved rats, *Biochem. J.* **94**:259.

Forsyth, G. W., Nagasawa, H. T., and Alexander, C. S., 1976, Ethanol metabolism by the rat heart and alcohol dehydrogenase activity, *Can. J. Biochem.* **54**:539.

Fraser, G. E., Anderson, J. T., Foster, N., Goldberg, D., Jacobs, D., and Blackburn, H., 1983, The effect of alcohol on serum high density lipoprotein (HDL), *Atherosclerosis* **46**:275.

French, S. W., Ihrig, T. J., and Morin, R. J., 1970, Lipid composition of red blood cell ghosts, liver mitochondria and microsomes of ethanol-fed rats, *Q. J. Stud. Alcohol* **31**:801.

French, S. W., Ihrig, T. J., Shaw, G. P., Tanaka, T. T., and Norum, M. L., 1971, The effect of ethanol on fatty acid composition of hepatic microsomes and inner and outer mitochondrial membranes, *Res. Commun. Chem. Pathol. Pharmacol.* **2**:567.

Fry, M. M., Spector, A. A., Connor, S. L., and Connor, W. E., 1973, Intensification of hyper-triglyceridemia by either alcohol or carbohydrate, *Am. J. Clin. Nutr.* **26**:798.

Fuchs, R., and Scheidt, S. S., 1983, Prevention of coronary atherosclerosis, Part I, *Cardiovasc. Rev. Rep.* **4**:671.

Goldberg, C. S., Tall, A. R., and Kumholz, S., 1984, Acute inhibition of hepatic lipase and increase in plasma lipoproteins after alcohol intake, *J. Lipid Res.* **25**:714.

Gordon, E. R., 1973, Mitochondrial functions in an ethanol-induced fatty liver, *J. Biol. Chem.* **248**:8271.

Gordon, E. R., 1984, Alcohol-induced tolerance in mitochondrial membranes, *Science* **223**:193.

Gordon, T., and Kannell, W. B., 1983, Drinking habits and cardiovascular disease: The Framingham Study, *Am. Heart J.* **105**:667.

Gordon, E. R., Rochman, J., Arai, M., and Lieber, C. S., 1982, Lack of correlation between hepatic mitochondrial membrane structure and functions in ethanol-fed rats, *Science* **216**:1319.

Goto, Y., Kikuchi, H., Abe, K., Nagawashi, Y., Ohira, S., and Kudo, H., 1974, The effects of ethanol on the onset of experimental atherosclerosis, *Tohoku J. Exp. Med.* **114**:35.

Gruchow, H. W., Hoffmann, R. G., Anderson, A. J., and Barboriak, J. J., 1982, Effects of drinking patterns on the relationship between alcohol and coronary occlusion, *Atherosclerosis* **43**:394.

Grunnet, N., and Kondrup, J., 1983, Effect of ethanol, noradrenaline and 3′,5′-cyclic AMP on oxidation of fatty acids and lipolysis in isolated rat hepatocytes, *Pharmacol. Biochem. Behav.* **18**(Suppl. 1):245.

Haagsman, H. P., DeHass, G. M., Geelen, M. J. H., and Van Golde, L. M. G., 1982, Regulation of triacylglycerol synthesis in the liver. Modulation of diacylglycerol acyltransferase activity *in vitro*, *J. Biol. Chem.* **257**:10593.

Haskell, W. L., Camargo, C., Williams, P. T., Vranizan, K. M., Krause, R. M., Lindgren, F. T., and Wood, P. D., 1984, The effect of cessation and resumption of moderate alcohol intake on serum high-density lipoprotein subfractions, *N. Engl. J. Med.* **310**:805.

Havel, R. J., 1982, Approach to the patient with hyperlipidemia, *Med. Clin. North Am.* **66**:319.

Havel, R. J., Kane, J. P., and Kashyap, M. L., 1973, Interchange of apolipoproteins between chylomicrons and high density lipoproteins during alimentary lipemia, *J. Clin. Invest.* **52**:32.

Heiss, G., Johnson, N. J., Reiland, S., Davis, C. E., and Tyroler, H. A., 1980, The epidemiology of plasma high-density lipoprotein cholesterol levels, *Circulation* **62**:116.

Hennekens, C. H., Willett, W., Rosner, B., Cols, D. S., and Mayrent, A. L., 1979, Effects of beer, wine and liquor in coronary deaths, *J. Am. Med. Assoc.* **242**:1973.

Hibbs, R. G., Ferrans, V. J., Black, W. C., Weilbaecher, D. G., Walsh, J. J., and Burch, G. E., 1965, Alcohol cardiomyopathy. An electron microscopic study, *Am. Heart J.* **69**:766.

Hron, W. T., Menahan, L. A., and Lech, J. J., 1978, Inhibition of hormonal stimulation of lipolysis in perfused rat heart by ketone bodies, *J. Mol. Cell. Cardiol.* **10**:161.

Jacqueson, A., Richard, J. L., Ducimetiere, P., Warnet, J. M., and Claude, J. R., 1983, High density lipoprotein cholesterol and alcohol consumption in a French male population, *Atherosclerosis* **48**:131.

Johansson, B. G., and Laurell, C. B., 1969, Disorders of serum α-lipoproteins after alcohol intoxication, *Scand. J. Clin. Lab. Invest.* **23**:231.

Joly, J. G., Feinman, L., Ishii, H., and Lieber, C. S., 1973, Effect of chronic ethanol feeding on hepatic microsomal glycerophosphate acyltransferase activity, *J. Lipid Res.* **14**:337.

Juhlin-Dannfelt, A., 1977, Ethanol effects of substrate utilization by the human brain, *Scand. J. Clin. Lab. Invest.* **37**:443.

Kaffarnik, H., and Schneider, J., 1970, Zur kurzfristigen Wirkung des Athylalkohols auf die Serumlipide gesunder fastender Versuchspersonen, *Z. Gesamte Exp. Med.* **152**:187.

Kagan, A., Yano, K., Rhoads, G. G., and McGee, D. L., 1981, Alcohol and cardiovascular disease: The Hawaiian experience, *Circulation* **64**(Suppl. III):27.

Kikuchi, T., and Kako, K. J., 1970, Metabolic effects of ethanol on the rabbit heart, *Circ. Res.* **26**:625.

Klatsky, A. L., Friedman, G. D., and Siegelaub, A. B., 1974, Alcohol consumption before myocardial infarction. Results from the Kaiser-Permanente epidemiological study of myocardial infarction, *Ann. Intern. Med.* **81**:294.

Klurfeld, D. M., and Kritchevsky, K., 1981, Differential effects of alcoholic beverages on experimental atherosclerosis in rabbits, *Exp. Mol. Pathol.* **34**:62.

Kondrup, J., Lundquist, F., and Damgaard, W. E., 1979, Metabolism of palmitate in perfused rat liver. Effect of low and high ethanol concentrations at various concentrations of palmitate in the perfusion medium, *Biochem. J.* **184**:83.

Kono, S., Ikeda, M., Ogata, M., Tokudome, S., Nishizumi, M., and Kuratsune, M., 1983, The relationship between alcohol and mortality among Japanese physicians, *Int. J. Epidemiol.* **12**:437.

Kozarevic, D., Demirovic, J., Gordon, T., Kaelber, C. G., McGee, D., and Zukel, W. J., 1982, Drinking habits and coronary heart disease, *Am. J. Epidemiol.* **116**:748.

Kudzma, D. J., and Schonfeld, G., 1971, Alcoholic hyperlipidemia: Induction by alcohol but not by carbohydrate, *J. Lab. Clin. Med.* **77**:384.

Lamb, R. G., Wood, C. K., and Fallon, H. J., 1979, The effect of acute and chronic ethanol intake on hepatic glycerolipid biosynthesis in the hamster, *J. Clin. Invest.* **63**:14.

Lange, L. G., 1982, Nonoxidative ethanol metabolism: Formation of fatty acid ethyl esters by cholesterol esterase, *Proc. Natl. Acad. Sci. USA* **79**:3954.

Lange, L. G., and Sobel, B. E., 1983, Myocardial metabolites of ethanol, *Circ. Res.* **52**:479.

Lange, L. G., Bergmann, S. R., and Sobel, B. E., 1981, Identification of fatty acid ethyl esters as products of rabbit myocardial ethanol metabolism, *J. Biol. Chem.* **256**:12968.

Levy, R. I., Brensike, J. F., Epstein, S. E., Kelsey, S. F., Passamani, E. R., Richardson, J. M., Loh, J. K., Stone, N. J., Aldreich R. F., Battaglini, J. W., Fisher, M. L., Friedman, L., Friedewald, W., and Detre, K. M., 1984, The influence of changes in lipid values induced by cholestyramine and diet on progression of coronary artery disease: Results of the NHLBI Type II Coronary Intervention Study, *Circulation* **69**:325.

Lieber, C. S., 1982, *Medical Disorders of Alcoholism: Pathogenesis and Treatment,* W. B. Saunders Co., Philadelphia.

Lieber, C. S., 1984, Metabolism and metabolic effects of alcohol, *Med. Clin. North Am.* **68**:3.

Lieber, C. S., and Schmid, R., 1961, The effect of ethanol on fatty acid metabolism: Stimulation of hepatic fatty acid synthesis *in vitro, J. Clin. Invest.* **40**:394.

Lieber, C. S. Jones, D. P., and DeCarli, L. M., 1965, Effects of prolonged ethanol intake: Production of fatty liver despite adequate diets, *J. Clin. Invest.* **44**:1009.

Lieber, C. S., Jones, D. P., Mendelson, J., and DeCarli, L. M., 1963, Fatty liver, hyperlipemia and hyperuricemia produced by prolonged alcohol consumption despite adequate dietary intake, *Trans. Assoc. Am. Physicians* **76**:289.

Lieber, C. S., Spritz, N., and DeCarli, L. M., 1966, Accumulation of triglycerides in heart and kidney after alcohol ingestion, *J. Clin. Invest.* **45**:1041.

Lieber, C. S., Lefevre, A., Spritz, N., Feinman, L., and DeCarli, L. M., 1967, Difference in hepatic metabolism of long- and medium-chain fatty acids. The role of fatty acid chain length in the production of the alcoholic fatty liver, *J. Clin. Invest.* **46**:1451.

Lindeneg, O., Mellemgaard, K., Fabricius, J., and Lundquist, F., 1964, Myocardial utilization of acetate, lactate and free fatty acids after ingestion of ethanol, *Clin. Sci.* **27**:427.

Lindros, K. O., 1970, Suppression by ethanol of acetate and hexanoate oxidation studied by non-recirculating liver perfusion, *Biochem. Biophys. Res. Commun.* **41**:635.

Lochner, A., Cowley, R., and Brink, A. J., 1969, Effect of ethanol on metabolism and function of perfused rat heart, *Am. Heart J.* **78:770.**

Losowsky, M. S., Jones, D. P., Davidson, C. S., and Lieber, C. S., 1963, Studies of alcoholic hyperlipemia and its mechanism, *Am. J. Med.* **35**:794.

Lumeng, L., and Davis, E. J., 1973, The oxidation of acetate by liver mitochondria, *FEBS Lett.* **29**:124.

Lundquist, F., 1975, Interference of ethanol in cellular metabolism, *Ann. NY Acad. Sci.* **252**:11.

Lundquist, F., Tygstrup, N., Winkler, E., Mellemgaard, K., and Munck-Petersen, S., 1962, Ethanol metabolism and production of free acetate in the human liver, *J. Clin. Invest.* **41**:955.

Majchrowicz, E., and Quastel, J. H., 1961, Effects of aliphatic alcohols and fatty acids on the metabolism of acetate by rat liver slices, *Can. J. Biochem. Physiol.* **39**:1895.

Marciniak, M., Gudbjarnason, S., and Bruce, T. A., 1968, The effect of chronic alcohol administration on enzyme profile and glyceride content of heart muscle, brain and liver, *Proc. Soc. Exp. Biol. Med.* **128**:1021.

Marmot, M. G., Rose, G., Shipley, M. J., and Thomas, B. J., 1981, Alcohol and mortality: A U-shaped curve, *Lancet* **1**:580.

Marth, E., Cazzolato, G., Bittolo, Bon, G., Avogaro, P., and Kostner, G. M., 1982, Serum concentrations of Lp(a) and other lipoprotein parameters in heavy alcohol consumers, *Ann. Nutr. Metab.* **26**:56.

Masarei, J. R. L., Fung, W. P., Prindiville, L. P., and Puddey, I. B., 1982, Lipids and lipoproteins in chronic alcohol abuse, *Ann. Acad. Med.* **11**:601.

Mayes, P. A., 1976, Control of hepatic triacylglycerol metabolism, *Biochem. Soc. Trans.* **4**:575.

Mayes, P. A., and Felts, J. M., 1967, Regulation of fat metabolism in the liver, *Nature* **215**:716.

McGarry, J. D., and Foster, D. W., 1971, The regulation of ketogenesis from oleic acid and the influence of antiketogenic agents, *J. Biol. Chem.* **246**:6247.

McGarry, J. D., and Foster, D. W., 1976, Ketogenesis and its regulation, *Am. J. Med.* **61**:9.

Menahan L. A., Ross, B. D., and Wieland, O., 1968, Studies on the mechanism of fatty acid and glucagon stimulated gluconeogenesis in the perfused rat liver, in *Stoffwechsel der isoliert perfundierten Leber* (W. Staib and R. Scholz, eds.), pp. 142–152, 3. Konferenz der Gesellschaft für Biologische Chemie, Springer-Verlag, Heidelberg.

Miller, N. E., Hammett, F., and Saltissi, S., 1981, Relation of angiographically defined coronary artery disease to plasma lipoprotein subfractions and apolipoproteins, *Br. Med. J.* **282**:1741.

Mishara, B. L., and Kastenbaum, R., 1980, *Alcohol and Old Age,* Grune and Stratton, Inc., New York.

Mogelson, S., and Lange, L. G., 1982, Altered myocardial triglyceride homeostasis: Mediation by ethanol and fatty acid ethyl esters, *Fed. Proc.* **41**:1922.

Nikkila, E. A., and Ojala, K., 1963, Role of hepatic L-α-glycerophosphate and triglyceride synthesis in the production of fatty liver by ethanol, *Proc. Soc. Exp. Biol. Med.* **113**:814.

Ontko, J. A., 1972, Metabolism of free fatty acids in isolated liver cells. Factors affecting the partition between esterification and oxidation. *J. Biol. Chem.* **247**:1788.

Ontko, J. A., 1973, Effects of ethanol on the metabolism of free fatty acids in isolated liver cells, *J. Lipid Res.* **14**:78.

Parker, B. M., 1974, The effects of ethyl alcohol on the heart, *J. Am. Med. Assoc.* **228**:741.

Pell, S., and D'Alonzo, C. A., 1973, A five year mortality study of alcoholics, *J. Occup. Med.* **15**:120.

Peuhkurinen, K. J., Kiviluoma, K. T., Hiltunen, J. K., Takala, T. E., and Hassinen, I. E., 1983, Effects of ethanol metabolites on intermediary metabolism in heart muscle, *Pharmacol. Biochem. Behav.* **18**(Suppl. 1):279.

Pikkukangas, A. H., Väänänen, R. A., Savolainen, M. J., and Hassinen, I. E., 1982, Precursor supply and hepatic enzyme activities as regulators of triacylglycerol synthesis in isolated hepatocytes and perfused liver, *Arch. Biochem. Biophys.* **217**:1982.

Pintar, K., Wolanskyj, B. M., and Buggay, E. R., 1965, Alcoholic cardiomyopathy, *Can. Med. Assoc. J.* **93**:103.

Poikolainen, K., and Simpura, J., 1983, One-year drinking history and mortality, *Prev. Med.* **12**:709.

Popham, R. E., Schmidt, W., and Israel, Y., 1983, Variation in mortality from ischemic heart disease in relation to alcohol and milk consumption, *Med. Hypotheses* **12**:321.

Pritchard, P. H., Bowley, M., Burditt, S. L., Cooling, J., Glenny, H. P., Lawson, N., Sturton, R. G., and Brindley, D. N., 1977, The effects of acute ethanol feeding and of chronic benfluorex administration on the activities of some enzymes of glycerolipid synthesis in rat liver and adipose tissue, *Biochem. J.* **166**:639.

Ragland, J. B., Heppner, C., and Sabesin, S. W., 1978, The role of lecithin-cholesterol acyltransferase deficiency in the apoprotein metabolism of alcoholic hepatitis, *Scand. J. Clin. Lab. Invest.* **38**(Suppl. 150):208.

Rawat, A. K., 1970, Effect of ethanol on glycerol metabolism in rat liver during different hormonal conditions, *Biochem. Pharmacol.* **19**:2791.

Rawat, A. K., and Lundquist, F., 1968, Influence of thyroxine on the metabolism of ethanol and glycerol in rat liver slices, *Eur. J. Biochem.* **5**:13.

Reboucas, G., and Isselbacher, K. J., 1961, Studies on the pathogenesis of the ethanol-induced fatty liver. I. Synthesis and oxidation of fatty acids by the liver, *J. Clin. Invest.* **40**:1355.

Regan, T. J., 1971, Ethyl alcohol and the heart, *Circulation* **44**:957.

Regan, T. J., Koroxenidis, G., Moschos, C. B., Oldewurtel, H. A., Lehan, P. H., and Hellems, H. K., 1966, The acute metabolic and hemodynamic responses of the left ventricle to ethanol, *J. Clin. Invest.* **45**:270.

Regan, T. J., Khan, M. I., Ettinger, P. O., Haider, B., Lyons, M. W., and Oldewurtel, H. A., 1974, Myocardial function and lipid metabolism in the chronic alcoholic animal, *J. Clin. Invest.* **54:**740.

Regan, T. J., Ettinger, P. O., Oldewurtel, H. A., and Haider, B., 1975, Heart cell responses to ethanol, *Ann. NY Acad. Sci.* **252:**250.

Rifkind, B. M., and Segal, P., 1983, Lipid research clinics program reference values for hyperlipidemia and hypolipidemia, *J. Am. Med. Assoc.* **250:**1869.

Rottenberg, H., Robertson, D. E., and Rubin, E., 1980, The effect of ethanol on the temperature dependence of respiration ATPase activities of rat liver mitochondria. *Lab. Invest.* **42:**318.

Rottenberg, H., Waring, A. J., and Rubin, E., 1981, Tolerance and cross-tolerance in chronic alcoholics: Reduced membrane binding of ethanol and other drugs, *Science* **213:**583.

Rottenberg, H., Waring, A., and Rubin, E., 1984, Alcohol-induced tolerance in mitochondrial membranes, *Science* **223:**193.

Rouach, H., Clement, M., Orfanelli, M. T., Janvier, B., and Nordmann, R., 1984, Fatty acid composition of rat liver mitochondrial phospholipids duirng ethanol inhalation, *Biochim. Biophys. Acta* **795:**125.

Rovinski, B., and Hosein, E. A., 1983, Adaptive changes in lipid composition of rat liver plasma membrane during postnatal development following maternal ethanol ingestion, *Biochim. Biophys. Acta* **735:**407.

Rubin, E., 1982, Alcohol and the heart: Theoretical considerations, *Fed. Proc.* **41:**2460.

Rubin, E., Beattie, D. S., and Lieber, C. S., 1970, Effects of ethanol on the biogenesis of mitochondrial membranes and associated mitochondrial functions, *Lab. Invest.* **23:**620.

Rudel, L. L., Leathers, C. W., Bond, M. G., and Bullock, B. C., 1981, Dietary ethanol-induced modifications in hyperlipoproteinemia and atherosclerosis in nonhuman primates, *Arteriosclerosis* **1:**144.

Sabesin, S. M., 1981, Lipid and lipoprotein abnormalities in alcoholic liver disease, *Circulation* **64**(Suppl. III):72.

Sabesin, S. M., and Weidman, S. W., 1984, Lipoprotein profiles in chronic alcoholics: Use of high density lipoprotein subspecies levels to differentiate subpopulations, *Hepatology* **4:**737.

Salonen, J. T., Puska, P., and Nissinen, A., 1983, Intake of spirits and beer and risk of myocardial infarction and death—a longitudinal study in eastern Finland, *J. Chronic Dis.* **36:**533.

Savolainen, M. J., 1977, Stimulation of hepatic phosphatidate phosphohydrolase activity by a single dose of ethanol, *Biochem. Biophys. Res. Comm.* **75:**511.

Savolainen, M. J., and Hassinen, I. E., 1978, Mechanisms for the effects of ethanol on hepatic phosphatidate phosphohydrolase, *Biochem. J.* **176:**885.

Savolainen, M. J., and Hassinen, I. E., 1980, Effect of ethanol on hepatic phosphatidate phosphodydrolase. Dose-dependent enzyme induction and its abolition by adrenalectomy and pyrazole treatment, *Arch. Biochem. Biophys.* **201:**640.

Savolainen, M. J., Baraona, E., Pikkarainen, P., and Lieber, C. S., 1984, Hepatic triacylglycerol synthesizing activity during progression of alcoholic liver injury in the baboon, *J. Lipid Res.* **25:**813.

Schaefer, E. J., Eisenberg, S., and Levy, R. I., 1978, Lipoprotein apoprotein metabolism, *J. Lipid Res.* **19:**667.

Schapiro, R. H., Drummey, G. D., Shimizu, Y., and Isselbacher, K. J., 1964, Studies on the pathogenesis of the ethanol-induced fatty liver. II. Effect of ethanol on palmitate-1-C$^{14}$ metabolism by the isolated perfused liver, *J. Clin. Invest.* **43:**1338.

Schapiro, R. H., Scheig, R. L., Drummey, G. D., Mendelson, J. H., and Isselbacker, K. J., 1965, Effect of prolonged ethanol ingestion on the transport and metabolism of lipids in man, *N. Engl. J. Med.* **272:**610.

Schonfeld, G., 1983, Disorders of lipid transport—Update 1983, *Prog. Cardiovasc. Dis.* **26:**89.

Sherrill, B. C., and Dietschy, J. M., 1978, Characterization of the sinusoidal transport processes responsible for uptake of chylomicrons by the liver, *J. Biol. Chem.* **253**:1859.

Smith, T. L., Vickers, A. E., Brender, K., and Yamamura, H. I., 1983, Influence of chronic ethanol treatment on alpha-adrenergic and vasopressin receptor-stimulted phosphotidylinositol synthesis in isolated rat hepatocytes, *Biochem. Pharmacol.* **32**:3059.

Srere, P. A., 1975, The enzymology of the formation and breakdown of citrate, *Adv. Enzymol.* **43**:57.

Stamford, B. A., Matter, S., Fell, R. D., Sady, S., Cresanta, M. K., and Papanek, P., 1984, Cigarette smoking, physical activity and alcohol consumption: Relationship to blood lipids and lipoproteins in premenopausal females, *Metabolism* **33**:585.

Stason, W. B., Neff, R. K., Miettinen, O. S., and Jick, H., 1976, Alcohol consumption and nonfatal myocardial infarction, *Am. J. Epidemiol.* **104**:603.

Steinberg, D., Pittman, R. C., Attie, A. D., Cares, T. E., Pangburn, S., and Weinstein, D., 1980, The role of the liver in LDL catabolism, in: *Atherosclerosis V* (A. M. Gotto, Jr., L. C. Smith, and B. Allen, eds.), pp. 800–803, Springer-Verlag, New York.

Sturton, R. G., Butterwith, S. C., Burditt, S. L., and Brindley, D. N., 1981, Effects of starvation, corticotropin injection and ethanol feeding on the activity and amount of phosphatidate phosphohydrolase in rat liver, *FEBS Lett.* **126**:297.

Thieden, H. I. D., and Lundquist, F., 1967, The influence of fructose and its metabolites on ethanol metabolism *in vitro, Biochem. J.* **102**:177.

Thompson, J. A., and Reitz, R. C., 1978, Effects of ethanol ingestion and dietary fat levels on mitochondrial lipids in male and female rats, *Lipids* **13**:540.

Thornton, J., Symes, C., and Heaton, K., 1983, Moderate alcohol intake reduces bile cholesterol saturation and raises HDL cholesterol, *Lancet* **2**:819.

Uthus, E. O., Skurdal, D. N., and Cornatzer, W. E., 1976, Effect of ethanol ingestion on choline phosphotransferase and phosphotidyl ethanolamine methyl transferase activities in liver microsomes, *Lipids* **11**:641.

Väänänen, R. A., Pikkukangas, A. H., Savolainen, M. J., and Hassinen, I. E., 1981, Cytosolic factor multiples the cotisol and ethanol effects on microsomal diacylglycerol acyltransferase, *Acta Pharmacol. Toxicol.* **49**(Suppl. IV):42.

Veech, R. L., Felver, M. E., Lakshmanan, M. R., Huang, M. T., and Wolf, S., 1981, Control of secondary pathway of ethanol metabolism by differences in redox state: A story of the failure to arrest the Krebs cycle for drunkenness, in: *Biological Cycles* (R. W. Estabrook and P. Srere, eds.), pp. 151–179, *Current Topics in Cellular Regulation*, Vol. 18, Academic Press, New York.

Waring, A. J., Rottenberg, H., Ohnishi, T., and Rubin, E., 1981, Membranes and phospholipids of liver mitochondria from chronic alcoholic rats are resistant to membrane disordering by alcohol, *Proc. Natl. Acad., Sci USA* **78**:2582.

Waring, A. J., Rottenberg, H., Ohnishi, T., and Rubin, E., 1982, The effect of chronic ethanol consumption on temperature-dependent physical properties of liver mitochondrial membranes, *Arch. Biochem. Biophys.* **216**:51.

Wiebe, T., and Belfrage, P., 1980, Effects of ethanol on fatty acid esterification and oxidation in isolated rat hepatocytes, *J. Stud. Alcohol* **41**:476.

Wieland, O. H., 1983, The mammalian pyruvate dehydrogenase complex: Structure and regulation, *Rev. Physiol. Biochem. Pharmacol.* **96**:124.

Wilhelmsen, L., Wedel, H., and Tibblin, G., 1978, Multivariate analysis of risk factors for coronary heart disease, *Circulation* **48**:950.

Willett, W., Hennekens, C. H., Siegel, A. J., Adner, M. M., and Castelli, W. P., 1980, Alcohol consumption and high density lipoprotein cholesterol in marathon runners, *N. Engl. J. Med.* **303**:1159.

Williams, E. S., and Li, T.-K., 1977, The effect of chronic alcohol administratiron on fatty acid metabolism and pyruvte oxidation of heart mitochondria, *J. Mol. Cell. Cardiol.* **9:**1003.

Williamson, J. R., Schulz, R., Browning, E. T., Thruman, R. G., and Fukami, M. H., 1969, Metabolic effects of ethanol in perfused rat liver, *J. Biol. Chem.* **244:**5044.

Wilson, E. E., Schreibman, P. H., Brewster, A. C., and Arky, R. A., 1970, The enhancement of alimentary lipemia by ethanol in man, *J. Lab. Clin. Med.* **75:**264.

Wing, D. R., Harvey, D. J., Hughes, J., Dunbar, P. G., McPherson, K. A., and Paton, W. D. M., 1982, Effects of chronic administration on the composition of membrane lipids in the mouse, *Biochem. Pharmacol.* **31:**3431.

Wing, D. R., Harvey, O. J., Belcher, S. J., and Paton, W. D. M., 1984, Changes in membrane lipid content after chronic ethanol administration with respect to fatty acyl compositions and phospholipid type, *Biochem. Pharmacol.* **33:**1625.

Wood, C. K., and Lamb, R. G., 1979, The effect of ethanol on glycerolipid biosynthesis by primary monolayer cultures of adult rat hepatocytes, *Biochim. Biophys. Acta* **572:**121.

Wu, A., and Windmueller, H. G., 1979, Relative contribution by liver and intestine to individual plasma apolipoproteins in the rat, *J. Biol. Chem.* **254:**7316.

Ylikahri, R. H., 1970, Ethanol-induced changes in hepatic $\alpha$-glycerophosphate and triglyceride concentrations in normal and thyroxine-treated rats, *Metabolism* **19:**1036.

Zakim, D., 1965, Effect of ethanol on hepatic acyl-coenzyme A metabolism, *Arch. Biochem. Biophys.* **111:**253.

Zieve, L., 1958, Jaundice, hyperlipemia and hemolytic anemia: A heretofore unrecognized syndrome associated with alcoholic fatty liver and cirrhosis, *Ann. Intern. Med.* **48:**471.

# 4

# Cellular Mechanisms Underlying Differences in Acute Ethanol Sensitivity

## Effects of Tolerance and Genetic Factors upon Neuronal Sensitivity to Ethanol


MICHAEL R. PALMER, THOMAS V. DUNWIDDIE, and
BARRY J. HOFFER


## 1.  INTRODUCTION

In this chapter, we will review studies of electrophysiological actions of ethanol on central neurons which may be correlated with the acute soporific effects of this drug. Since there have been several excellent reviews surveying the actions of ethanol on neuronal excitability at many levels of the neuraxis (Berry and Pentreath, 1980; Deitrich and Spuhler, 1984; Himwich and Callison, 1972; Klemm, 1979; Seiger et al., 1983; Siggins and Bloom, 1980), we should like to focus this review on those aspects of neuronal sensitivity to the effects of ethanol that may arise primarily as a result of genetic selection for acute sensitivity to ethanol-induced ataxia.


MICHAEL R. PALMER AND BARRY J. HOFFER ● Department of Pharmacology, University of
Colorado Health Sciences Center, Denver, Colorado 80262. THOMAS V. DUNWID-
DIE ● Medical Research Service, Denver Veterans Administration Hospital, Denver, Colorado
80220.

## 2.  DIFFERENCES IN ACUTE ETHANOL SENSITIVITIES OF SELECTIVELY INBRED LINES OF MICE

Previous neurophysiological investigations have demonstrated that the cerebellum is sensitive to the acute effects of systemically administered ethanol (Eidelberg et al., 1971; Kalant, 1974; Klemm and Stevens, 1974; Mitra, 1977; Rogers et al., 1980; Sinclair and Lo, 1980, 1981). Similarly, ethanol alters the firing rates of cerebellar Purkinje cells in rats when applied locally by micropressure-ejection and electro-osmosis (Siggins and Bloom, 1980; Siggins and French, 1979). The firing rates of single cerebellar Purkinje neurons from mice are also slowed by the micropressure-ejection application of ethanol from a second barrel of double-barrel recording pipette (Palmer et al., 1980; Sorensen et al., 1980).

The development of long-sleep (LS) and short-sleep (SS) mouse lines, which differ markedly in their ataxic responses to acute ethanol administration (Heston et al., 1973, 1974; McClearn et al., 1970), offered a unique opportunity to study the relationship between neurophysiological and behavioral effects of this drug. In LS mice the duration of the loss of righting response in response to a standard test dose of ethanol is over an order of magnitude longer than in SS mice. This differential behavioral sensitivity is paralleled by a similar difference in Purkinje neuron sensitivity to the inhibitory actions of ethanol. Purkinje neurons of LS mice require approximately a 30-fold lower dose of locally applied ethanol to elicit an equivalent inhibition of spontaneous neuronal discharge than do Purkinje cells from SS mice (Fig. 1). The heterogeneous stock (HS) of mice, which constitute the parental stock from which the LS and SS mice were bred, expressed Purkinje neuron sensitivities to ethanol intermediate to those found for LS and SS mice. This differential neuronal sensitivity to ethanol in LS and SS mice is apparently brain region-specific since Sorensen and colleagues (1981*b*) reported that no differences existed between LS and SS mice in terms of either the sensitivity of hippocampal pyramidal neuron firing rates to micropressure-ejected ethanol *in vivo* (Fig. 2) or the sensitivity of the hippocampal population spike to ethanol superfused in *in vitro* brain slices (Fig. 3).

The actions of ethanol applied locally to cerebellar Purkinje cells, as in the experiments described above, are most likely to reflect direct effects upon cerebellar neurons in the vicinity of the pipette tip. However, ethanol administered by this method might also alter the activity of nearby nerve terminals or other cell types in the same region, and previous investigations have shown that some of the effects of systemic ethanol may be mediated via actions on input pathways to the cerebellum (Rogers et al., 1980; Siggins and Bloom, 1980; Sinclair and Lo, 1980, 1981; Sorensen et al., 1981*a*). On the other hand, a direct action of ethanol in the cerebellum is supported by the work of Forney and Klemm (1976) who reported that the electrophysiological effects of systemic ethanol persist in cerebella that have been surgically isolated from the rest of the brain. In addition,

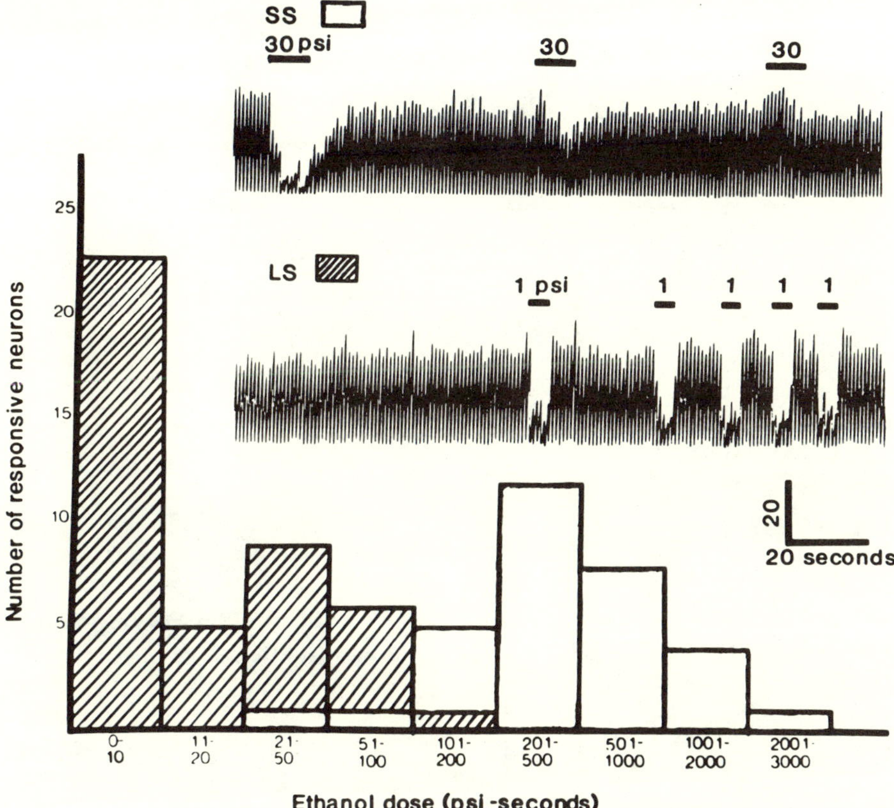


Figure 1. Sensitivity of cerebellar Purkinje neurons from LS and SS mice to the depressant effects of acute ethanol applied by micropressure-ejection. The number of neurons responding is indicated on the ordinate and the average dose to cause half-maximal inhibitions in pounds per square inch (psi) × seconds is plotted on the abscissa. The pressure ejection dose required to depress the firing rate of Purkinje neurons from LS mice (29 ± 6 psi-sec; hatched bars) was significantly lower ($p < 0.001$) than that required to depress SS Purkinje cells (888 ± 147 psi-sec; open bars). Inset, representative ratemeter records showing the effects of pressure-ejected ethanol (pressures indicated in psi above trace) on the spontaneous firing rates of single Purkinje neurons from SS (above) and LS (below) mice. In this and succeeding ratemeter records, the calibration bars indicate action potentials per second on the ordinate and time in seconds on the abscissa. The SS cell illustrated (above) shows rapid acute tolerance to a 30 psi local application of ethanol. (Adapted from Sorensen et al., 1980.)

ethanol perfused *in vitro* alters the firing rates of neurons in explants of cerebellum (Seil et al., 1977).

In order to determine whether the differences between LS and SS mice reflected direct or indirect actions of ethanol, we characterized the ethanol sensitivity of Purkinje neurons in intraocular brain grafts of LS and SS cerebella

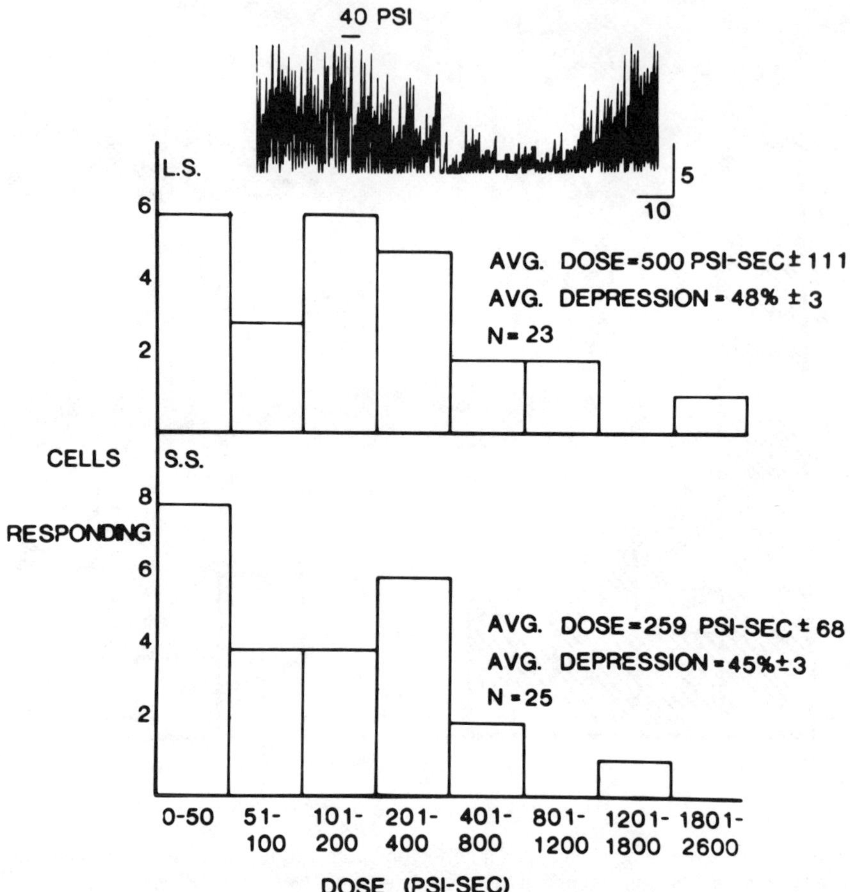



Figure 2. Effects of locally applied ethanol on hippocampal pyramidal neurons *in situ*. Bars indicate the number of neurons manifesting a 30–70% depression of spontaneous activity to micropressure-ejected ethanol for LS (upper) and SS (lower) cells. Typical ratemeter record for an LS neuron is shown in the inset above the bar graphs; calibration represents 5 spikes/sec, 10 sec. There were no significant differences in the ethanol sensitivities of hippocampal pyramidal neurons between LS and SS mice. (From Sorensen et al., 1981*b*.)

(Palmer et al., 1982). Cerebellar buds were obtained from 11- to 12-day fetuses of LS and SS mice and were homologously transplanted into the anterior eye chamber of adult mice (Olson and Malmfors, 1970; Olson et al., 1983), where they were allowed to mature prior to physiological investigations (Hoffer et al., 1974). Fetal cerebellar tissue was transplanted both within and across mouse lines. The cerebellar grafts attached to the anterior surface of the iris and de-

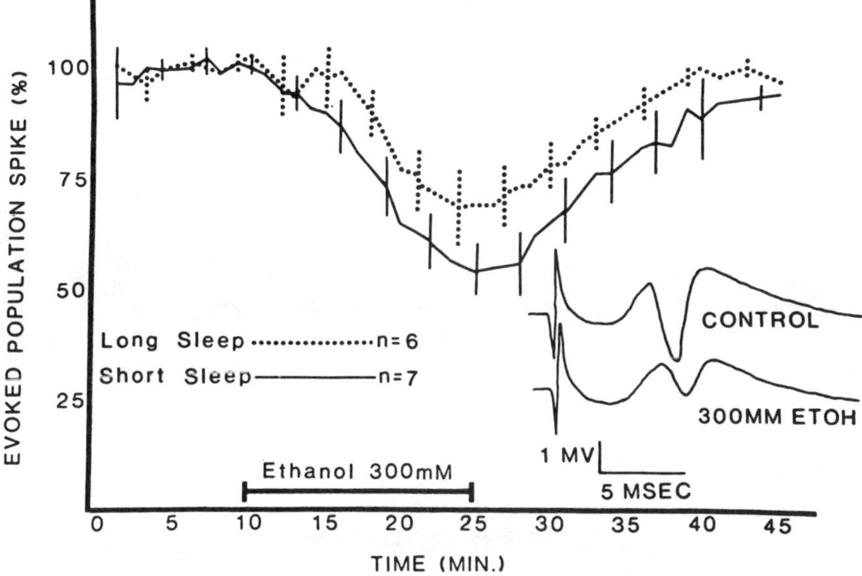


Figure 3. Effects of superfusion of 300 mM ethanol on the size of the pyramidal neuron population spike evoked by stimulation of Schaffer and commissural afferents in the *in vitro* hippocampal slice. Curves show mean normalized spike amplitude with SEM displayed for every fourth point. Stimuli were pairs of shocks at 5-sec intervals delivered once per minute. Typical responses during control and ethanol perfusion are inset in the lower right portion of the figure. The first deflection is the stimulus artifact; the negatively sloping (downward) part of the waveform with the roughly 5-msec latency to onset is the population spike. This spike results from the simultaneous firing of a population of pyramidal neurons in the vicinity of the electrode in response to stimulation of the afferent pathway and the size of the spike reflects the number of neurons responding to the stimulation. There were no significant differences in the effects of ethanol for decreasing population spike size between LS and SS mice. (From Sorensen et al., 1981*b*.)

veloped a trilaminar organization typical of the *in situ* cerebellum (Palmer et al., 1982). When such grafts are superfused *in oculo* with ethanol, characteristic dose-dependent depressions of Purkinje neuron firing rates are observed (Fig. 4). Purkinje neurons from LS grafts were about an order of magnitude more sensitive to the depressant effects of perfused ethanol than were Purkinje cells from SS grafts (Fig. 5). Because the *in oculo* transplants are not innervated directly by the central nervous system, the observed ethanol sensitivity difference could not be due to indirect actions on central afferents. However, ethanol effects on adrenergic and cholinergic axons from the host iris, which are known to innervate cerebellar grafts *in oculo* (Taylor et al., 1980), cannot be ruled out.

We also found that the intrinsic ethanol sensitivity of Purkinje cells is

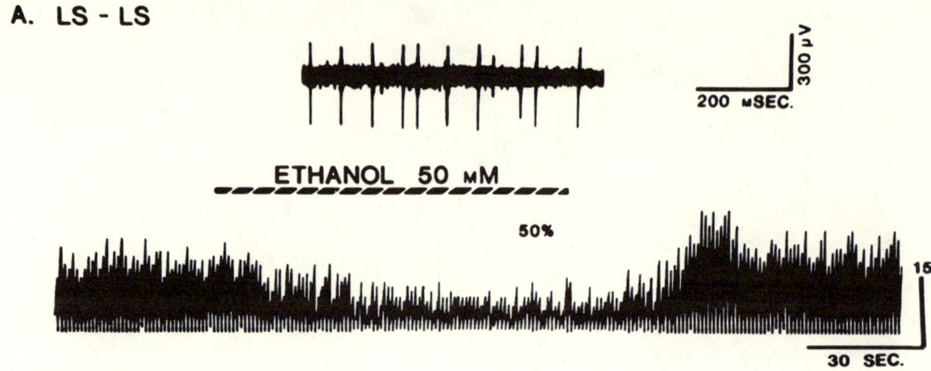



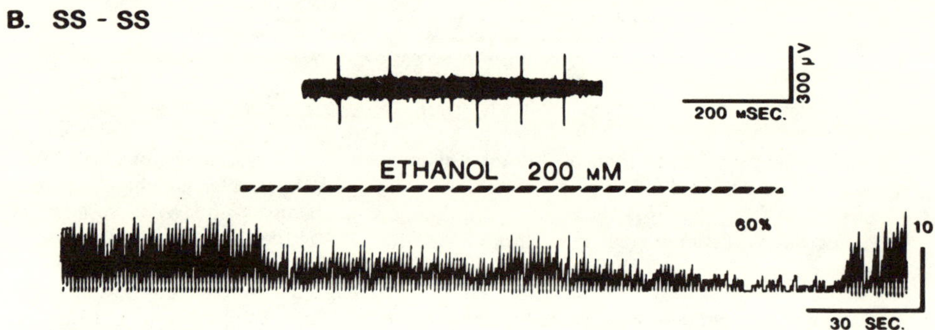



Figure 4. (A) Response of a Purkinje cell in a LS graft transplanted *in oculo* to an LS host (50 mM ethanol). (B) Response from a Purkinje cell in an SS graft transplanted *in oculo* to an SS host (200 mM ethanol). The action potential insets illustrate the spontaneous firing of the Purkinje cells prior to ethanol application. The values above the ratemeter record indicate the percent depression of firing rate during the ethanol application.

maintained in cerebellar grafts transplanted between mouse lines (Fig. 5). Thus, Purkinje neurons in grafts derived from LS fetal brain and transplanted to SS host mice exhibited ethanol sensitivities similar to neurons from LS cerebellum transplanted to LS host mice (Fig. 6; cf. Fig. 4). Likewise, the ethanol sensitivity of Purkinje neurons from SS donors were similar regardless of whether they were transplanted to LS or SS host mice. Since the LS and SS transplants maintain the ethanol sensitivity of the mouse line from which they were derived, the depressant effects of ethanol on grafted cerebella do not appear to be markedly influenced by the constitutional influences of the host mouse line. These results suggest that the primary determinants of ethanol sensitivity in cerebellar Purkinje

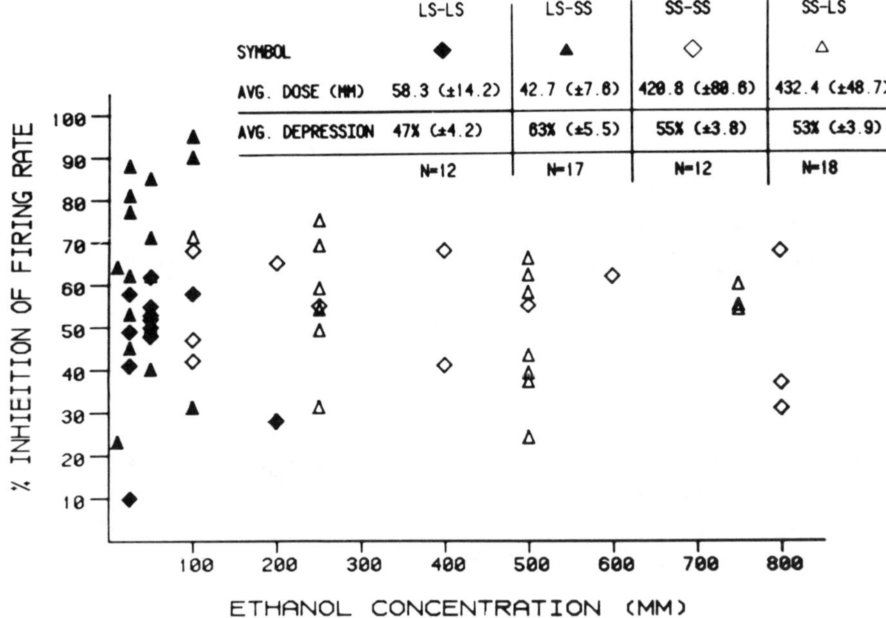



Figure 5. A scattergram indicating ethanol sensitivity of neurons from the four types of intra-ocular transplants. The abscissa shows the concentration of ethanol superfused over the transplant, the ordinate the neuronal response to ethanol administration measured as the percentage of inhibition of spontaneous discharge. The inset table shows the population means (± SEM) for the dose of ethanol administered and the percent inhibition of firing rate in the four transplant types. (From Palmer et al., 1982.)

neurons of LS and SS mice are genetically determined factors that are intrinsic to the fetal grafts.

Consistent with the hypothesis that genotypic differences between Purkinje neurons from LS and SS lines of mice determine the cerebellar ethanol sensitivities of these animals, Basile et al. (1983) have reported that Purkinje neurons in slices of LS cerebella are fivefold more sensitive to the depressant effects of ethanol superfused *in vitro* than are neurons from slices of SS cerebella (Fig. 7). As is the case *in situ,* the ethanol sensitivity of HS Purkinje neurons is intermediate to that found in the LS and SS lines of mice. These experiments do not demonstrate directly that the differences between these lines is intrinsic to Purkinje neurons per se, since much of the local circuitry of the cerebellum remains. However, because many of the afferents to the cerebellum have been truncated in the preparation of cerebellar slices, there is relatively little spontaneous synaptic activity in this preparation that could be modulated by ethanol. Furthermore, the differential ethanol sensitivity is maintained in medium con-

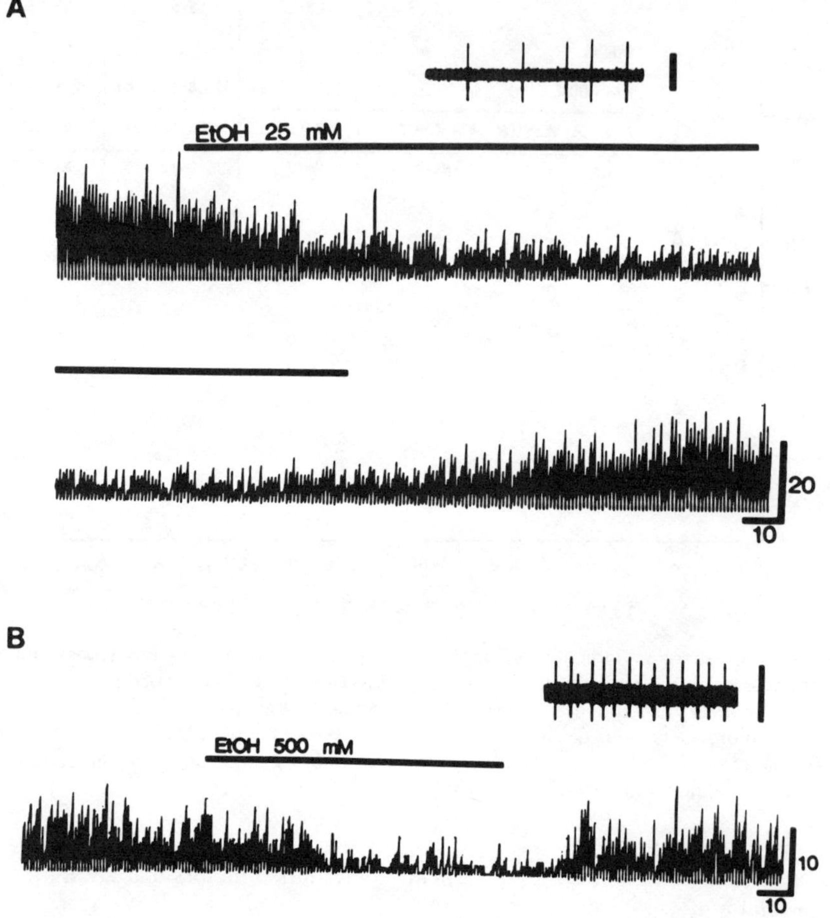



Figure 6.　(A) Ratemeter records of the response to perfusion of 25 mM ethanol from a Purkinje cell in an LS graft transplanted *in oculo* to an SS host mouse. (B) Ratemeter responses to the perfusion of 500 mM ethanol from a Purkinje cell in an SS graft transplanted *in oculo* to an LS host mouse. The insets illustrate the spontaneous firing rates of these Purkinje neurons prior to ethanol application; sweep duration is (A) 250 msec and (B) 1000 msec. Calibration bar: (A) 300 μV and (B) 400 μV. (From Palmer et al., 1982.)

taining reduced calcium and high magnesium, which nearly eliminates the modest synaptic activity normally found in the slice. Thus, although ethanol may affect Purkinje cell activity by a variety of mechanisms in the intact animal (including indirect actions), the physiological actions of ethanol described in these *in vitro* experiments must be attributed to direct actions, most probably on the Purkinje neurons themselves.

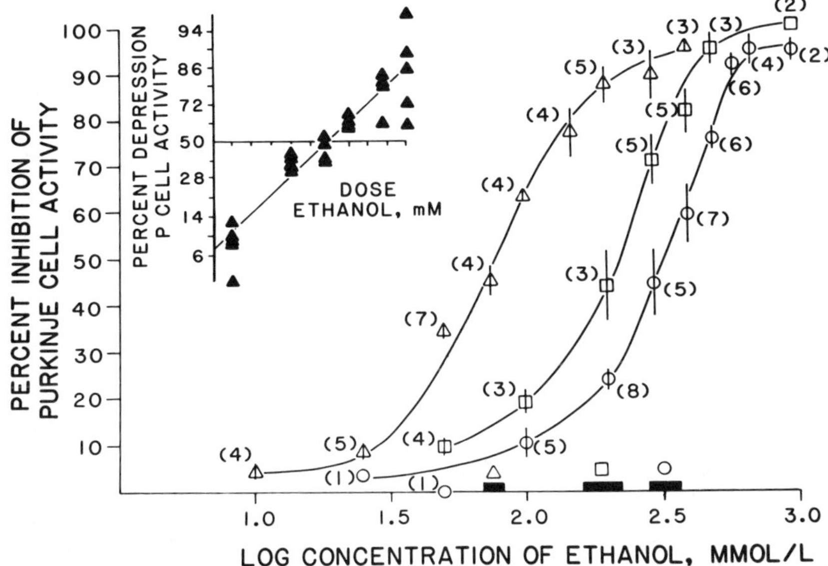



Figure 7. Log dose–response curves for the depression of spontaneous Purkinje neuron discharge rates during ethanol superfusion in cerebellar slices from LS (triangles), HS (squares), and SS (circles) mice. The $EC_{50}$ for the ethanol-induced depressions was calculated using Hill plots, an example of which is shown for the LS mice in the upper left corner of the figure (concentration range 1–500 mM). The horizontal bars on the abscissa represent the 95% confidence limits for the $EC_{50}$ for LS, HS, and SS mice. There were significant differences between all three mean values ($p < 0.05$). Numbers in parentheses represent the number of cells comprising each data point. These cells were taken from a total of 13 LS, 8 HS, and 20 SS mice. (From Basile et al., 1983.)

## 3. ETHANOL TOLERANCE IN LS AND SS MICE

We have recently studied the electrophysiological activity of cerebellar Purkinje neurons in LS and SS mice tolerant to the effects of ethanol (Palmer et al., 1985). After 1 to 4 weeks of feeding on a liquid ethanol diet, mice of both lines were less sensitive to the sedative and ataxic effects of parenteral ethanol than were controls. Measurements of blood ethanol levels confirmed that such tolerance was not purely metabolic, but was at least partially the result of an altered CNS sensitivity to ethanol. In terms of the electrophysiological experiments, cerebellar Purkinje cells in ethanol-fed LS and SS mice were found to be less responsive than those in controls to the depressant effects of ethanol applied either via bath perfusion *in vitro* (Fig. 8) or via local application *in vivo* (Fig. 9). Tolerance to the electrophysiological effects of ethanol was apparent

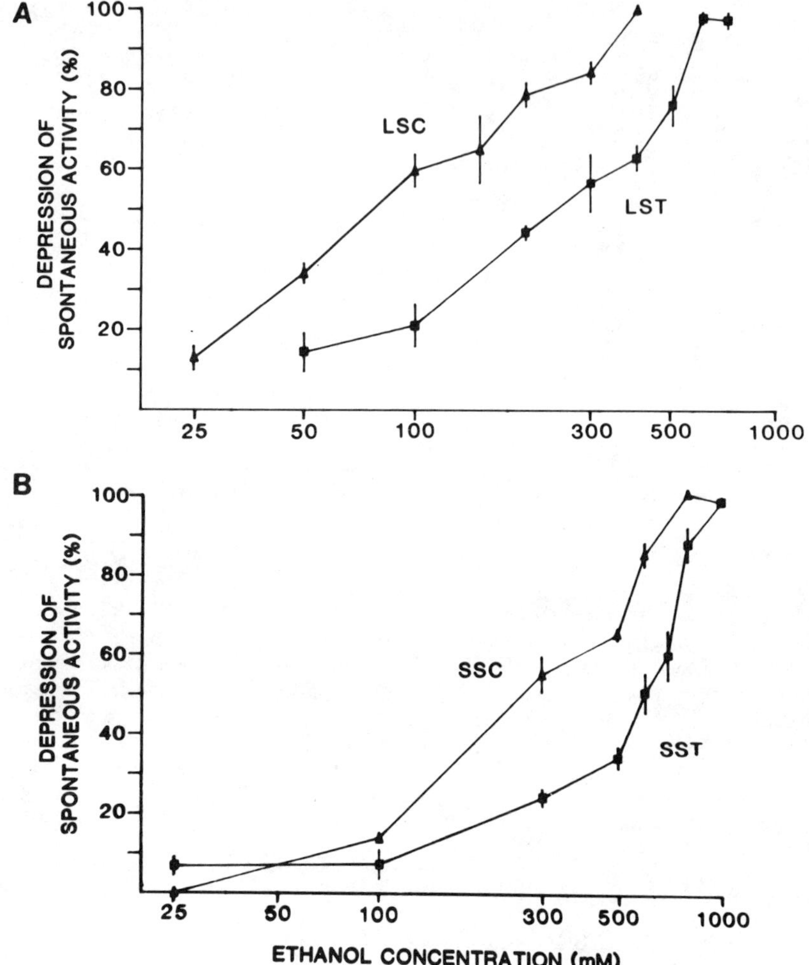


Figure 8. Log dose–response curves for ethanol-induced depressions of the spontaneous activity of Purkinje neurons *in vitro*. The values indicate results from tolerant (squares) and pair-fed control (triangles) (A) SS mice and (B) LS mice. $EC_{50}$ values calculated from tolerant and pair-fed control curves were significantly different ($p < 0.05$) for both mouse lines. All mice were fed liquid diet for 2 weeks prior to testing. Each data point represents the average $\pm$ SEM of responses from 3 to 6 neurons.

within 7 to 9 days after initiating the ethanol diet, and the degree of tolerance did not increase significantly in either mouse line during an additional 1–3 weeks of ethanol administration. In addition, the differential ethanol sensitivities of naive mice (LS > SS) were maintained in tolerant mice as well. Thus, tolerance to both the cellular and behavioral effects of ethanol can be observed after chronic

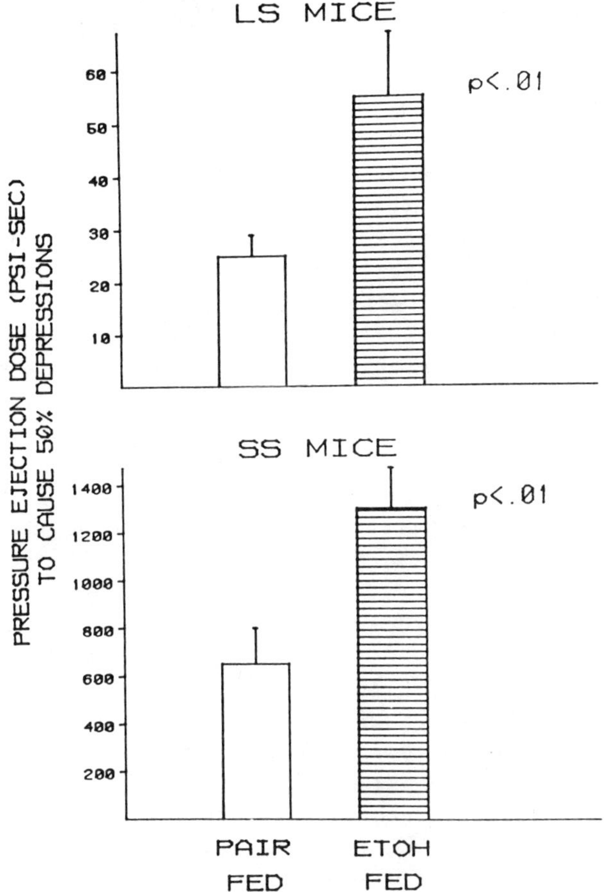



Figure 9.  Effects of chronic ethanol administration on sensitivity to local pressure ejection of ethanol in *in vivo* cerebellar Purkinje neurons of LS mice (upper graph) or SS mice (lower graph). Each pair of bars indicates the dose of ethanol (mean ± SEM) to produce a 50% inhibition of spontaneous neuronal activity in ethanol- (hatched bars) or pair-fed (open bars) mice.

administration of ethanol in LS and SS mice, and there are no significant differences in the degree of tolerance developed by these mice. Like the genetic differences in ethanol sensitivity, the differences in ethanol sensitivity that come about as a result of chronic ethanol administration appear to be intrinsic to the Purkinje neurons; however, the lack of interaction between tolerance and genetic sensitivity to ethanol suggests that these differences arise by independent mechanisms.

## 4. DIFFERENCES IN ETHANOL SENSITIVITY OF INBRED MOUSE AND RAT STRAINS

The association between the sensitivity of Purkinje neurons to ethanol and the behavioral effects of ethanol, documented by the studies described above, support the hypothesis that the ethanol sensitivities of certain areas of the CNS may be directly related to behavioral responses to this drug. However, it is also possible that this association of behavioral and neurophysiological phenotypic differences in the LS and SS mice may be fortuitous; perhaps this phenomenon is related to the occurrence of genetic drift or inbreeding during the selection experiments. To test for a genetic correlation of the two phenotypes, we compared the phenotypic differences in behavioral and neurophysiological responses to acute ethanol administration among the eight inbred mouse strains used to generate the HS mice (Spuhler et al., 1982). In these experiments, the estimates of genetic variation in ethanol-induced behavioral ataxia, as measured by loss of righting response (sleep time), and in the depressant response of cerebellar Purkinje neurons to ethanol applied locally by micropressure-ejection were significant. All eight strains showed prominent depressant effects of ethanol on Purkinje neuron activity as is illustrated for four of the mouse strains in Figure 10. A highly significant estimate of genetic correlation of behavioral and cerebellar sensitivity to ethanol is evident from the bivariate plot of the observed mean behavioral and neuronal sensitivities (Fig. 11).

It is noteworthy that points for the LS and SS mice, taken from Sorensen et al. (1980), would fall well beyond the extremes of this linear distribution. This is consistent with the fact that the selected lines have been bred for at least 20 generations for a specific combination of gene complexes that express the extreme phenotypic responses in ethanol sensitivity. The inbred strains, on the other hand, have had no selection pressure applied that would either increase or decrease their ethanol sensitivity. Indeed, the observed distribution of ethanol sensitivities might be expected for the inbred strains if the fixation of gene complexes for ethanol sensitivity within the strains during their development was a chance event, consistent with random genetic drift.

From the analysis of covariance, the estimate of genetic correlation of ethanol-induced sleep time and Purkinje neuron ethanol sensitivity was quite high; this suggests that either common or closely linked genes influence both behavioral and neuronal sensitivity to this depressant agent. Further experiments, using recombinant inbred strains (Bailey, 1971; Eleftheriou and Elias, 1975), might clarify this point. However, recent data suggest that the genetic correlation of behavioral and neuronal ethanol sensitivities is not as high in rats as in mice (Spuhler et al., 1984). This appears to be due to a greater role for differential metabolism in determining the behavioral sensitivity to ethanol in rats than in mice. Moreover, it is apparent that the variation in expression of behavioral

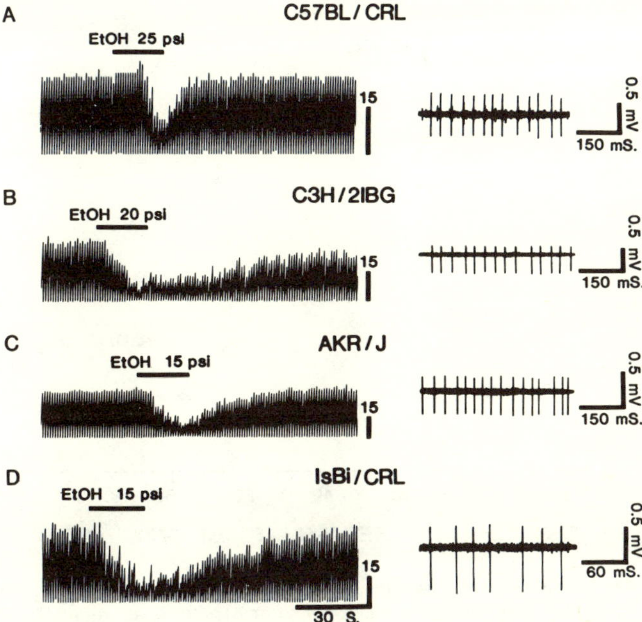



Figure 10. Sample ratemeter records illustrating the effects of micropressure-ejected ethanol on the *in situ* firing rates of Purkinje neurons (left) from four of the eight mouse strains studied in the genetic correlation analysis. Filtered action potential records from these cells are shown to the right of the ratemeter records. This figure shows ethanol-induced responses of neurons from (A) C57BL/Crgl, (B) C3H/2Ibg, (C) AKR/J, and (D) ISBI/Crgl inbred strains of mice. These neurons fire in a pattern of rapid, regular spontaneous discharge, which was slowed by the local pressure ejection (5–30 psi) of ethanol. (From Spuhler et al., 1982.)

sensitivity to ethanol is under polygenic control. Even so, the consistent relationship of Purkinje neuron sensitivity and behavioral sensitivity to ethanol suggests that major genes regulate the ethanol sensitivity of certain areas of the CNS which, in turn, influence behavioral responsiveness to this depressant drug.

## 5. EFFECTS OF NEONATAL CEREBELLECTOMY ON ETHANOL-INDUCED SEDATION

Because of the high degree of correspondence between Purkinje neuron sensitivity to ethanol and behavioral sensitivity, we wished to examine the alterations in behavioral sensitivity that would occur as a result of neonatal cerebellectomy. More specifically, we characterized sensitivity to ethanol-induced depression of the righting response by determining sleep times of LS and SS

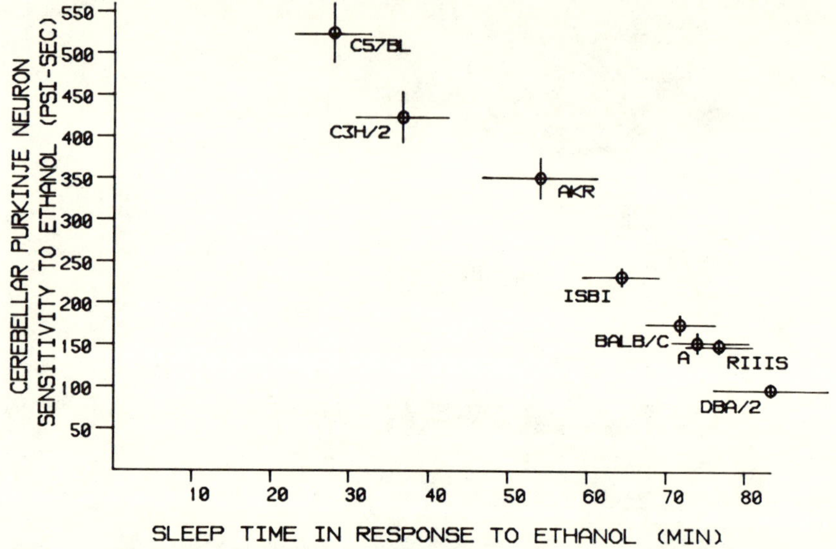



Figure 11.   The correlation of sleep time and Purkinje neuron sensitivity to acute ethanol among eight inbred strains of mice, five mice per strain. The ordinate represents the pressure ejection dose (in psi × seconds) applied from a two-barrel micropipette, that caused approximately 50% inhibition of Purkinje neuron firing rates *in situ*. Each point in the plot represents the mean sensitivity of 25 neurons, pooled over five mice within each strain. The abscissa represents the length of time (in minutes) that the animals lost the righting response (sleep time) after receiving 3.3 g/kg IP ethanol. Each point shows the average of five sleep time scores, one for each animal studied. The vertical and horizontal bars associated with each point represent SEM for those measurements. (From Spuhler et al., 1982.)

mice in which the cerebella had been lesioned (Palmer et al., 1984; Seiger et al., 1983). Mice of both strains were cerebellectomized within ten days of birth and behavioral testing was performed several weeks later. Postmortem investigations showed that the degree of cerebellectomy of animals from both lines ranged from 0 to 100%. The histological observations for any one animal agreed well with the results of cerebellar function tests administered prior to measurement of sleep times. The results of these studies indicated that neonatal cerebellectomy caused SS mice to become more sensitive to the ataxic effects of ethanol while the sleep times of LS mice were essentially unchanged (Fig. 12). Furthermore, the ethanol-induced sleep times of cerebellectomized LS mice were still fourfold to fivefold longer than those observed for cerebellectomized SS mice. While the proposed explanations for these results are complex, it can be reasonably concluded that cerebellar Purkinje cells are not the only neuronal elements involved in determining the sensitivity of LS and SS mice to the ataxic

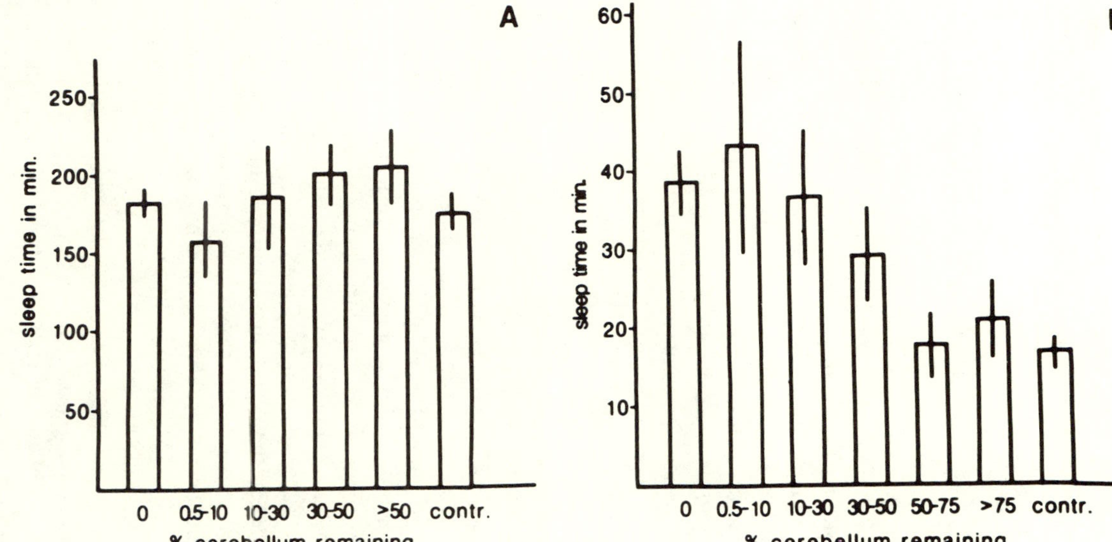


Figure 12.   The effect of cerebellectomy on the ethanol-induced sleep time (± SEM) of 3- to 6-week old (A) LS and (B) SS mice which were partially or totally cerebellectomized shortly after birth. The abscissa represents the percentage of cerebellum remaining as assessed by gross morphology at autopsy and confirmed by histological examination at a later date. Cerebellectomy appears to have a significant effect on the sleep times of SS, but not LS mice. (From Palmer et al., 1984.)

effects of ethanol. These data suggest that populations of neurons in other brain areas could express the differential sensitivity to the depressant effects of ethanol in LS and SS mice that we originally described for cerebellar Purkinje neurons. Nevertheless, it is also apparent from the work of Sorensen and colleagues (1981*b*) that neurons in at least some brain regions do not show this differential ethanol sensitivity.

## 6.  ENDOGENOUS PURINERGIC MECHANISMS IN LS AND SS MICE

With the exception of ethanol, the differences between the LS and SS in responsiveness to centrally acting drugs such as anesthetics, neurotransmitter receptor agonists and antagonists, and so forth have been relatively small. However, we have recently reported that these two lines do differ in terms of their behavioral sensitivity to drugs that interact with adenosine receptors in the brain. Agonists at these receptors (so-called purinergic agonists) have been previously reported to be sedative, anticonvulsant, and to be prohypnotic, in the sense that they may promote sleep without being able to directly induce a hypnotic state, at least in rodents (Haulica et al., 1973; Vapaatalo et al., 1975; Snyder et al., 1981; Dunwiddie and Worth, 1982). Purinergic antagonists such as caffeine and theophylline have an alerting action upon the CNS and increase spontaneous motor activity. Using several types of behavioral tasks, Proctor and Dunwiddie (1984) demonstrated that LS mice are generally fourfold to sixfold more sensitive to the actions of phenylisopropyl adenosine (a nonmetabolized adenosine receptor agonist) than are the SS mice (Fig. 13). The degree of differential sensitivity was quite similar to that observed with the sedative effects of subanesthetic doses of ethanol in these two lines. In a subsequent paper, Proctor et al. (1985) have also shown that the behavioral effects of the adenosine antagonist theophylline are also greater in the LS mice than they are in the SS line (Fig. 13) and that there are no differences in the blood and brain drug levels in the two lines of mice following systemic injections. Taken together, these data suggest that an important difference between these lines of mice may be in terms of some aspect of purinergic systems in the CNS, specifically in terms of the behavioral responses to drugs that affect adenosine receptors.

Putting these results into the larger context of ethanol actions, it is unclear at this point whether these differences in purinergic sensitivity reflect an important role for adenosine as a mediator in responses to ethanol, or whether they may provide an explanation for *differences* in ethanol sensitivity, without reference to the actual mechanism of ethanol action. If ethanol exerts its actions via alterations in adenosine receptors, then one might expect parallel sensitivity to ethanol and adenosine agonists and antagonists. However, it is not necessary to postulate such a direct interaction to explain the present results.

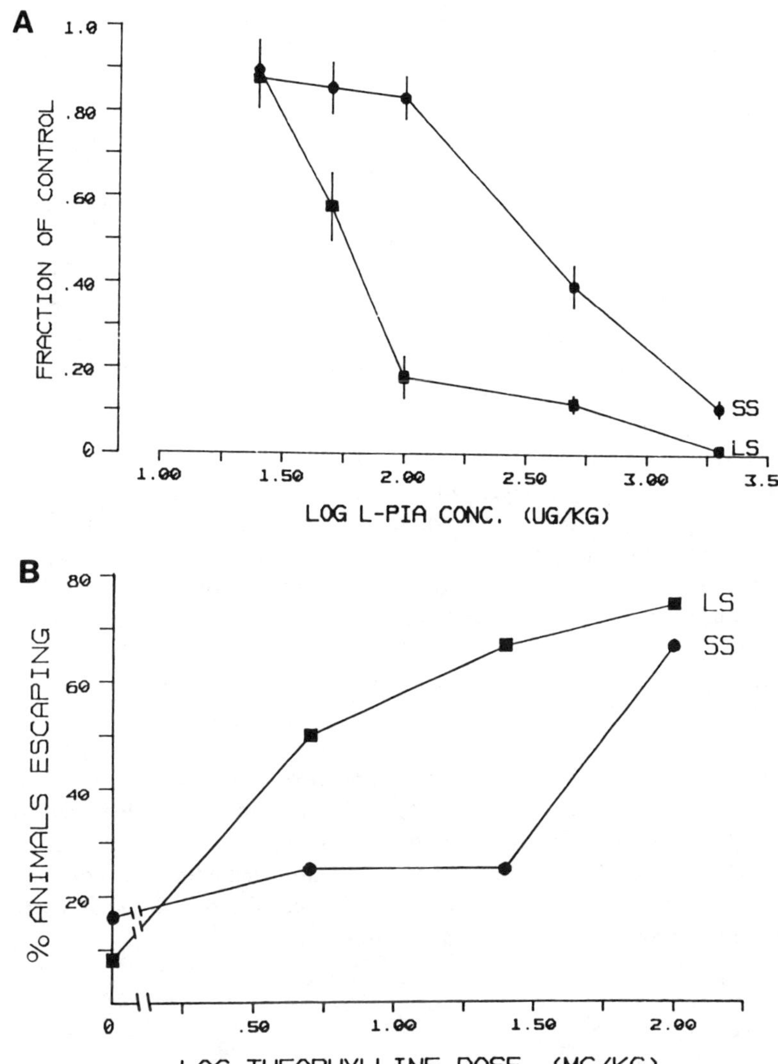



Figure 13.   Behavioral actions of PIA and theophylline. (A) The sedative effect of PIA in LS and SS mice is illustrated. The behavioral measure was the number of escape attempts from an elevated platform during a 2-min test period. LS and SS mice did not differ in their control responses on this test (following saline injection), but they did differ in their sensitivity to the suppressant effect of IP injection of PIA on the number of escape attempts. (B) Mice were tested for successful escape from an elevated platform as in (A). Control mice usually did not escape during a 2-min test period, but theophylline increased the frequency of successful escape in a dose-dependent fashion. LS mice were significantly more sensitive to this activating effect of theophylline than were SS mice.

One way in which to test the relationship between ethanol and adenosine is to characterize the extent of interaction between adenosine receptor agonists and antagonists and the behavioral effects of ethanol. It has been known for some time that adenosine antagonists such as caffeine can reduce but not completely block many of the behavioral actions of ethanol (Linnoila and Mattila, 1981; Seppala et al., 1979), suggesting that the interactions of these two agents are not direct. Dar et al. (1983) have shown that theophylline can reduce ethanol-induced sleep time, whereas an adenosine reuptake inhibitor (dipyridamole) markedly increases sleep time. In addition, these authors showed that chronic treatment with ethanol appeared to increase adenosine receptor number, whereas withdrawal decreased receptor number. Again, it is difficult to determine whether such changes indicate a common site of action for the two types of drugs, or merely a situation in which changes induced by ethanol may result in indirect alterations in purinergic systems.

In preliminary studies, de la Garza and Dunwiddie (1985) have approached this issue by comparing the degree of cross-tolerance between ethanol and PIA when given chronically. Mice made tolerant to the effects of ethanol by chronic administration of a liquid ethanol-containing diet did not show any alteration in their response to PIA when compared with pair-fed controls. Conversely, mice chronically administered PIA showed tolerance to the effects of an acute injection of PIA, but did not show a change in their responses to acute ethanol administration. These results would seem to provide relatively strong evidence that these drugs act at different sites in the nervous system to produce their behavioral responses.

Figure 14 illustrates some of the potential ways in which ethanol and adenosine might interact in the nervous system. A direct interaction between these agents would appear to be inconsistent with the present results, but models such as those presented in Fig. 14B or C, in which ethanol and adenosine both appear to work via a common output pathway, would be consistent with what has been presented regarding the actions of these drugs.


## 7.  CONCLUSIONS

In summary, groups of neurons have been described in the cerebella of LS and SS mice that show differential responses to locally applied ethanol that are consistent with the sensitivity of these animals to the ataxic effects of this drug. This differential ethanol sensitivity is not found in all brain areas, and the expression of this phenotype by Purkinje neurons is an intrinsic property of the cerebellum. There is, in addition, a high genetic correlation between the behavioral and Purkinje neuron sensitivity to ethanol among inbred mouse strains. However, neurons in other brain regions are probably also important in mediating

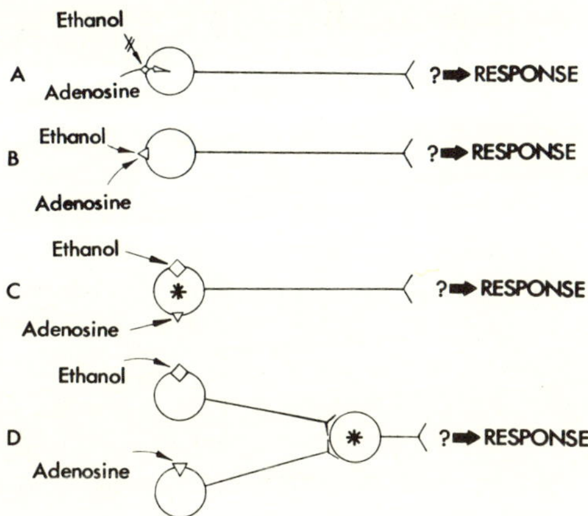


Figure 14.   Hypothetical interactions between ethanol and adenosine. Some potential inter-actions between ethanol and adenosine are schematized in this drawing. If ethanol interferes with the release, uptake (illustrated in A), or metabolism of adenosine, at least some of the effects of ethanol might be exerted indirectly through actions on purinergic receptors. However, in this case, the effects of ethanol should be completely reversed by adenosine antagonists, and by and large this is not the case. (B) Alternatively, ethanol and adenosine might act upon some common site, such as a receptor. However, most if not all of the effects of PIA involve actions at highly specific membrane receptors, and ethanol does not appear to directly affect the interaction between adenosine receptor ligands and the binding site (Fredholm, personal communication). If adenosine and ethanol act upon (C) different sites on the same neuron or (D) on different neurons that converge on a final common pathway involved in the the behavior, this would be consistent with the present results. In this situation there would be little if any cross-tolerance, and physiological but not pharmacological interactions between the two classes of drugs. Furthermore, if the neurons in (C) and (D) marked with the asterisk differ in the LS and SS mice, this would provide a mechanism by which both purinergic drugs and ethanol could elicit differential behavioral responses in these two lines, without requiring a common site of action. Although these latter two hypotheses appear to be most consistent with the experimental data, further work will be required to determine which of these is the case.

the differential ethanol sensitivity of LS and SS mice to the ataxic effects of this depressant agent. Finally, although many neurochemical properties are similar in LS and SS mice, there are significant differences in adenosine sensitivities that parallel those to ethanol. The extent to which differences in puriner-gic systems underlie the differences in ethanol sensitivity must await future experimentation.

Despite the utility of these electrophysiological and behavioral approaches to the question of ethanol action, they have not provided any specific answers regarding mechanisms (such actions upon specific ionic conductances) by which ethanol affects neuronal activity. Moreover, if the primary action of ethanol is to alter some cellular parameter such as membrane fluidity that cannot be directly measured using these techniques, it is unlikely that they will ever provide any such answers by themselves. However, by demonstrating in an unequivocal fashion that ethanol sensitivity can be regulated by intrinsic cerebellar determinants, these experiments have "narrowed the field" in terms of the search for substrates of ethanol action. Whether these prove ultimately to be changes in membrane structure or function, direct effects upon cellular proteins, or as yet unspecified changes in cellular function remains to be determined.

ACKNOWLEDGMENTS

This work was supported by USPHS grants #AA05915, AA 03527, DA 02702, and by the Veterans Administration Medical Research Service.

# REFERENCES

Bailey, D. W., 1971, Recombinant-inbred strains: An aid to finding identity, linkage, and function of histocompatibility and other genes, *Transplantation* **11**:325.

Basile, T., Hoffer, B., and Dunwiddie, T., 1983, Differential sensitivity of cerebellar Purkinje neurons to ethanol in selectively outbred lines of mice: Maintenance *in vitro* independent of synaptic transmission, *Brain Res.* **264**:69.

Berry, M. S., and Pentreath, V. W., 1980, The neurophysiology of alcohol, in: *Psychopharmacology of Alcohol* (M. Sandler, ed.), pp. 43–72, Raven Press, New York.

Dar, M. S., Mustafa, S. J., and Wooles, W. R., 1983, Possible role of adenosine in the CNS effects of ethanol, *Life Sci.* **33**:1363.

Deitrich, R. A., and Spuhler, K., 1984, Genetics of alcoholism and alcohol actions, in: *Research Advances in Alcohol and Drug Problems,* Vol. 4 (Y. Israel, F. Glaser, H. Kalant, R. Popham, W. Schmidt, and R. Smart, eds.), pp. 47–98, Plenum Press, New York.

de la Garza, R., and Dunwiddie, T. V., 1985, Tolerance to the sedative and hypothermic effects of R-PIA and lack of cross-tolerance to ethanol, *Fed. Proc.* **44**:723.

Dunwiddie, T. V., and Worth, T., 1982, Sedative and anticonvulsant effects of adenosine analogs in mouse and rat, *J. Pharmacol. Exp. Ther.* **220**:70.

Eidelberg, E., Bond, M., and Kelter, A., 1971, Effects of alcohol on cerebellar and vestibular neurons, *Arch. Int. Pharmacodyn. Ther.* **192**:213.

Eleftheriou, B. E., and Elias, P. K., 1975, Recombinant inbred strains: A novel genetic approach for psychopharmacogeneticists, in: *Psychopharmacogenetics* (B. E. Eleftheriou, ed.), pp. 43–71, Plenum Press, New York.

Forney, E., and Klemm, W. R., 1976, Effect of ethanol on impulse activity in isolated cerebellum, *Res. Commun. Chem. Pathol. Pharmacol.* **15**:801.

Haulica, I., Ababei, L., Branisteanu, D., and Topoliceanu, F., 1973, Preliminary data on the possible hypnogenic role of adenosine, *J. Neurochem.* **21**:1019.

Heston, W., Anderson, S., Erwin, V., and McClearn, G., 1973, A comparison of the actions of various hypnotics in mice selectively bred for the effects of alcohol, *Behav. Genet.* **3**:402.

Heston, W., Erwin, V., Anderson, S., and Robbins, H., 1974, A comparison of the effects of alcohol on mice selectively bred for differences in ethanol sleep time, *Life Sci.* **14**:356.

Himwich, H. E., and Callison, D. A., 1972, The effects of alcohol on evoked potentials of various parts of the central nervous system of the cat, in: *The Biology of Alcoholism,* Vol. 2 (B. Kissin and H. Begleiter, eds.), pp. 67–84, Plenum Press, New York.

Hoffer, B., Seiger, A., Ljungberg, T., and Olson, L., 1974, Electrophysiological and cytological studies of brain homografts in the anterior chamber of the eye: Maturation of cerebellar cortex *in oculo, Brain Res.* **79**:165.

Kalant, H., 1974, Ethanol and the nervous system: Experimental neurophysiological aspects, *Int. J. Neurol.* **9**:111.

Klemm, W. R., 1979, Effects of ethanol on nerve impulse activity, in: *Biochemistry and Pharmacology of Ethanol*, Vol. 2 (E. Majchrowicz and E. P. Noble, eds.), pp. 243–267, Plenum Press, New York.

Klemm, W. R., and Stevens, III, R. E., 1974, Alcohol effects on EEG and multiple unit activity in various brain regions of rats, *Brain Res.* **70**:351.

Linnoila, M., and Mattila, M. J., 1981, How to antagonize ethanol-induced inebriation, *Pharmacol. Ther.* **15**:99.

McClearn, G. E., Wilson, J. R., and Meredith, W., 1970, The use of isogenic and heterogenic mouse stocks in behavioral research, in: *The Mouse as a Prototype* (G. Lindzey and D. O. Theissen, eds.), pp. 3–22, Appleton-Century-Crofts, New York.

Mitra, J., 1977, Differential effects of ethanol on unit activity in cerebellum and other brain areas in the rat, *Soc. Neurosci. Abstr.* **3**:268.

Olson, L., and Malmfors, T., 1970, Growth characteristics of adrenergic nerves in the adult rat. Fluorescence histochemical and $^3$H-noradrenaline uptake studies using tissue transplantations to the anterior chamber of the eye, *Acta. Physiol. Scand. Suppl.* **348**:1.

Olson, L., Seiger, A., and Strömberg, I., 1983, Intraocular transplantation in rodents: A detailed account of the procedure and examples of its use in neurobiology with special reference to brain tissue grafting, in: *Advances in Cellular Neurobiology* (S. Fedoroff and A. Herz, eds.), pp. 407–442, Academic Press, New York.

Palmer, M. R., Wuerthele, S. M., and Hoffer, B. J., 1980, Physical and physiological characteristics of micropressure ejection of drugs from multibarreled pipettes, *Neuropharmacology* **19**:931.

Palmer, M. R., Sorensen, S., Freedman, R., Olson, L., Hoffer, B. J., and Seiger, A., 1982, Differential ethanol sensitivity of intraocular cerebellar grafts in long-sleep and short-sleep mice, *J. Pharmacol. Exp. Ther.* **222**:480.

Palmer, M. R., Olson, L., Dunwiddie, T. V., Hoffer, B. J., and Seiger, A., 1984, Neonatal cerebellectomy alters ethanol-induced sleep time of short sleep but not long sleep mice, *Pharmacol. Biochem. Behav.* **20**:153.

Palmer, M. R., Basile, A. S., Proctor, W. R., Baker, R. C., and Dunwiddie, T. V., 1985, Ethanol tolerance of cerebellar Purkinje neurons from selectively outbred mouse lines: *In vivo* and *in vitro* electrophysiological investigations, *Alcoholism Clin. Exp. Res.* **9**:291.

Proctor, W. R., and Dunwiddie, T. V., 1984, Behavioral sensitivity to purinergic drugs parallels ethanol sensitivity in selectively bred mice, *Science* **224**:519.

Proctor, W. R., Baker, R. C., and Dunwiddie, T. V., 1985, Differential CNS sensitivity to PIA and theophylline in long-sleep and short-sleep mice, *Alcohol* **2**:287.

Rogers, J., Siggins, G. R., Schulman, J. A., and Bloom, F. E., 1980, Physiological correlates of ethanol intoxication, tolerance and dependence in rat cerebellar Purkinje cells, *Brain Res.* **916**:183.

Seiger, A., Sorensen, S. M., and Palmer, M. R., 1983, Cerebellar role in the differential ethanol sensitivity of long sleep and short sleep mice, *Pharmacol. Biochem. Behav.* **18**:495.

Seil, F. J., Leiman, A. L., Herman, M. M., and Fisk, R. A., 1977, Direct effects of ethanol on central nervous system cultures: An electrophysiological and morphological study, *Exp. Neurol.* **55**:390.

Seppala, T., Linnoila, M., and Mattila, M., 1979, Drugs, alcohol and driving, *Psychopharmacology* **56**:185.

Siggins, G. R., and Bloom, F. E., 1980, Alcohol-related electrophysiology, *Pharmacol. Biochem. Behav.* **13**(Suppl. 1):203.

Siggins, G. R., and French, E., 1979, Central neurons are depressed by iontophoretic and micropressure application of ethanol and tetrahydropapaveraline, *Drug Alcohol Depend.* **4**:239.

Sinclair, J. G., and Lo, G. F., 1980, The effects of ethanol on cerebellar Purkinje cells in naive and alcohol-dependent rats, *Can. J. Physiol. Pharmacol.* **56**:668.

Sinclair, J. G., and Lo, G. F., 1981, The effects of ethanol on cerebellar Purkinje cell discharge pattern and inhibition evoked by local surface stimulation, *Brain Res.* **204**:465.

Snyder, S., Katims, J. J., Annau, Z., Bruns, R. F., and Daly, J., 1981, Adenosine receptors and the behavioral actions of methylxanthines, *Proc. Natl. Acad. Sci.* **78**:3260.

Sorensen, S., Palmer, M., Dunwiddie, T., and Hoffer, B., 1980, Electrophysiological correlates of ethanol-induced sedation in differentially sensitive lines of mice, *Science* **210**:1143.

Sorensen, S., Carter, D., Marwaha, J., Baker, R., and Freedman, R., 1981*a*, Disinhibition of rat cerebellar Purkinje neurons from noradrenergic inhibition during rising blood ethanol, *J. Stud. Alcohol* **42**:908.

Sorensen, S., Dunwiddie, T., McClearn, G., Freedman, R., and Hoffer, B., 1981*b*, Ethanol-induced depressions in cerebellar and hippocampal neurons of mice selectively bred for differences in ethanol sensitivity: An electrophysiological study, *Pharmacol. Biochem. Behav.* **14**:227.

Spuhler, K., Hoffer, B., Weiner, N., and Palmer, M., 1982, Evidence for genetic correlation of hypnotic effects and cerebellar Purkinje neuron depression in response to ethanol in mice, *Pharmacol. Biochem. Behav.* **17**:569.

Spuhler, K., Fossom, L., Palmer, M., and Deitrich, R. A., 1984, Differences in acute sensitivity to the effects of ethanol in rat inbred strains, *Alcoholism* **8**:121.

Taylor, D., Seiger, A., Freedman, R., Olson, L., and Hoffer, B., 1980, Electrophysiological analysis of reinnervation of transplants in the anterior chamber of the eye by the autonomic ground plexus of the iris, *Proc. Natl. Acad. Sci. USA* **75**:1009.

Vapaatalo, H., Onken, D., Neuvonen, P. J., and Westermann, E., 1975, Stereospecificity in some central and circulatory effects of phenylisopropyladenosine (PIA), *Arzneim.-Forsch.* **25**:407.

5

# Acute Alcohol Amnesia
## What Is Remembered and What is Forgotten

SHAHIN HASHTROUDI and ELIZABETH S. PARKER

## 1. INTRODUCTION

Alcohol is a drug with potent effects on human emotions and cognition. Drinking enables individuals to alter their emotional state, their cognitive functioning, and their interactions with their environment. Certain cognitive responses to alcohol seem to be the idiosyncratic response of a particular individual to a particular dose of alcohol. There are, however, other cognitive responses that have a more orderly relationship to alcohol and reflect a basic effect of the drug. It appears that one such response is the disruption in memory produced by modest doses of alcohol in both alcoholics and social drinkers (e.g., Parker et al., 1974; Rosen and Lee, 1976). Episodes of impaired memory can be viewed as isolated instances of dysfunction, yet they are bound to disrupt the drinker's continuity of experience between past and future. Memory is the remarkable capacity that records both interactions with the environment and mental events. These records of past experience shape and mold later behavior, thereby connecting internal representations of past, present, and future.

In this chapter we focus on acute alcohol amnesia. Acute alcohol amnesia is defined as memory impairment that occurs while alcohol is in the bloodstream. It is the most prevalent form of mnemonic disruption related to alcohol. Significant memory impairment has been reliably documented in studies of social drinkers with alcohol levels of about 0.08 g/100 ml (e.g., Carpenter and Ross,

---

SHAHIN HASHTROUDI ● Department of Psychology, George Washington University, Washington, D. C. 20052. ELIZABETH S. PARKER ● Alcohol Research Center, Department of Psychiatry and Behavioral Sciences, UCLA School of Medicine, Los Angeles, California 90024.

179

1965; Jones and Jones, 1980; Hashtroudi et al., 1983; Birnbaum et al., 1978). Anyone who has consumed 5–6 drinks over the course of several hours has probably experienced a measurable degree of memory loss for that drinking event. Unlike the alcoholic blackout (Goodwin, 1977), which is a total amnesia for events and experiences, most instances of acute alcohol amnesia are more subtle and go unnoticed by the drinker. It has been demonstrated that there is an orderly relation between the amount of alcohol in the blood and memory function; higher blood alcohol levels result in greater memory deficits (Kalin, 1964; Carpenter and Ross, 1965; Parker et al., 1974, 1976, 1983). Another consistent finding is that alcohol has particularly disruptive effects on encoding new information. During intoxication, subjects have difficulty forming new memories, yet they are able to access memories that were formed before drinking (Birnbaum et al., 1978; Miller et al., 1978). Most of the research on acute alcohol amnesia is based on studies of male college students, with doses of alcohol these students often consume (e.g., 1.0 ml or 0.8 g absolute alcohol per kilogram of body weight, or 4–5 standard drinks for a 70-kg person).

The main theme of this chapter is that only certain forms of memory are impaired by alcohol, while other forms of memory appear to be resistant to the effects of the drug. We propose a framework for investigating acute alcohol amnesia that distinguishes between alcohol-sensitive and alcohol-resistant forms of memory. These separate forms of memory involve different cognitive processes and are characterized on the basis of these processes. In contrast to a stage model that separates memory according to storage versus retrieval (Birnbaum and Parker, 1977; Gerrein and Chechile, 1977), this new framework separates memory according to mental activities that can be differentiated from one another but which can operate simultaneously at both encoding and retrieval.

The notion of different forms of memory can be illustrated by examples. One form of memory is involved in recollecting discrete events such as "Was wine served at the last party you attended?" Another form of memory is involved in gaining access to one's general knowledge of the world as in answering the question, "What alcoholic beverage is made by fermenting grapes?" Perhaps another form of memory is involved in shaping our behavior based on recent events. If you were asked, "Name a type of alcoholic beverage," outside the context of reading this chapter you might have responded with a variety of beverages. In the context of the preceding sentences, however, there is a high probability that you would say "wine." Alcohol, it appears, selectively impairs memory for particular events (e.g., was wine served at the last party you attended?), while leaving other forms of memory relatively uninterrupted.

New questions about alcohol and its impact on the drinker are raised by the possibility that intoxication affects only some of the simultaneously operating memory processes. Does alcohol disrupt the memory processes involved in conscious recollection? Is alcohol impairing the formation of memories that allow us to know that something happened? Are alcohol-resistant forms of memory those that guide behavior, without the subject necessarily knowing or remembering why? What is the nature of the experience of heavy drinkers if they record and remember their world on the basis of the alcohol-resistant memory processes? What is the cognitive set of someone who repeatedly dampens memories for particular events by drinking? In what way could this serve an adaptive function for the drinker? Could this be the basis for the disinhibiting effects or the reduction in self-awareness attributed to alcohol by Hull and his associates (Hull, 1981; Hull et al., 1983)?

In the rest of the chapter we will review recent research from neuropsychological studies of the amnesic syndrome as well as normal human memory which supports the view that memory is composed of dissociable parts. We then will review research on alcohol-sensitive memory processes. We propose that alcohol disrupts the active cognitive processes that organize and integrate new information with preexisting knowledge. Thus, for example, if one was intoxicated at the last party one attended, it would be difficult to recall whether or not wine was served. To recollect this information one needs to have processed that fact in an organized way, such as noticing that the label was of a superior quality, or constructing an image of the attractive person who was drinking wine. When intoxicated, the active, reflective processes that provide a rich representation of experience are impaired.

The spared forms of memory are more difficult to characterize since much less research has been addressed to this topic. In the final section we will review this research. There is scant but compelling evidence that intoxicated subjects can activate preexisting knowledge (Rosen and Lee, 1976; Hashtroudi et al., 1983). For example, intoxicated subjects are able to remember what alcoholic beverage is made by fermenting grapes. This could simply reflect the fact that well-learned information is resistant to disruption. What is perhaps more striking is that intoxicated subjects can acquire new information. We will suggest that these forms of memory do not require active cognitive processing. Intoxicated subjects would be as likely as sober subjects to be biased to name wine as a type of alcoholic beverage if this particular beverage had just been mentioned.

This is a selective review and certain important topics on alcohol and human memory are not covered. We do not review studies on alcohol-related memory loss in the *sober* individual which would include Korsakoff's syndrome (Butters and Cermak, 1980), detoxified alcoholic patients (Ryan and Butters, 1983), or social drinkers (Parker and Parker, 1982). Nor do we address state-dependent retrieval with alcohol since this has been reviewed by Eich (1980). Similarly, retrograde enhancement of predrinking events by alcohol has recently been reviewed (Parker and Weingartner, 1984).

## 2.   FORMS OF MEMORY AND AMNESIA

The proposed framework of alcohol-sensitive and alcohol-resistant forms of memory is supported by recent research on the amnesic syndrome and normal memory. Traditionally, human memory has been characterized as a single unified system. Functions as varied as learning and retention of motor skills, acquisition of language, recollection of past events, and recognition of faces are assumed to reflect the operation of the same memory system, and research has been directed at finding general laws that account for all learning and memory phenomena (McGeogh and Irion, 1952).

Recently, however, researchers with different persuasions have challenged the unified memory concept and have proposed the existence of separate forms of memory (Tulving, 1983, 1984*a,b;* Johnson, 1983; Jacoby and Witherspoon, 1982; Warrington and Weiskrantz, 1982; Cohen and Squire, 1980). How many forms of memory exist and how to best conceptualize them are issues of considerable debate in memory research today. Some theorists refer to these different forms of memory as different memory *systems* that are represented in the brain by different structures or neural mechanisms (Tulving, 1983; Cohen, 1984; Warrington and Weiskrantz, 1982). Others characterize different forms of memory as separate memory subsystems (Johnson, 1983) or simply in terms of separate processes (Jacoby and Witherspoon, 1982). Our framework differentiates between alcohol-sensitive and alcohol-resistant forms of memory on the basis of separate cognitive processes. This requires the fewest assumptions about the independence of different forms of memory and about their underlying neural representations.

The major impetus for formulation of separate processes or systems has been the findings of dissociations in performance of various memory tasks observed in amnesic patients, especially alcoholic Korsakoff patients. There is now substantial evidence suggesting a distinction between impaired and preserved memory functions in these patients. A particularly striking example was reported by Cohen and Squire (1980). These investigators demonstrated that amnesic patients learned a cognitive skill (reading inverted texts) as easily as normal subjects and retained this skill for three months. However, amnesics had no recollection of the learning experience and they could not distinguish the learned words from new words even after 20 exposures. In a classic series of studies, Warrington and Weiskrantz (1970) showed that performance of amnesic patients was indistinguishable from normal controls when they were asked to identify degraded versions of words they had just seen. Yet the same patients had severely impaired recall and recognition for the words. Jacoby and Witherspoon (1982) had amnesic subjects hear homophones (e.g., Read/Reed) in the context of sentences that biased the less common interpretations of the words (e.g., Name a musical instrument that employs a reed). Later, the amnesic subjects spelled the homophone consistent with its recently biased interpretation (e.g., Reed), but they did not remember whether they had heard that homophone before. Schacter and Tulving (1982) showed that amnesic subjects can learn and retain new facts (e.g., Who holds the world's record for shaking hands?) even though they do not recall the experience of learning the fact or the source of information.

Warrington and Weiskrantz (1982, p. 233) have fully summarized the capabilities of amnesic patients:

> It has become evident in recent years that amnesic patients can demonstrate long-term learning and retention remarkably well under certain laboratory conditions and with certain kinds of test material: these include motor-skill learning, the recall and identification of words and pictures with partial information, semantically and phonemically related paired-associate learning, the facilitation of perception of depth random-dot stereograms, the retention of the McCulloch colour-grating illusion, the retention of jigsaw-puzzle problems and word-anagram problems, and eyelid conditioning. Tasks like these, and the list is not exhaustive, stand in contrast to those on which the amnesic performs disastrously badly, such as multiple-choice recognition and yes/no recognition of words, pictures and faces; paired-associate learning; story recall; free recall; and memory for past remote events.

Several theoretical accounts of these results have been proposed. These accounts converge on the idea that the memory deficit is not global or all-encompassing. Rather, there are specific memory processes or specific systems that underlie the amnesic syndrome. For example, Schacter and Tulving (1982) have described amnesia as a failure of episodic memory. According to this view, there are at least two propositional memory systems that serve independent but parallel functions. Semantic memory is generic and represents knowledge of language, facts, and concepts, whereas episodic memory is unique to the individual and it preserves information about the temporal and spatial context in which an event occurs. Episodic memory includes autobiographical and personal reference information and it requires an integration of the incoming information with the personal past of the rememberer. It is this conscious and well-integrated system that is affected in amnesia (Schacter and Tulving, 1982; Tulving, 1983, 1984a,b).

Other investigators (Cohen and Squire, 1980) have proposed a distinction between procedural and propositional knowledge represented by the corresponding memory systems. Procedural memory mediates skill acquisition and it can only be expressed through action; its contents are not accessible to consciousness. Propositional memory, on the other hand, records verbal and symbolic information and requires conscious processing. According to Cohen and Squire, procedural knowledge (knowing how) is intact in amnesics, whereas verbal or propositional expression (knowing that) is severely impaired.

In her multiple-entry, modular memory system (MEM), Johnson (1983) suggests a distinction between the sensory and perceptual memory subsystems, on the one hand, and the reflection subsystem on the other. The first two subsystems preserve information about external events. The reflection subsystem, however, records the consequences of self-generated activities, such as organization, elaboration, and integration of information. According to Johnson, "Reflection involves the conscious construction of patterns and relationships; it is our 'commentary' on what is happening to us." (Johnson et al., 1985, p. 22). The reflection subsystem gives our memories a personal quality and allows us to voluntarily recall the past events. Johnson (1983) proposes that it is the deficit in these reflective conscious processes that accounts for the amnesic syndrome. The sensory and perceptual processes seem to be relatively intact.

Other theorists have described impaired and preserved forms of memory in terms of a dissociation between learning requiring cognitive mediational strategies and stimulus–response (S-R) learning (Warrington and Weiskrantz, 1982), a dissociation between conscious remembering that involves a perceptual-like act and another memory system that does not depend on such processes (Poulos and Wilkinson, 1984), a dissocation between vertical associations involving higher units and horizontal or S-R type associations (Wicklegren, 1979), a dissociation between elaborative and perceptual processing (Graf et al., 1984), and finally, a dissociation between remembering with and without awareness (Jacoby and Witherspoon, 1982).

It is clear that theorists do not yet agree on the terminology or explanations for different forms of retention, particularly with regard to the new learning that amnesics express. There is, however, a consensus that the impaired form of memory in amnesics requires conscious analysis and reflective processing on the part of the subject. There is also some agreement that the spared memory functions do not depend on these activities either during initial encoding or the subsequent memory test (see Johnson, 1983; Jacoby and Witherspoon, 1982, for a detailed discussion of this point).

In this chapter we will extend this characterization of the impaired and the preserved memories to acute alcohol amnesia. We will demonstrate that alcohol impairs the same memory processes that are disrupted in amnesic patients. As is the case with the amnesic syndrome, our characterization of the spared processes will be incomplete. Since very few studies have investigated the memory processes that are resistant to disruption during intoxication, perhaps the most useful approach is to examine the effect of intoxication on a number of different memory tasks. Research on amnesics indicates that we should focus on those memory tasks that are stimulus- or data-driven (Jacoby, 1983a) and that do not rely on reflective processing (Johnson, 1983).


## 3.  ALCOHOL-SENSITIVE MEMORIES: CHARACTERISTICS OF IMPAIRED FUNCTIONS

In this section, we present evidence suggesting that alcohol disrupts those cognitive activities that require conscious deliberate processing and that profoundly improve retention in normal situations. These activities are of a reflective* nature (Johnson, 1983) and include organization of to-be-remembered items, elaboration of the items on the basis of what subjects already know, comparison of past and present events, and placing new information in the context of ideas about the future. Under normal conditions, subjects form a rich representation of new information by integrating it with their preexisting knowledge structures, by relating the items to the experimental context, and by forming relationships among items. Under alcohol, however, there seems to be a deficit in these reflective cognitive activities (Craik, 1977; Birnbaum et al., 1980; Hashtroudi et al., 1983, 1984).

We will examine the disruption in cognitive activities by reviewing alcohol-induced changes in organization and elaboration. Organization refers to formation of interitem relationships and elaboration refers to relating the item to other facts that we know about the item. Organization and elaboration, of course, rely on activating preexisting semantic knowledge. Alcohol-induced changes in organizational and elaborative processes, therefore, could arise from an inability to activate these semantic structures. Activation of knowledge structures, however, is a necessary but not a sufficient condition for reflective processing. In the next section, we will review the evidence suggesting that alcohol does not disrupt semantic activation. Intoxicated subjects can sort words into appropriate categories, produce reasonable associates to words, and comprehend prose material as well as sober subjects. It appears that memory deficits under alcohol are not due to qualitative changes in semantic activation. Rather, we propose that alcohol interferes with the additional reflective processing that occurs after semantic activation. Alcohol disrupts those processes that integrate the products of semantic activation with new information to create a unique representation of a particular event. Thus, intoxicated subjects are able to decide that a list of words, (e.g., dog, chipmunk, skunk, cat, etc.) belongs to a particular category (e.g., animal). To be able to recall these words, however, they must carry out additional processing beyond simple classification. This additional processing involves forming interrelations among items or integrating the item with the experimental context (e.g., the word "dog" occurred on a list in the psychology experiment). It is these additional activities that are impaired by alcohol.

## Deficits in Organizational Processes and Free Recall

Organization refers to the use of strategies for grouping items on the basis of common physical or conceptual categories. In human memory research, or-

---

* We have borrowed the term reflection from Johnson (1983). It should be noted, however, that Johnson has defined reflection as one of three subsystems in her MEM model. Here, we have defined reflection as a process and have used it in a more general sense. For a precise definition of the reflective subsystem, the reader is referred to Johnson's work (Johnson, 1983).

ganization is typically studied in a free recall test. In this task, subjects are given a list of items to learn and are subsequently asked to recall the items in any order. The list can be presented and tested once or over several trials. The most popular type of organization studied is known as *category clustering*. In this situation, subjects are given a list of items that are instances of several different well-defined categories. For example, subjects may be presented with doctor, maple, elm, violin, lawyer, oak, flute, and engineer, which belong to categories of professions, trees, and musical instruments. The use of organization (category clustering) is inferred when items from the same category are recalled adjacently. In another type of organization known as *subjective organization,* subjects build structure into a list. Organization is inferred on the basis of consistency in grouping of items over many trials.

Several studies have shown correlations between organization and recall (Klatzky, 1980). It is generally assumed that subjects engage in organizational activities in free recall and that improvements in recall are related to these activities. Thus, if alcohol impairs organization, free recall of intoxicated subjects should be lower than recall of sober subjects and examination of their recall should show less evidence of structuring.

Numerous studies have shown that acute alcohol intoxication severely disrupts free recall and that free recall might be the most sensitive task to the detrimental effect of alcohol. Impairments in free recall have been demonstrated with lists of categorized items (Parker et al., 1974; Rosen and Lee, 1976; Birnbaum et al., 1978; Peterson, 1977; Poulos et al., 1981) as well as unrelated items (Baddeley, 1976; Jones, 1973; Jones and Jones, 1977, 1980; Hashtroudi et al., 1984; Miller et al., 1978; Peterson, 1977), and in multitrial free recall (Rosen and Lee, 1976; Parker et al., 1974; Birnbaum et al., 1978; Rundell and Williams, 1977, 1980) as well as in single-trial free recall (Jones, 1973; Hashtroudi et al., 1984; Miller et al., 1978, Exp. 1; Weingartner et al., 1976). Also, in free recall under alcohol have been observed with a wide range of retention intervals (Jones, 1973; Miller et al., 1978).

Aside from the well-documented decrements in free recall, which indirectly implicate impairments in organization, more direct measures of category clustering clearly reveal deficits in organizational processes under alcohol. Three different studies (Parker et al., 1974; Rosen and Lee, 1976; Birnbaum et al., 1978) have examined category clustering with intoxicated subjects and a fourth study has focused on associative clustering where subjects are presented with pairs of associated words rather than instances of a category (Rundell and Williams, 1977).

In the first study, Parker et al. (1974) assessed the effect of alcohol on free recall and category clustering scores of 24 subjects (12 alcoholics and 12 matched nonalcoholic controls). Each subject participated in three sessions: sober (no alcohol), moderate dose (.67 ml/kg), and high dose (1.33 ml/kg). The sessions were about 1 week apart and dose was counterbalanced across sessions. In each session, subjects learned a different list of 30 words consisting of five taxonomic categories with six words per category. The list was presented for four study-test trials and the items were given in a different random order on each trial. The results demonstrated that free recall decreased with increasing doses of alcohol. More importantly, the $Z$-score measure of clustering* showed significantly less organization in intoxicated subjects' recall than in sober subjects' recall.

Similar results were obtained by Rosen and Lee (1976). In this study, 24 subjects were tested first in a sober session and then in an intoxicated (blood alcohol level of 100 mg%) session on two consecutive days. In each session, subjects learned two lists of 20 words from four taxonomic categories in four study-test trials. One of these lists was tested with a cued recall procedure where subjects heard the category names before and after list presentation. The other list was tested with a free recall procedure where category names were given only before the list and subjects were free to recall the lists in any order. In agreement with the findings of Parker et al., intoxicated subjects recalled significantly fewer words than sober subjects in both free and cued recall tests. The $Z$-score measure also showed that intoxicated subjects displayed less organization than sober subjects.

A third study by Birnbaum et al. (1978, Exp. 1) also supported the finding of lower recall and lower clustering under alcohol. In this study, two different groups of 24 subjects each were tested in placebo and alcohol (1.0 ml/kg) conditions. The list comprised six words from each of ten taxonomic categories and was presented in a blocked fashion (i.e., instances of a given category presented adjacently). Subjects were given three study-test trials and a fourth trial in which they received the category names as retrieval cues. Once again, intoxicated subjects were impaired in both free recall and cued recall tests and they also showed less evidence of category clustering.

Finally, Rundell and Williams (1977) extended these findings by demonstrating a disruption in associative clustering under alcohol. Eighteen intoxicated (1.0 gm 95% ethanol per kilogram) and 18 sober subjects received two lists of ten associated word pairs. List 1 consisted of word pairs with high associative value and list 2 contained word pairs with low associative value. The items of the word pairs were separated from each other and presented in a random fashion in six study-test free recall trials. The results were consistent with the studies on category clustering. Alcohol impaired free recall and clustering; intoxicated subjects were less likely than sober subjects to recall the items of the word pairs adjacently. Furthermore, an index of sequential organization called intertrial

---

* The $Z$-score measure of clustering developed by Frankel and Cole (1971) is based on the number of runs of the same-category items in recall.

repetitions (ITR) (Bousefield and Bousefield, 1966) was lower for intoxicated subjects than for sober subjects. Thus Rundell and Williams provided some evidence that alcohol reduces stability of recall across trials. Similar results were reported in a later study (Rundell and Williams, 1980).

Each of these studies indicates that intoxicated subjects are less likely to engage in organizational activities than sober subjects. Moreover, Rundell and Williams (1977, 1980) have demonstrated that the deficits in organization are not limited to clustering but can be extended to sequential organization. Hence, it seems reasonable to suggest that the severe impairment observed in free recall under alcohol is due, in part, to a deficit in using organizational strategies.

Another line of evidence for alcohol-induced organizational deficits in free recall is provided by studies that have examined improvement in recall over trials. Presumably, this improvement is a result of increased organization or increased clustering. Thus, if alcohol disrupts organizational processes, alcohol-related differences in free recall should increase over trials. In fact, the studies reviewed above have all reported significant state (sober vs. alcohol) by trials interactions, indicating that improvements over trials were slower for intoxicated subjects than for sober subjects. Likewise, Miller et al. (1978), using a list of unrelated words, showed that alcohol impaired free recall and that impairment was greater on trial 2 than on trial 1. Finally, Rundell and Williams (1977) reported significant interactions between state and trials on two measures of organization (Z-scores and ITR), thereby providing direct evidence for less efficient development of organization under alcohol.

In summary, three lines of evidence clearly indicate that alcohol impairs organizational processes. First, free recall, a memory task that depends heavily on organization, is severely disrupted by alcohol. Second, direct measures of organization (categorical or associative clustering) are significantly reduced under alcohol. Finally, intoxicated subjects show a slower rate of improvement in free recall and clustering than do sober subjects.

## Deficits in Elaborative Processing

A second cognitive activity of considerable importance is elaboration. Elaboration refers to the mental operations that integrate incoming information with prior knowledge and clarify new information in light of what subjects already know (Birnbaum et al., 1980; Stein and Bransford, 1979). Elaboration and organization are similar in that they both involve semantic processing and require relational and integrative activities. There is, however, a difference. In organization the emphasis is on the formation of relationships among the items within a list. In elaboration the integrative processing involves forming a relationship between an individual item and other facts known about the item. Elaboration can vary in quality or type of processing (e.g., Stein and Bransford, 1979), or it can vary in quantity and extensiveness of processing (e.g., Craik and Tulving, 1975).

Several studies have demonstrated that elaborative processing profoundly improves retention in normal subjects (e.g., Craik and Tulving, 1975; Jacoby and Craik, 1979; Bransford et al., 1979). This has led some investigators to suggest that a failure to engage in elaborative processing is a critical source of alcohol-related memory impairments (Craik, 1977; Birnbaum et al., 1980; Hashtroudi et al., 1983). According to the present framework, elaborative processing should be particularly disrupted by alcohol since it involves an integration between new information and the semantic structures activated by the information. This integrative processing, of course, requires reflective activities on the part of subjects. Intoxicated subjects can activate preexisting semantic structures, but they may fail to engage in additional reflective processing necessary for integration.

The first study that examined the effect of alcohol on elaboration was conducted by Hartley et al. (1978). These investigators varied the amount (quantity) of semantic elaboration by changing the complexity of sentence frames (Craik and Tulving, 1975). Intoxicated (1.0 ml/kg) and sober subjects received sentence frames that had one missing word and they had to decide whether a particular word fitted the sentence. There were three levels of complexity of sentence frames, simple (e.g., She cooked the _______ ), medium (e.g., The _______ frightened the children), and complex (e.g., The great bird swooped down and carried off the struggling _______ ). After completion of the sentence frame task, a surprise free recall test for words was given followed by a cued recall test in which sentence frames served as cues. The results showed that intoxicated subjects recalled fewer words than sober subjects, but there was no significant interaction between groups and sentence complexity in either free or cued recall. Thus, increasing the amount of semantic elaboration and providing adequate retrieval cues (sentence frames) did not differentially affect the level of recall for intoxicated subjects.

There was, however, a hint in the data from this experiment that alcohol might interfere with the ability of intoxicated subjects to benefit from complexity of elaboration. This is shown in free recall of words that were processed in appropriate sentences (Hartley et al., 1978, Exp. 3, Table 3, p. 644). Recall for sober subjects was nearly four times higher with complex sentence frames than with simple sentences (mean recall was 1.62 and 0.46, respectively). In the alcohol groups, however, recall improved only half as much with complex compared with simple sentences (mean recall was 0.75 and 0.38, respectively). The authors noted that the very low recall performance might have interfered with the identification of significant interactions involving alcohol. Thus, although the differential benefit from complex elaboration between alcohol and sober groups of subjects was not significant, it certainly is suggestive.

A series of three studies by Birnbaum et al. (1980), using a quite different approach to elaboration, clearly showed that intoxicated subjects differentially benefitted from provision of elaborative contexts. In these studies, intoxicated (1.0 ml/kg) and sober subjects were given essentially meaningless sentences such as "The notes went sour when the seam split." For one group of subjects, the sentences were preceded by elaborative context words, such as "bagpipes," which made the sentence meaningful. Another group of subjects did not receive these context words. All subjects were given a recognition memory test in which distractors were constructed by re-pairing components of the study sentences (e.g., The notes went sour when the fire got too hot). The major finding of this study was that without the context words (elaborative schemas) intoxicated subjects had lower recognition memory than sober subjects. Presenting the context words reduced the difference in recognition. Birnbaum et al. concluded that in the absence of elaborative contexts, intoxicated subjects fail to engage in spontaneous production of schemas for comprehending sentences. However, they can use elaborators when they are provided by the experimenter. Alcohol-induced memory impairment could be attributed to deficits in elaborative processing.

Thus, the results of the Birnbaum et al. (1980) studies did indicate that alcohol disrupts elaborative processing. Furthermore, intoxicated subjects seemed to be able to use elaborators that made essentially meaningless material meaningful. The results of the Hartley et al. (1978) experiment, however, suggested that intoxicated subjects might not be able to use elaborators that were related to more refined aspects of meaning. In fact, providing the complex elaborators seemed to reduce the level of free recall under alcohol. In view of these findings, we decided to conduct a systematic study on the role of elaboration in acute alcohol amnesia. We focused on subtle but powerful elaborators that facilitate normal retention.

Stein and Bransford (1979) have distinguished between precise and imprecise elaborators. For a given "base" sentence such as "The fat man read the sign," a precise elaborator would be a continuation such as "warning about the thin ice" and an imprecise elaborator would be "that was two feet high." Both sentences are meaningful. However, only precise elaborators improve recall in normal subjects. Stein and Bransford (1979) proposed that precise elaborators are effective because they specify the potential significance and relevance of target words (e.g., fat). Imprecise elaborators, on the other hand, simply put the words in the context of semantically congruous knowledge.

We used Stein and Bransford's procedure (1979, Exp. 1) to examine the effect of alcohol on elaborative processing. Independent groups of intoxicated (1.0 ml/kg) and sober subjects were tested in four different elaboration conditions. Subjects in condition 1 heard a simple "base" sentence without instructions to elaborate. In conditions 2 and 3, subjects received precise or imprecise elaborators for the base sentences. In all three conditions they were instructed to rate the comprehensibility of each sentence. In condition 4, subjects were instructed to produce their own meaningful continuations for base sentences. After presentation of 12 sentences, all subjects heard the base sentences with the target word missing (e.g., the adjective "fat") and they were asked to recall the target word.

Condition 4 was included to permit a direct examination of the quality of elaborators produced during intoxication. This allowed us to examine two different sources of impairment in elaboration under alcohol. First, alcohol may alter the quality of semantic interpretation. Intoxicated subjects might not activate preexisting knowledge structures as effectively as sober subjects or their elaborative schemas may be activated but more idiosyncratic (Birnbaum et al., 1980; Weingartner and Murphy, 1977). These semantically impoverished elaborators could interfere with later retrieval of the material (Birnbaum et al., 1980; Hartley et al., 1978). If alcohol amnesia is due to a deficiency in activating semantic structures, intoxicated subjects should generate imprecise or less-refined elaborators. Second, as noted earlier, alcohol may not disrupt semantic interpretation. Rather, sober subjects engage in additional reflective processes that enhance the effective utilization of elaborators during later recall, whereas intoxicated subjects do not. This explanation would be supported if intoxicated subjects produced reasonable elaborators, but they still showed poor recall performance. By rating the generated elaborators for precision, we could distinguish between these two alternatives.

The recall data are shown in Table 1. There was a marked deficit in using semantic elaborators. Intoxicated subjects could not use normally effective but subtle elaborators. The advantage of precise over imprecise elaborators was much greater for sober subjects than for intoxicated subjects. Furthermore, compared with the no-elaboration (base) condition, precise elaborators substantially enhanced the recall for sober subjects but did not improve the performance of the intoxicated subjects.

Table 1. Mean Recall of Target Words[a]

|                         | State |             |
| ----------------------- | ----- | ----------- |
| Elaboration condition   | Sober | Intoxicated |
| No elaboration (base)   | 6.81  | 3.68        |
| Imprecise elaboration   | 4.50  | 2.25        |
| Precise elaboration     | 9.31  | 3.37        |

[a] From Hashtroudi et al., 1983.

Table 2. Proportion of Elaborators Generated
for Sober and Intoxicated Groups[a]

|  | State | |
| --- | --- | --- |
| Type of elaborator | Sober | Intoxicated |
| Precise | .58 | .48 |
| Imprecise | .42 | .52 |

[a] From Hashtroudi et al., 1983.

The generation data shown in Table 2 indicate that the quality of elaborators produced by intoxicated subjects are strikingly similar to those generated by sober subjects. Alcohol intoxication did not disrupt activation of semantic structures and did not lead to production of more idiosyncratic elaborators. Moreover, a comparison of recall performance for precise-generated versus imprecise-generated elaborators indicated that intoxicated subjects did not benefit from precise elaborators even when they produced these elaborators.

Evidently, intoxicated subjects can activate preexisting memory structures as effectively as sober subjects, but mnemonic enhancement from this activation requires an additional processing step. In our experiment, the additional step may involve the integration between the target word and the meaningful elaboration. Sober subjects perceive the significance of semantic relatedness and form an integrated unit that enhances the accessibility of any part during later recall. Intoxicated subjects fail to engage in additional activities necessary for the unitization and integration processes. The precise nature of these activities cannot be specified at this point. It appears though that these are self-initiated activities that require expenditure of effort and result in formation of new connections among items. Thus, simple tasks such as generation of elaborators or sorting items into categories (Rosen and Lee, 1976) which rely on automatic activation of preexisting knowledge structure may not reveal the deficit in integrative processing. Future experiments should focus on tasks that tap the recently-acquired information and require deliberate processing on the part of subjects. Examination of overt rehearsal patterns and manipulations of the complexity and novelty of stimulus materials might help to clarify the underlying processes involved in integration and unitization.

The proposed deficit in reflective processes under alcohol may explain why the depth-of-processing interpretation of alcohol amnesia has not been supported (Hartley et al., 1978; Williams and Rundell, 1984). Depth of processing is similar to elaboration in that the memory trace is the product of the type of analysis performed on incoming information. However, the key distinction in the depth-of-processing view is between deep (e.g., semantic) processing, which results in a durable memory trace, and shallow (e.g., physical) processing, which does not (Craik and Lockhart, 1972; Craik and Tulving, 1975). In elaboration, the emphasis is on variations of processing within the semantic domain.

Craik (1977) suggested that the depth-of-processing view could explain the effects of acute alcohol intoxication on memory. Intoxicated subjects might fail to encode the meaning of the item as deeply as sober subjects and the encoding deficit may lead to poor retention. Thus, if depth of processing could be equated for sober and intoxicated subjects by presenting semantic orienting tasks, the memory decrement under alcohol should be eliminated or reduced. Following Craik's (1977) suggestion, several studies have provided subjects with physical (e.g., Does this word contain the letter R? Beer) or semantic (e.g., Does this word represent a living thing? Student) orienting tasks (Hartley et al., 1978; Williams and Rundell, 1984). The results have not been supportive of the depth notion. Intoxicated subjects show improvements from performing the semantic tasks but they do not benefit more than sober subjects.

In the context of the present framework, it seems obvious that activation of a category (living vs. nonliving) does not tap the cognitive processes that are impaired by alcohol. This semantic orienting task can be accomplished in a fairly automatic fashion. Hence, intoxicated subjects may have no difficulty in performing this task. Alcohol seems to disrupt the more complex operations that are used by sober subjects. These additional activities may involve ". . . elaboration of a network of semantic associations between elements of a list . . ." (Hartley et al., 1978, p. 646) and integration of the item with preexisting knowledge structures. Nominally, intoxicated subjects may engage in semantic processing, but functionally their encodings may be less elaborate and extensive than that of sober subjects (cf., Craik and Simon, 1980).

In summary, the studies reviewed in this section clearly demonstrate that intoxicated subjects have a marked deficit in using semantic elaborators and organizational activities that profoundly improve normal retention. This deficit, however, cannot be attributed to the inability to activate preexisting knowledge structures. Alcohol interferes with deliberate processing, and this seems to be a primary source of acute alcohol amnesia.


## 4.  ALCOHOL-RESISTANT MEMORIES: CHARACTERISTICS OF PRESERVED FUNCTIONS

If alcohol disrupts memories based on reflective processing, what forms of memory proceed in a relatively normal manner during intoxication? In this section we will review studies in which subjects have been shown to accomplish certain memory tasks at levels of intoxication that disrupt other forms of memory. It is the juxtaposition of affected and unaffected forms of memory at the same doses of alcohol that is particularly striking. At moderate levels of intoxication, there is some alteration in the delicate interplay between different forms of memory. With very high doses, this interplay may become a complete dissociation leaving the drinker with only those forms of memory that continue to function.

Although research on spared memory processes under alcohol is just beginning, certain consistent results have been obtained. Doses of alcohol that significantly impair free recall do not produce a state of bizarre, idiosyncratic thinking. While very high doses in alcoholic subjects may induce such a state (see Weingartner and Faillace, 1971, Exp. 1 results for alcoholic subjects only), this is not a consequence of moderate intoxication in social drinkers. Activating well-learned facts and concepts about the world is automatic and does not require reflective operations. Thus a subject who, when sober, knows that Toronto is a city in southern Ontario is not, after drinking, going to think that Toronto is in northern Quebec. Drinking and parties can create an atmosphere that permits silly behavior, but research indicates that such behavior is not a direct result of the drug.

In addition to being able to activate and utilize semantic structures, which could simply reflect the fact that practiced or well-learned information and skills are resistant to disruption, there is now evidence that new information can be acquired under alcohol (Hashtroudi et al., 1984; Parker et al., 1983). The acquisition of new information is revealed when memory is tested indirectly by, for example, asking the subject to complete word fragments. It also appears that the results of this new learning can be expressed on recognition tasks under particular conditions. Some new memories are formed under alcohol and they can be detected as long as the experimenter queries memory indirectly or by tasks that do not require well-integrated encodings.

## Activation and Utilization of Preexisting Structures

The acquisition of new information depends, in large part, on the activation of abstract organized knowledge structures. Any changes in semantic activation could lead to substantial changes in the way information is encoded or retrieved (e.g., Burke and Light, 1981). Indeed, a popular hypothesis about alcohol's amnesic effects is that it disrupts the quality of semantic activation and leads to poor recall (Birnbaum et al., 1980; Weingartner and Murphy, 1977). Several lines of research, nonetheless, suggest that organization of semantic memory seems to remain relatively intact during intoxication and that alcohol does not alter the nature of activated structures.

Changes in semantic organization could lead to changes in classification or sorting tasks. For example, intoxicated subjects might be less likely to sort a group of items on the basis of taxonomic categories (e.g., grouping doctor with professions). Rosen and Lee (1976), however, demonstrated that intoxicated subjects had no difficulty in sorting the items into the appropriate taxonomic categories. Yet, as noted before, these subjects had significantly lower free recall and clustering scores than did sober subjects. Rosen and Lee concluded that ". . . there is no reason to suppose that this [the deficit in organization and recall] is a result of an inability to organize materials by semantic category per se." (1976, p. 315).

Changes in semantic organization should also be reflected in the quality of elaborators or free associations generated during intoxication. There are, however, very few studies on this issue and the results are seemingly contradictory. Our recent study, described above, directly examined the nature of elaborators (Hashtroudi et al., 1983) and showed that the quality of elaborators generated by intoxicated and sober subjects were remarkably similar. In contrast, two widely cited earlier studies (Weingartner and Faillace, 1971; Hartocollis and Johnson, 1956) reported that alcohol increases the production of idiosyncratic associations. Similarly, Wallgren and Barry (1970) in their review of literature on alcohol and verbal abilities concluded that intoxicated subjects generate superficial and inappropriate associations. A closer examination of the data, however, reveals that many investigators, including ourselves, might have been less than cautious in accepting this conclusion.

In the first study (Weingartner and Faillace, 1971, Exp. 1), six alcoholic and six nonalcoholic subjects were tested in intoxicated and sober states. The dose of alcohol was administered incrementally in 1-ounce steps over several sessions. Alcoholic subjects received 2 ounces of 95% alcohol on the first day and 6 ounces on the fifth day. Nonalcoholic subjects started with 1 ounce of alcohol and could not tolerate more than 3 ounces. All subjects were asked to produce free associations to noun stimuli selected from Palermo and Jenkins (1964) free association norms. The responses were categorized as common or idiosyncratic on the basis of normative data. The major findings were that during intoxication, alcoholic subjects produced less frequent and more idiosyncratic associations. But the quality of free associations produced by nonalcoholic subjects did not differ in intoxicated and sober states, although free recall was significantly impaired.

The common interpretation of the Weingartner and Faillace results is that alcohol modifies the nature of semantic organization (Birnbaum et al., 1980; Weingartner and Murphy, 1977; Hashtroudi et al., 1983). This interpretation, however, is not demanded by the data. The production of bizarre associations was demonstrated only with alcoholic subjects who had consumed extremely large amounts of alcohol. It appears that with the doses discussed throughout this chapter, the quality of associations remains unchanged during intoxication.

The results of a second study (Hartocollis and Johnson, 1956), which have been interpreted as evidence for changes in associations, are also equivocal. In this study, intoxicated (blood alcohol concentration, or BAC, of .10%) and sober subjects performed four different verbal fluency tasks, one of which involved naming members of common categories such as birds, trees, and so forth (semantic production). It was expected that intoxicated subjects would produce fewer words than sober subjects. This was the case for the other fluency tasks but not for the semantic production task. There was no difference between sober and intoxicated subjects in the number of words produced from common categories. The authors suggested that even though the number of words was not reduced, an impairment in quality of associations counteracted the impairment in quantity. For example, they mentioned that intoxicated subjects produced such responses as "big tree" and "Christmas tree." It should be noted, however, that this study was a test of verbal fluency and only the number of correct response was recorded. No systematic method for measuring the quality of responses was developed. The few examples were merely offered as an explanation for the absence of an effect of alcohol on verbal fluency in the semantic production task.

Taken together, the two above-mentioned studies do not provide any evidence for changes in the nature of semantic associates. Additional studies using better techniques for recording the quality of responses and controlling the doses of alcohol are needed to clarify the effect of alcohol on semantic associations.

Changes in semantic organization could also affect language comprehension. Comprehension of words and connected discourse is, of course, dependent on activation of appropriate preexisting structures. Changes in the pattern of activation or initial interpretation of information, therefore, might result in poor comprehension. On the basis of several studies, it appears that comprehension and inferential activities are relatively resistant to acute alcohol amnesia. For example, in our study on elaborative processing, intoxicated and sober subjects gave the same comprehensibility ratings to the base, precise, and imprecise elaborators (Hashtroudi et al., 1983). Furthermore, in two other studies we have found no differences in prose comprehension between intoxicated and sober subjects (E.S. Parker, S. Hashtroudi, and L. Yablick, in preparation).

In the first study, we used a procedure developed by Cohen (1979) for assessing subjects' ability to comprehend facts and draw inferences. Cohen reported that older subjects made more inferential errors than younger subjects. In contrast, we found that intoxicated (1.0 ml/kg) subjects drew inferences as well as sober subjects. Evidently, alcohol does not impair comprehension and inferential processing.

In a second experiment, intoxicated (1.0 ml/kg) and sober subjects were presented with simple stories and were asked to recall these stories. According to current theories of comprehension (e.g., Kintsch and van Dijk, 1978; Mandler and Johnson, 1977; Thorndyke, 1977), these stories fit a particular structure or schema that determines how the events in the story are organized. For example, Mandler and Johnson (1977) suggest that stories are hierarchically organized.

We can study the underlying structure or the story grammar by representing the story as consisting of various categories such as settings, beginnings, outcomes, attempts, reactions, and endings (for a detailed description of the story grammar, see Mandler and Johnson, 1977). A good story begins with a setting that introduces the major characters and the time and place of the story. It then proceeds through a series of events that describe causes of behavior, and attempts and reactions of the characters. Finally, the story ends by describing the outcomes and the closing statements.

We used Mandler and Johnson's stories and analysis to test the effect of alcohol on comprehension and recall. We reasoned that poorer overall recall under alcohol would indicate that the underlying structure of the story is unaffected. On the other hand, differential recall from various categories (e.g., settings, reactions, outcomes) would provide some evidence that intoxicated subjects have difficulty in activating a story schema and using the schema to comprehend the stories. The results were consistent with the first outcome. Although recall was lower under alcohol, the size of the alcohol difference was similar for all categories. In short, the studies on comprehension show that activation of schemas and sensitivity to the semantic structure of prose seem to be unaffected by alcohol.

Overall, the available research on classification, production of elaborators, and comprehension clearly shows that the *nature* of activated semantic structures (semantic organization) is relatively resistant to acute alcohol amnesia. However, there is some, albeit inconsistent, evidence that speed of access to these structures may be affected by alcohol.

Only a few experiments have examined the speed of activation for complex semantic information. In one study, Baddeley (1976) used a modified version of a semantic verification task (e.g., canaries have wings), and he reported that alcohol significantly impaired speed of access to semantic structures. Baddeley's task, however, was different from a typical semantic memory task (e.g., Collins and Quillian, 1969) in that the reaction time (RT) was not recorded after each sentence. Rather, a large number of sentences were typed on paper and subjects were allotted 3 min to check correct and incorrect responses. Moreover, the study was conducted in a natural setting and the blood alcohol level measured by an alcometer varied from .05 to .20%. Although the experiment can be considered as evidence for impairment in speed of access to semantic memory, several procedural aspects (e.g., variability in BAC levels and measurement of RT) complicate this interpretation.

In contrast to Baddeley's findings, Hartley et al. (1978, Exp. 3) failed to find differences in RT for deciding whether a word fitted sentences of various complexity. Median response latencies and error rates did not differ for sober and intoxicated subjects. Similarly, several other experiments (e.g., Hartley et al., 1978; Williams and Rundell, 1984) comparing physical and semantic processing have reported no reliable differences in response latencies. We suspect that the discrepancy among these studies arises from major procedural differences. The fact that the stimulus materials are complex and vary greatly among experiments makes it even harder to draw any conclusions about speed of access to semantic memory.

Other studies that employ simpler materials, nevertheless, provide some evidence for a general slowing in semantic access under alcohol. In a series of studies, Moskowitz and his co-workers have shown a reduction in processing speed in naming numbers (Moskowitz and Burns, 1973) and pictures of objects (Moskowitz and Roth, 1971). For instance, Moskowitz and Roth (1971) presented their subjects with outline drawings of objects and used a voice-activated electronic timer to measure the speed of naming the picture. The objects were selected on the basis of frequency of occurrence of their names in the English language; there were six frequency groupings varying from rare to common words. All subjects were tested in intoxicated (.52 g/kg) and sober states. The major finding was that alcohol increased naming latency, thereby indicating that speed of access to semantic information is reduced. It is interesting to note, however, that alcohol did not differentially increase naming latency for various frequency groups. There appears to be an overall cognitive slowing under alcohol, but this slowing seems to be independent of changes in semantic organization. Sensitivity to frequency of occurrence of words is not diminished.

Changes in speed of access to semantic information have also been demonstrated with verbal fluency tests. Fluency measures the number of words that can be written or spoken in a given period of time. In general, intoxicated subjects generate fewer responses on fluency tasks than do sober subjects, suggesting that it takes them longer to access particular words. For instance, Hartocollis and Johnson (1956) reported that intoxicated subjects (BAC .10) produced fewer words on a variety of fluency tasks ranging from simple naming of words to naming of words that begin with a specific letter. The reduction in frequency did not increase with semantic loading. In fact, subjects under alcohol gave fewer words when they were simply asked to name words for 3 min (low semantic load), but they produced as many words as sober subjects when they were required to name words from semantic categories (high semantic load).

In summary, the studies on verbal fluency and naming latency indicate that speed of access to semantic information is reduced under alcohol. Yet, the experiments on sorting, production of elaborators, and language comprehension show that the quality of semantic response remains unchanged during intoxication. Moderate intoxication does not alter the pattern of semantic activation, although it might induce a general slowing in the rate of activation. How the speed of access affects memory and whether it might ultimately influence interpretation of stimuli and quality of responses are questions that remain to be elucidated by future research.

## Acquisition and Retention of New Information

In our brief reference to the literature on the amnesic syndrome, we noted that despite severe impairments in recall and recognition, amnesic patients show relatively normal retention of new information when their memory is tested in an implicit fashion which does not require reflective processing or conscious recollection of the learning episode. Exposure to an item on a study list facilitates later processing of the item in a variety of memory tasks, such as identification of degraded words (Warrington and Weiskrantz, 1970), spelling of homophones (Jacoby and Witherspoon, 1982), and deciding whether a string of letters is a word or nonword (Moscovitch, 1982). Warrington and Weiskrantz (1970), for example, showed that amnesic subjects did not differ from control subjects when they were asked to identify visually degraded versions of words on a recently studied list even though they could not recall or recognize these words. Since neither amnesics nor controls were able to identify these words before the study trial, facilitation in performance was attributed to learning as a result of prior exposure.

Tulving et al. (1982) extended these results to normal subjects. After a single exposure to a long list of words, subjects were given two types of memory tests for studied as well as unstudied items. In the recognition test, subjects had to indicate whether they had seen the words before. In the fragment completion task, subjects had to complete word fragments such as A__A__IN by filling in the missing letters (ASSASSIN). Although subjects could complete some of the words on the basis of their general knowledge, they could complete more after exposure to these words on the study trial. Furthermore, recognition performance declined over a one-week interval but fragment completion did not. Evidently, even a single exposure to the item facilitates later performance and this facilitation can be independent of the ability to recall or recognize the item (see also Jacoby and Witherspoon, 1982).

There is little agreement among theorists about the characterization of this facilitation effect. Consequently, identification, fragment completion, and other similar tasks have been variously described as tests of perceptual memory (Jacoby and Dallas, 1981), sensory memory (Johnson, 1983), or priming tests which reflect the operation of a third and as yet unknown memory system (Tulving et al., 1982). Despite the differences, there is a consensus that facilitation in performance represents a new memory record that is formed in a specific situation. It does not result from simple reactivation of preexisting knowledge structures or a temporary increase in semantic activation (Schacter, 1984; Tulving, 1983; Jacoby and Witherspoon, 1982). The hallmark of this form of memory is that it can occur in the absence of integrative and reflective processing. Identifying or completing items does not require comparison with other items or conscious recollection of the learning episode (Johnson, 1983; Jacoby and Witherspoon,

1982; Schacter and Tulving, 1982). Therefore, according to the framework presented here, performance on identification of degraded stimuli, fragment completion, and similar tasks should be resistant to acute alcohol amnesia. The results of two recent studies have supported this prediction.

In the first experiment (Hashtroudi et al., 1984), 48 intoxicated (1.0 ml/kg) and 48 sober subjects studied a list of 29 words. The words were presented in a nondegraded (i.e., normal) fashion on a single trial. After a 5-min interval filled with a distractor task, one-third of the subjects in each group were tested in one of the three different retention tests: free recall, a modified version of an identification task used by Warrington and Weiskrantz (1970), and yes/no recognition. The free recall test, which clearly requires integrative processing, was included as a sensitive measure of acute alcohol amnesia.

The identification test was selected as an ideal task for examination of memory processes that might be resistant to alcohol's amnesic effects. This test consisted of presentation of the studied words in a visually degraded form, starting with the most degraded (80% degraded) and presenting increasingly more complete versions of the words (65% and 50% degraded; see Fig. 1). To measure the benefit from studying the items, we modified Warrington and Weiskrantz's task (1970) by testing the subjects on the studied items as well as new or "distractor" items that were degraded in the same fashion. Warrington and Weiskrantz excluded words that subjects could guess before the study trial, but they did not include distractors at test. Given alcohol's negative effect on speed of access to words and/or sensorimotor performance, we expected an overall deficit in identification by intoxicated subjects. Comparison of performance on old and new items was, therefore, essential to reveal the benefit from studying the items.

The third memory measure was a yes/no recognition test. Jacoby (1983b) has suggested that recognition tasks are intermediate between recall tasks and identification tasks since recognition can rely either on elaborative information or on familiarity based on perceptual information (see also Mandler, 1980). Some test conditions may promote a recognition decision that is based largely on the perceptual characteristics of the item. In these situations, the item seems to "jump out" from the page or the screen, and the subjects decide that they have seen the item before. In other situations, however, recognition of the item

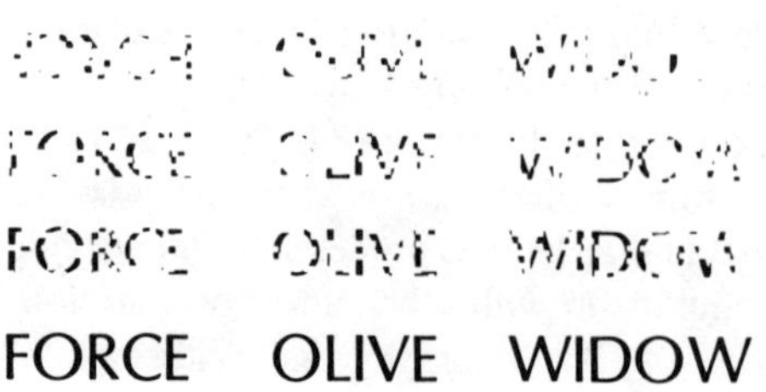

FORCE  OLIVE  WIDOW

Figure 1.  Examples of degraded words. (From Hashtroudi et al., 1984.)

Table 3. Proportions of Responses in the
Test Conditions[a]

|  | State | |
| Measure | Sober | Intoxicated |
| --- | --- | --- |
| Recall | .21 | .09 |
| Identification | | |
|   Old[b] | .34 | .23 |
|   New[c] | .25 | .12 |
|   Corrected | .09 | .11 |
| Recognition | | |
|   Old | .78 | .72 |
|   New[d] | .18 | .13 |
|   Corrected | .60 | .59 |
| $d'$[e] | 1.90 | 1.89 |

[a] From Hastroudi et al., 1984.
[b] Number of old words identified at the 80% level divided by total number of old words (29).
[c] Number of new words identified at the 80% level divided by total number of new words (29).
[d] Proportion of false alarms.
[e] Average of $d'$ values calculated for individual subjects.

may require more deliberate processing and reflection on prior experience. In a somewhat similar vein, Johnson (1983) has proposed that recognition can draw on reflective processes, but in many circumstances it is more heavily dependent on nonreflective or perceptual processing. Thus, it is difficult to predict the effect of alcohol on recognition. In fact, previous experiments on recognition have yielded mixed results. Some studies (e.g., Parker et al., 1976; Ryback et al., 1970; Wicklegren, 1975; Williams and Rundell, 1984) have reported impaired recognition under alcohol, whereas others have not (Goodwin et al., 1969, 1973; T.H. Taylor, M.K. Johnson, and I.M. Birnbaum, 1980, unpublished manuscript). In view of these unexplained mixed results, it seemed important to include a recognition task in this study.

The results of this experiment are shown in Table 3. The corrected recognition score is the difference between the proportion of hits (old words correctly recognized as old) and false positives (new words incorrectly recognized as old). The corrected identification measure is the difference in proportion of old and new items correctly identified at the most degraded level (80%). The proportion of new items identified, therefore, reflects the difficulty in gaining access to words, and the corrected identification score provides a measure of improvement from having studied the words (i.e., a measure of memory independent of the rate of access to words).

The data clearly support the differential effect of alcohol on free recall and identification. Free recall was severely impaired under alcohol, but identification performance was not. In fact, the benefit from prior presentation (i.e., corrected identification score) was almost the same for sober and intoxicated subjects. It is clear that studying the items facilitates later identification of these items and that this facilitation is not affected by alcohol. The absolute level of identification (proportion of old and new items correctly identified) was lower for intoxicated subjects, and was attributed to the effect of alcohol on speed of access to words or to slowing of sensorimotor performance.

Recognition performance measured by corrected recognition and $d'$ (a signal detection statistic) was not affected by alcohol. As noted earlier, alcohol does not seem to have a consistent effect on recognition. It is possible that the disparate results arise from the different types of information that can be used for recognition. The demand for reflective processes may vary in different situations (Johnson, 1983). The detrimental effect of alcohol on recognition may occur when recognition judgments are heavily dependent on reflective processes. We will return to this issue later.

Overall, the dissociation between recall and identification tasks in this experiment supports our suggestion that alcohol may have different effects on different forms of memory. Intoxicated subjects seem to learn and retain new information, but this information is not tapped by tasks that require deliberate recollection.

Further evidence that certain new memories can be acquired under alcohol comes from a study by Parker et al. (1983). This study examined the relationship between blood alcohol level and mnemonic performance. Intoxicated subjects studied materials over the course of a protracted 7-hr drinking episode, before they began to drink and on the rising and falling limbs of the blood alcohol curve. A week later, they were tested for these materials before and after drinking. The control (placebo) group participated in identical procedures during the acquisition and the test sessions.

A total of 64 male college students were randomly divided into two groups. Thirty-two subjects received alcohol both days and 32 received placebo. Alcohol was administered in a dose of 0.8 g/kg that was consumed over a 1.5-hr period and produced an average peak breath alcohol reading of 0.08 g/100 ml. All subjects studied a variety of different materials during the experiment. The results of only one task are discussed here.

The task of interest was a modified version of the Tulving et al. (1982) fragment completion task. During study, subjects were shown words with three or four missing letters and were asked to fill in the letters. Each word was presented first in its most incomplete form (3–4 missing letters), then with one more letter at a time up to the whole word (e.g., _ _RT _ _ I, M _ RT _ _ I, MART _ _I, MARTI_I, MARTINI). The next week, during the test session, word fragments of targets (study words) were shown mixed with distractors. As at study, subjects were presented with progressively more complete versions of words up to the whole word. At the point subjects complete a word fragment, they were asked whether they remembered having seen the word the week before. Fragment completion and recognition performance were examined separately.

The results showed that memory expressed through word fragment completion was not affected by alcohol. We know that memories were formed because the proportion of target words completed at their most degraded (least complete cue) was larger on the test session than on the study session. Neither alcohol at study nor alcohol at test affected this form of memory: The probability of completing words with the most incomplete fragment was .118 for the alcohol group and .124 for the placebo group. Furthermore, the difference in completion between targets and distractors was almost identical for intoxicated and sober subjects (.12 for intoxicated subjects and .14 for sober subjects).

In contrast to apparently normal memory under alcohol, as measured by fragment completion, recognition of the same words was affected. The ability to remember whether or not a word had been studied a week before was impaired when subjects had been intoxicated at the time they studied the material. Recognition performance was not affected by having alcohol at the time of test. This is consistent with other studies that have found that alcohol particularly impairs encoding (Parker et al., 1976; Birnbaum et al., 1978).

Recognition performance was closely linked to BAC at study: The probability of correctly recognizing a word as one that had been studied decreased as BAC increased. Recognition was most impaired at the peak BAC of 0.08 g/100 ml and recovered as alcohol was metabolized. This orderly relation between BAC and recognition performance stands in marked contrast to word fragment completion performance which was unaffected by alcohol.

Taken together, these two studies showed that indirect tests of memory (e.g., identification of visually degraded words and word fragment completion) are resistant to acute alcohol amnesia. Despite major procedural differences, the magnitude of improvement in memory for intoxicated subjects in the identification of visually degraded words (.11) compared favorably with the level of memory in word fragment completion (.12).

Recognition, however, was unaffected by alcohol in the first experiment, but it was impaired in the second. Once again, alcohol seems to have an inconsistent effect on recognition. We have previously suggested that this discrepancy may be due to the different types of information used in a recognition test. It is, therefore, important to identify the procedural differences between the recognition tests in the two studies. Aside from differences in the duration of delay (5 min vs. 1 week) and in the methods of study (presenting the whole word vs. presenting progressively complete versions), the two experiments differed in one major respect. In the first study, recognition was tested with an independent group of subjects and it did not follow identification of degraded stimuli. In contrast, in the second study, recognition followed word fragment completion.

We suspect that recognition differences between the two studies might arise from this procedural difference. Two other pieces of evidence provide some support for this notion. First, in an unpublished study, we presented 16 intoxicated and 16 sober subjects with the same list of words used in the Hashtroudi et al. (1984) study. During the test, subjects were first required to identify the visually degraded words (80%, 65%, and 50% degraded). At the point of identification, they were asked to indicate whether they had seen the item before. Thus, recognition followed identification. In agreement with the results of Parker et al. (1983), alcohol disrupted recognition in this situation. Second, in one of the tasks used in the Parker et al. (1983) study, recognition was unaffected by alcohol when it occured by itself, without following another task. Although these data are preliminary, it seems that presentation of a prior task may be an important determinant of recognition performance under alcohol.

It is possible that when recognition occurs by itself, there is greater reliance on perceptual characteristics of the items. Some items may simply be recognized on the basis of their perceptual familiarity and without much reflection on the prior presentation. On the other hand, when recognition follows another task, such as word completion, subjects might be more likely to notice the difference between the two tasks and they might use a more deliberate and reflective strategy in recognition. Clearly, there is much to be learned about the recognition response under alcohol. The framework proposed in this chapter may provide a basis for systematic study of this issue. If alcohol impairs recognition based on reflection, then increasing subjects' reliance on deliberate processes and/or decreasing the reliance on perceptual information should result in impairment of recognition during intoxication.

To summarize, indirect tests of memory such as fragment completion and identification, which focus on the task at hand and do not require subjects to reflect on prior experience, seem to be resistant to acute alcohol amnesia. To the extent that recognition can draw on perceptual information, it may also proceed normally during intoxication. It is clear that this nonreflective form of memory guides intoxicated subjects' behavior and plays an important role in interpretation of later events.


## 5.  CONCLUSIONS

We have presented a framework for interpreting the acute effect of alcohol on memory. The value of such a framework lies in the extent to which it helps to explain problematic results and guides future research in a meaningful direction. The framework is, of course, preliminary and will have to be further developed and refined. Nevertheless, we believe that it has been useful in accounting for seemingly disparate results. For example, we have suggested that the inconsistent effects of alcohol on recognition can be explained by the extent to which a particular recognition test involves alcohol-sensitive and alcohol-resistant forms of memory. When reflective, deliberate processes are required to perform a recognition task, then alcohol should have detrimental effects. When recognition is based on alcohol-resistant forms of memory, such as a vague feeling of familiarity or the perceptual characteristics of the item, then no recognition impairment should be seen.

We have also suggested that the depth-of-processing framework could not account for acute alcohol amnesia because it did not distinguish between reflective and nonreflective processes. The earlier version of this theory (Craik and Lockhart, 1972) focused on variations between different domains of encoding (e.g., physical vs. semantic features). Since semantic activation is not affected by alcohol at moderate levels of intoxication, there is no particular reason to describe acute alcohol amnesia as a failure to engage in semantic analysis. Rather, the framework we propose points to those cognitive processes that follow initial semantic activation as the source of alcohol's amnesic effects.

The present framework also explains why free recall is particularly vulnerable to disruption by alcohol. In a free recall task, subjects typically integrate the information and recollect it in a conscious, deliberate manner. Thus, the key processes underlying alcohol amnesia are usually required in free recall. It should be noted, however, that these processes are not unique to free recall and can be called upon for other memory tasks as well.

Finally, we should consider one important finding about memory and alcohol that has not been incorporated into the alcohol-sensitive/alcohol-resistant framework. A number of studies have found that alcohol has its primary amnesic actions on encoding new information rather than on retrieving the information that was acquired before drinking (Parker et al., 1976; Birnbaum et al., 1978; Rosen and Lee, 1976; Miller et al., 1978). Since both encoding and retrieval can involve reflective processes, the particular susceptibility of encoding seems paradoxical. The present framework leads us to the hypothesis that the conditions under which retrieval has been tested have not depended upon deliberate processes. In contrast, studies that have demonstrated encoding impairments by alcohol have used tasks that rely on these processes. This account is, of course, speculative and requires empirical corroboration.

Perhaps the most important issue for future research will be the characterization of the forms of memory that are unimpaired during intoxication. So far only two studies have examined the new memories that can be acquired under alcohol (Hashtroudi et al., 1984; Parker et al., 1983). Several questions regarding the nature of spared memories remain unanswered. For example, can motor skills be acquired under alcohol? Are emotionally-based memories recorded during intoxication? Can preferences be formed under alcohol even though the subject may have no conscious recollection of them later? And, finally, can information on memories that are formed during intoxication help us to understand why people drink? These questions provide exciting new opportunities for understanding the role of cognitive processes in drinking, and particularly heavy drinking.


Acknowledgments

We are grateful to Virginia Rappold and Lauri Yablick for their helpful suggestions on earlier drafts of this chapter. We would like to thank Linda Matthews and Vee Richardson for their able assistance in preparing this manuscript.


# REFERENCES


Baddeley, A. D., 1976, Information processing and alcohol: A field experiment, paper presented at the First Scandinavian Symposium on Information Processing and Performance, Uppsala.

Birnbaum, I. M., and Parker, E. S., 1977, Acute effects of alcohol on storage and retrieval, in: *Alcohol and Human Memory* (I. M. Birnbaum and E. S. Parker, eds.), pp. 99–108, Lawrence Erlbaum, Hillsdale, New Jersey.

Birnbaum, I. M., Parker, E. S., Hartley, J. T., and Noble, E. P., 1978, Alcohol and memory: Retrieval processes, *J. Verb. Learn. Verb. Behav.* **17**:325–335.

Birnbaum, I. M., Johnson, M. K., Hartley, J. T., and Taylor, T. H., 1980, Alcohol and elaborative schemas for sentences, *J. Exp. Psychol. Hum. Learn. Mem.* **6**:293–300.

Bousfield, A. K., and Bousfield, W. A., 1966, Measurement of clustering and of sequential constancies in repeated free recall, *Psychol. Rep.* **19**:935–942.

Bransford, J. D., Franks, J. J., Morris, C. D., and Stein, B. S., 1979, Some general constraints on learning and memory research, in: *Levels of Processing in Human Memory* (L. S. Cermak and F. I. M. Craik, eds.), pp. 331–354, Lawrence Erlbaum, Hillsdale, New Jersey.

Burke, D. M., and Light, L. L., 1981, Memory and aging: The role of retrieval processes, *Psychol. Bull.* **90**:513–546.

Butters, N., and Cermak, L. S., 1980, *Alcoholic Korsakoff's Syndrome: An Information Processing Approach to Amnesia,* Academic Press, New York.

Carpenter, J. A., and Ross, B. M., 1965, Effect of alcohol on short-term memory, *Q. J. Stud. Alcohol* **26**:561–579.

Cohen, G., 1979, Language comprehension in old age, *Cognitive Psychol.* **11**:412–429.

Cohen, N. J., 1984, Preserved learning capacity in amnesia: Evidence for multiple memory systems, in: *Neuropsychology of Memory* (L. R. Squire and N. Butters, eds.), pp. 83–103, The Guilford Press, New York.

Cohen, N. J., and Squire, L. R., 1980, Preserved learning and retention of pattern analyzing skill in amnesia: Dissociation of knowing how and knowing that, *Science* **210**:207–210.

Collins, A. M., and Quillian, M. R., 1969, Retrieval time from semantic memory, *J. Verb. Learn. Verb. Behav.* **8**:240–247.

Craik, F. I. M., 1977, Similarities between the effects of aging and alcoholic intoxication on memory performance construed within a "levels-of-processing" framework, in: *Alcohol and Human*

*Memory* (I. M. Birnbaum and E. S. Parker, eds.), pp. 9–21, Lawrence Erlbaum, Hillsdale, New Jersey.

Craik, F. I. M., and Lockhart, R. S., 1972, Levels of processing: A framework for memory research, *J. Verb. Learn. Verb. Behav.* **11**:671–684.

Craik, F. I. M., and Simon, E., 1980, Age differences in memory: The roles of attention and depth of processing, in: *New Directions in Memory and Aging: Proceedings of the George Talland Memorial Conference* (L. W. Poon, J. L. Fozard, L. S. Cermak, D. Arenberg, and L. W. Thompson, eds.), pp. 95–112, Lawrence Erlbaum, Hillsdale, New Jersey.

Craik, F. I. M., and Tulving, E., 1975, Depth of processing and the retention of words in episodic memory, *J. Exp. Psychol. Gen.* **104**:268–294.

Eich, J. E., 1980, The cue dependent nature of state-dependent retrieval, *Mem. Cognit.* **8**:157–173.

Frankel, F., and Cole, M., 1971, Measures of category clustering in free recall, *Psychol. Bull.* **76**:39–44.

Gerrein, J. R., and Chechile, R. A., 1977, Storage and retrieval processes of alcohol-induced amnesia, *J. Abnorm. Psychol.* **86**:285–294.

Goodwin, D. W., 1977, The alcoholic blackout and how to prevent it, in: *Alcohol and Human Memory* (I. M. Birnbaum and E. S. Parker, eds.), pp. 177–183, Lawrence Erlbaum, Hillsdale, New Jersey.

Goodwin, D. W., Powell, B., Bremer, D., Hoine, H., and Stern, J., 1969, Alcohol and recall: State-dependent effects in man, *Science* **163**:1358–1360.

Goodwin, D. W., Hill, S. Y., Powell, B., and Viamontes, J., 1973, Effect of alcohol on short-term memory in alcoholics, *Br. J. Psychiatry* **122**:93–94.

Graf, P., Squire, L. R., and Mandler, G., 1984, The information that amnesic patients do not forget, *J. Exp. Psychol. Learn. Mem. Cog.* **10**:164–178.

Hartley, J. T., Birnbaum, I. M., and Parker, E. S., 1978, Alcohol and storage deficits: Kind of processing? *J. Verb. Learn. Verb. Behav.* **17**:635–647.

Hartocollis, P., and Johnson, D. M., 1956, Differential effects of alcohol on verbal fluency, *Q. J. Stud. Alcohol* **17**:183–189.

Hashtroudi, S., Parker, E. S., DeLisi, L. E., and Wyatt, R. J., 1983, On elaboration and alcohol, *J. Verb. Learn. Verb. Behav.* **22**:164–173.

Hashtroudi, S., Parker, E. S., DeLisi, L. E., Wyatt, R. J., and Mutter, S. A., 1984, Intact retention in acute alcohol amnesia, *J. Exp. Psychol. Learn. Mem. Cog.* **10**:156–163.

Hull, J. G., 1981, A self-awareness model of causes and effects of alcohol consumption, *J. Abnorm. Psychol.* **90**:586–600.

Hull, J. G., Levenson, R. W., Young, R. D., and Sher, K. J., 1983, Self-awareness-reducing effects of alcohol consumption, *J. Pers. Soc. Psychol.* **44**:461–473.

Jacoby, L. L., 1983a, Remembering the data: Analyzing interactive processes in reading, *J. Verb. Learn. Verb. Bahav.* **22**:485–508.

Jacoby, L. L., 1983b, Perceptual enhancement: Persistent effects of an experience, *J. Exp. Psychol. Learn. Mem. Cog.* **9**:21–38.

Jacoby, L. L., and Craik, F. I. M., 1979, Effects of elaboration of processing at encoding and retrieval: Trace distinctiveness in recovery of initial context, in: *Levels of Processing in Human Memory* (L. S. Cermak and F. I. M. Craik, eds.), pp. 1–21, Lawrence Erlbaum, Hillsdale, New Jersey.

Jacoby, L. L., and Dallas, M., 1981, On the relationship between autobiographical memory and perceptual learning, *J. Exp. Psychol. Gen.* **110**:306–340.

Johnson, M. K., 1983, A multiple-entry, modular memory system, in: *The Psychology of Learning and Motivation*, Vol. 17 (G. H. Bower, ed.), pp. 81–123, Academic Press, New York.

Johnson, M. K., Kim, J. K., and Risse, G., 1985, Do alcoholic Korsakoff's syndrome patients acquire affective reactions? *J. Exp. Psychol. Learn. Mem. Cog.* **11**:22–36.

Jones, B. M., 1973, Memory impairment on the ascending and descending limbs of the blood alcohol curve, *J. Abnorm. Psychol.* **82**:24–32.

Jones, B. M., and Jones, M. K., 1977, Alcohol and memory impairment in male and female social drinkers, in: *Alcohol and Human Memory* (I. M. Birnbaum and E. S. Parker, eds.), pp. 127–138, Lawrence Erlbaum, Hillsdale, New Jersey.

Jones, M. K., and Jones, B. M., 1980, The relationship of age and drinking habits to the effects of alcohol on memory in women, *J. Stud. Alcohol* **41**:179–186.

Kalin, R., 1964, Effects of alcohol on memory, *J. Abnorm. Soc. Psychol.* **69**:635–641.

Kintsch, W., and van Dijk, T. A., 1978, Toward a model of text comprehension and production, *Psychol. Rev.* **85**:363–394.

Klatzky, R. L., 1980, *Human Memory: Structures and Processes,* W. H. Freeman, San Francisco.

Mandler, G., 1980, Recognizing: The judgment of previous occurrence, *Psychol. Rev.* **87**:252–271.

Mandler, J. M., and Johnson, N. S., 1977, Remembrance of things parsed: Story structure and recall, *Cog. Psychol.* **9**:111–151.

McGeogh, J. A., and Irion, A. L., 1952, *The Psychology of Human Learning,* Longmans Green, New York.

Miller, M. E., Adesso, V. J., Fleming, J. P., Gino, A., and Lauerman, R., 1978, Effects of alcohol on storage and retrieval processes of heavy social drinkers, *J. Exp. Psychol. Hum. Learn. Mem.* **4**:245–255.

Moscovitch, M., 1982, Multiple dissociations of function in amnesia, in: *Human Memory and Amnesia* (L. S. Cermak, ed.), pp. 337–370, Lawrence Erlbaum, Hillsdale, New Jersey.

Moskowitz, H. A., and Burns, M., 1973, Alcohol effects on information processing time with an overlearned task, *Percept. Mot. Skills* **37**:835–839.

Moskowitz, H. A., and Roth, S., 1971, Effect of alcohol on response latency in object naming, *Q. J. Stud. Alcohol* **32**:969–975.

Palermo, D. S., and Jenkins, J. J., 1964, *Word Association Norms: Grade School through College,* University of Minnesota Press, Minneapolis, Minnesota.

Parker, E. S., and Parker, D. A., 1982, Towards an epidemiology of cognitive deficits among alcohol consumers, in: *Cerebral Deficits in Alcoholism* (D. A. Wilkinson, ed.), pp. 21–46, Addiction Research Foundation, Toronto.

Parker, E. S., and Weingartner, H., 1984, Retrograde facilitation of human memory by drugs, in: *Memory Consolidation: Psychobiology of Cognition* (H. Weingartner and E. S. Parker, eds.), pp. 231–251, Lawrence Erlbaum, Hillsdale, New Jersey.

Parker, E. S., Alkana, R. L., Birnbaum, I. M., Hartley, J. T., and Noble, E. P., 1974, Alcohol and the disruption of cognitive processes, *Arch. Gen. Psychiatry* **31**:824–828.

Parker, E. S., Birnbaum, I. M., and Nobel, E. P., 1976, Alcohol and memory: Storage and state dependency, *J. Verb. Learn. Verb. Behav.* **15**:691–702.

Parker, E. S., Schoenberg, R., Schwartz, B. L., and Tulving, E., 1983, Memories on the rising and falling blood alcohol curve, paper presented at the twenty-fourth annual meeting of the Psychonomic Society.

Peterson, R. C., 1977, Retrieval failures in alcohol state-dependent learning, *Psychopharmacology* **55**:141–146.

Poulos, C. X., and Wilkinson, D. A., 1984, A process theory of remembering: Its application to Korsakoff amnesia and a critique of context and episodic-semantic theories, in: *Neuropsychology of Memory* (L. R. Squire and N. Butters, eds.), pp. 67–82, The Guilford Press, New York.

Poulos, C. X., Wolff, L., Zilm, D. H., Kaplan, H., and Cappell, H., 1981, Acquisition of tolerance to alcohol-induced memory deficits in humans, *Psychopharmacology* **73**:176–179.

Rosen, L. J., and Lee, C. L., 1976, Acute and chronic effects of alcohol use on organizational processes in memory, *J. Abnorm. Psychol.* **85**:309–317.

Rundell, O. H., and Williams, H. L., 1977, Effects of alcohol on organizational aspects of human memory, in: *Currents in Alcoholism,* Vol. 2 (F. Seixas, ed.), pp. 175–186, Grune and Stratton, New York.

Rundell, O. H., and Williams, H. L., 1980, Alcohol and organizational processes in free recall, *Alcohol Tech. Rep.* **9:**12–19.

Ryan, C., and Butters, N., 1983, Cognitive deficits in alcoholics, in: *The Pathogenesis of Alcoholism: Biological Factors* (B. Kissin and H. Begleiter, eds.), pp. 485–538, Plenum Press, New York.

Ryback, R. S., Weinert, J., and Fozard, J. L., 1970, Disruption of short-term memory in man following consumption of ethanol, *Psychon. Sci.* **20:**353–354.

Schacter, D. L., 1984, Priming of old and new knowledge in amnesic patients and normal subjects, *Ann. N.Y. Acad. Sci.* **444:**41–53.

Schacter, D. L., and Tulving, E., 1982, Memory, amnesia, and the episodic/semantic distinction, in: *The Expression of Knowledge* (R. L. Isaacson and N. E. Spear, eds.), pp. 33–65, Plenum Press, New York.

Stein, B. S., and Bransford, J. D., 1979, Constraints on effective elaboration: Effects of precision and subject generation, *J. Verb. Learn. Verb. Behav.* **18:**769–777.

Thorndyke, P. W., 1977, Cognitive structures in comprehension and memory of narrative discourse, *Cog. Psychol.* **9:**77–110.

Tulving, E., 1983, *Elements of Episodic Memory,* Oxford University Press, New York.

Tulving, E., 1984*a*, Multiple learning and memory systems, in: *Psychology in the 1990's* (K. M. J. Lagerspetz and P. Niemi, eds.), pp. 163–184, North-Holland, Amsterdam.

Tulving, E., 1984*b*, Precis of elements of episodic memory, *Behav. Brain Sci.* **7:**223–268.

Tulving, E., Schacter, D. L., and Stark, H. A., 1982, Priming effects in word-fragment completion are independent of recognition memory, *J. Exp. Psychol. Learn. Mem. Cog.* **8:**336–342.

Wallgren, H., and Barry, H., 1970, *Actions of Alcohol: Biochemical, Physiological, and Psychological Aspects,* Elsevier, Amsterdam.

Warrington, E. K., and Weiskrantz, L., 1970, Amnesic syndrome: Consolidation or retrieval? *Nature* **228:**628–630.

Warrington, E. K., and Weiskrantz, L., 1982, Amnesia: A disconnection syndrome? *Neuropsychologia* **20:**233–248.

Weingartner, H., and Faillace, L. A., 1971, Alcohol state-dependent learning in man, *J. Nerv. Ment. Dis.* **153:**395–406.

Weingartner, H., and Murphy, D. L., 1977, State-dependent storage and retrieval of experience while intoxicated, in: *Alcohol and Human Memory* (I. M. Birnbaum and E. S. Parker, eds.), pp. 159–173, Lawrence Erlbaum, Hillsdale, New Jersey.

Weingartner, H., Adefris, W., Eich, J. E., and Murphy, D. L., 1976, Encoding-imagery specificity in alcohol state-dependent learning, *J. Exp. Psychol. Hum. Learn. Mem.* **2:**83–87.

Wicklegren, W. A., 1975, Alcoholic intoxication and memory storage dynamics, *Mem. Cognit.* **3:**385–389.

Wicklegren, W. A., 1979, Chunking and consolidation: A theoretical synthesis of semantic networks, configuring in conditioning, S-R versus cognitive learning, normal forgetting, the amnesic syndrome, and the hippocampal arousal system, *Psychol. Rev.* **86:**44–60.

Williams, H. L., and Rundell, O. H., 1984, Effect of alcohol on recall and recognition as functions of processing levels, *J. Stud. Alcohol* **45:**10–15.

6

# Experimental Social Psychology and the Causes and Effects of Alcohol Consumption

JAY G. HULL and RONALD R. VAN TREUREN

## 1. INTRODUCTION

Recently research has begun to focus on the social causes and consequences of alcohol consumption. While some of this research has been correlational in nature (e.g., investigating the relationship between peer pressure to drink and self-reports of alcohol consumption), a sizable body of research has accumulated that adopts a strictly experimental approach. The hallmark of this tradition is the random assignment of subjects to different experimental conditions. Research that employs such an experimental approach has examined the effects of alcohol on such diverse topics as aggression, sexual arousal, moods and emotion, self-disclosure, risk taking, assertiveness, moral judgment, social influence, helping, and social interaction. Experimental research has also examined the effects of a variety of social settings on the consumption of alcohol. It is the purpose of this chapter to provide a broad overview of this experimental literature on the social causes and consequences of alcohol consumption.

## 2. THE BEHAVIORAL CONSEQUENCES OF ALCOHOL CONSUMPTION: RESEARCH DESIGNS

Early experimental research on the social consequences of alcohol consumption adopted an experimental design in which subjects were randomly as-

---

JAY G. HULL and RONALD R. VAN TREUREN ● Department of Psychology, Dartmouth College, Hanover, New Hampshire 03755.

signed to an alcohol or placebo beverage condition. Typically, subjects were told that their beverages contained alcohol. The actual content of the beverage was disguised by a variety of procedures designed to insure that the beverages could not be distinguished on the basis of taste. Differences in behavior as a function of beverage content were attributed to the pharmacological properties of the drug.

In contrast to this interest in the pharmacological properties of alcohol, recent research has also explored the hypothesis that alcohol has effects by altering subjects' cognitive expectancies (cf. Marlatt and Rohsenow, 1980). Thus, it is possible that the cognition that one has consumed alcohol, irrespective of drink content, may be sufficient to alter subsequent behavior. Research has explored this hypothesis using a "balanced placebo" design in which beverage content and expected beverage content are factorially crossed.

There are at least two advantages of a balanced placebo design over the more traditional design in which beverage content is manipulated while expectancy is held constant (i.e., all subjects expect alcohol). First, the balanced placebo design allows one to investigate the independent effects of expectation. Second, by including additional control conditions, the design allows for a test of the pharmacological effects of alcohol that is not confounded with subjects' expectations. Thus, it is possible that the pharmacological effects of alcohol are only evident when subjects think they are consuming the drug. One cannot directly test this hypothesis within the traditional design since there are no conditions in which subjects consume alcohol but think that they are consuming a placebo. This deficiency is corrected by the balanced placebo design.

## 3. THE BEHAVIORAL CONSEQUENCES OF ALCOHOL CONSUMPTION: THEORETICAL APPROACHES

Different theoretical traditions have developed to explain the effects of alcohol consumption and alcohol expectancies. Current thinking on the effects of alcohol consumption adopts the view that alcohol has pharmacological effects on social behaviors by virtue of its detrimental effects on perception and cognition (Hull, 1981; Pernanen, 1976; Steele & Southwick, 1985; Zeichner & Pihl, 1979). According to these cognitive impairment models, alcohol intoxication leads to behavioral disinhibition by disrupting the processing of information that is necessary for effective self-regulation. To take the concrete example of aggression, the general logic of such models is as follows: Alcohol impairs information processing; as a consequence, it impairs both interpretation of the other's behavior and recognition of situational and personal standards for ones' own behavior; therefore, if one is challenged when intoxicated, one is more likely to perceive the other's behavior as being arbitrary and less likely to consider internal and external standards for one's own conduct; the ultimate result is that one is more likely to respond aggressively. Pernanen (1976) argued that the effect of alcohol responsible for this pattern of behavior involves a narrowing of the perceptual field that effectively reduces the number of cues available to the intoxicated person. Hull (1981) argued that the effect responsible for this pattern of behavior involves a deficit in higher-order encoding strategies used in information storage and self-regulation with respect to internal and external standards of appropriate conduct. Steele and Southwick (1985) endorse both points of view:

> Negative consequences of the response must be conceptualized, standards of conduct must be accessed and evaluated in relation to the response, and potentially inhibiting cues (again, both external and internal) must be perceived and their inhibitory significance understood. Alcohol's damage to perception, however, makes it harder for one to notice peripheral inhibiting cues. For cues that are noticed, its damage to cognition makes it harder for one to grasp their inhibitory meaning. (p. 19)

A different theoretical tradition has developed to explain the effects of alcohol expectancies. Current thinking on the effects of expectancies adopts the view that the cognition that one has consumed alcohol can function as an attributional excuse for deviant behavior (Lang et al., 1975; Marlatt and Rohsenow, 1980). To take the example of aggression, alcohol is again proposed to disinhibit aggressive behavior, but in this case as a consequence of alcohol expectancies since ". . . aggression may be attributed to the effects of alcohol, thus reducing the individual's own responsibilities for his actions" (Lang et al., 1975, p. 517).

As will be seen in the review of the experimental literature, there is evidence that both alcohol consumption and expectancy have independent effects on a variety of social behaviors. In addition, the two variables rarely interact. In a comprehensive review of research employing the balanced placebo design, Hull and Bond (1986) have calculated that the frequency of alcohol by expectancy interactions is virtually identical to what one would expect by chance (5.1%). This would suggest that the fear that the effects of alcohol consumption vary as a consequence of one's expectancies is unfounded. At the same time, the two variables do appear to have independent effects on behavior.

Over three-quarters of the experimental studies on the social consequences of alcohol have been done in the areas of aggression, sexual arousal, and mood. The remaining studies have been done in a number of other areas (self-disclosure, risk taking, assertiveness, moral judgment, social influence, helping, and social interaction). We will begin by considering research on the major topics. Since more research has been done on the effects of alcohol on aggression than on any other single topic, we will spend a considerable amount of time reviewing this research. Furthermore, we will use this literature to illustrate the strengths and weaknesses of a number of different theoretical approaches to the effects of alcohol on social behavior.

## 4. THE BEHAVIORAL CONSEQUENCES OF ALCOHOL CONSUMPTION: EXPERIMENTAL LITERATURE

### Alcohol and Aggression

The correlation of alcohol consumption and aggression is relatively well-established. For example, a number of studies have shown a link between alcohol and a variety of crime statistics. This body of literature has been adequately reviewed elsewhere (e.g., Pernanen, 1976; Taylor and Leonard, 1983) and will not be the major focus here. The observed correlation between alcohol and aggression led to an interest in conducting controlled experiments that investigated the causal relationship between alcohol and aggression.

The Effects of Alcohol Consumption on Aggression. In one of the first experiments on the effects of alcohol on aggression, Bennett et al., (1969) used the teacher–learner paradigm to measure shock intensity and duration as a function of various doses of alcohol. The teacher–learner paradigm is a common experimental design for the study of aggression. Subjects are told that one person in the experiment will have to teach another person a list of paired associates. The teacher is instructed to administer an electric shock as punishment for incorrect responses by the learner. The shock level and duration is left to the discretion of the teacher. Subjects are then assigned to the role of teacher or learner. Through a rigged drawing, subjects are always assigned the role of teacher and a confederate of the experimenter is assigned to the role of learner. Aggression is measured by the level and duration of shock the teacher is willing to apply to the learner for incorrect responses. The study by Bennett et al. had subjects consume an alcohol or placebo beverage prior to the teacher–learner task. The study involved a four-cell design with a placebo control group and three experimental groups who consumed various dosages of alcohol. Somewhat surprisingly, there were no significant effects involving alcohol.

In a subsequent study by Shuntich and Taylor (1972), the effects of alcohol on aggression were examined within a reaction time–competition paradigm. In this design, subjects compete against an opponent (actually a confederate) in responding as quickly as possible to a stimulus. The person responding most quickly is the winner of that trial. Prior to each trial, the players set a shock level. If they win the trial, the opponent receives a shock at the level they had set. If they lose, they receive a shock at the level the opponent had set. Typically subjects are allowed to win one-half of the trials. In the experiment by Shuntich and Taylor (1972), subjects were randomly assigned to consume either an alcoholic beverage, a placebo beverage, or no beverage prior to the competition. The results of the study indicated that subjects in the alcoholic beverage condition set significantly higher shock intensities than the placebo or no beverage control groups, who did not significantly differ from each other.

Two methodological differences were initially identified as potential sources of the discrepancy between the findings of Shuntich and Taylor (1972) and Bennett et al. (1969). First, Bennett et al. manipulated intoxication by having subjects consume a beverage that contained vodka, whereas Shuntich and Taylor used bourbon. Perhaps the higher congener content of bourbon could have been responsible for the increased aggression observed by Shuntich and Taylor. Subsequent research by Taylor and Gammon (1975) eliminated this alternative by finding no significant difference between the two beverages in their ability to elicit aggression.

The second methodological difference between Bennett et al. (1969) and Shuntich and Taylor (1972) concerned the experimental task. In the case of the teacher–learner paradigm used by Bennett et al., the target of the shocks was a passive victim, incapable of retaliatory responses. In the reaction time–competition paradigm used by Shuntich and Taylor, the partner was clearly an opponent in a competitive situation that involved the real possibility of retaliation.

The perception of threat or provocation is an important element in aggressive social situations and appears to distinguish the Bennett et al. from Shuntich and Taylor experiment. Taylor et al. (1976) conducted a direct test of the hypothesis that the effect of alcohol on aggression depends on the level of threat. These authors used a paradigm similar to Shuntich and Taylor (1972). Threat and alcohol were treated as independent variables. As predicted, there was a significant interaction of alcohol and level of threat such that intoxicated subjects initiated higher levels of shock than nonintoxicated subjects, but only under conditions of threat. Taylor et al. (1976) concluded that aggression induced by alcohol is a function of the interaction between alcohol consumption and degree of perceived threat inherent in a particular social situation.

Although the studies by Bennett et al. (1969), Shuntich and Taylor (1972), and Taylor et al. (1976) took place under strict "laboratory" conditions, a study by Boyatzis (1974) adopted a more naturalistic approach. Boyatzis (1974) collected data from "experimental parties" designed to simulate conditions under which drinking normally takes place. He was also interested in assessing the differences between types of alcohol ingested. There were three primary party conditions: one in which distilled spirits were served; one in which beer was served; and a third in which soft drinks and juice were served. In each case, party-going subjects participated in a number of competitive games. The behavior at the parties was video-taped at regular intervals, and the tape was later analyzed for aggression. The men in both the distilled spirits and beer conditions showed significantly more aggressive behavior than the soft drink group. In addition, distilled spirits drinkers were more aggressive than beer drinkers.* Boyatzis concluded that due to the naturalistic setting of the experiment, research showing

---

* This latter effect is similar to a finding by Pihl et al. (1984) that individuals who consumed or thought they were consuming beer were less aggressive than individuals who consumed or thought they were consuming distilled spirits.

increased aggression as a function of alcohol consumption could be generalized to social-drinking situations outside the laboratory.

Two theoretical accounts of alcohol and aggression appear to be eliminated on the basis of the above results. According to a pharmacological disinhibition model (cf. Taylor and Gammon, 1975), alcohol impairs cortical functions responsible for the inhibitory control of behavior. As a consequence, behavior comes under the control of more primitive subcortical centers in the brain. The result is the release of innate aggressive tendencies. Since such a model assumes that alcohol invariably increases aggressive tendencies, it would appear to be contradicted by evidence associating the effect of alcohol with the provocative character of the stimulus (Bennett et al., 1969; Taylor et al., 1976).

As an alternative to the pharmacological disinhibition model, some theorists have proposed a cognitive expectation model (cf. Lang et al., 1975; Marlatt and Rohsenow, 1980). According to this analysis, the knowledge that one has consumed alcohol rather than the pharmacological effect of the drug is responsible for aggressive behavior. This model will be reviewed in a later section of the present chapter at which time it will be seen that cognitions about one's level of intoxication in fact have effects on social behaviors that are independent of the pharmacological effects of the drugs. At the same time, the model has difficulty accounting for the findings on aggression reviewed above. Thus, increasing the actual alcohol content of a beverage has been shown to increase aggression in response to provocation even when expectancies have been controlled such that all subjects think they have consumed alcohol. Such a finding suggests that alcohol intoxication has effects independent of subjects' cognitive expectations.

Additional research would appear to eliminate a third theoretical account of the effects of alcohol on aggression. According to this analysis, alcohol may enhance the effects of salient contextual cues. When such cues indicate threat, alcohol increases the probability of aggression. Such an attentional or salience analysis has been adopted by a number of theorists (Pernanen, 1976; Taylor and Leonard, 1983; Zeichner, et al., 1982). As stated by Taylor and Leonard (1983):

> It is the current authors' contention that . . . the intoxicated person might simply focus on the most salient aspects of the situation. According to this position, when instigative cues are dominant, an intoxicated person would be more likely to focus on these dominant cues and, therefore, be more likely to react aggressively. On the other hand, in a situation in which instigative cues are minimal, aggressive behavior would not be expected to occur. (p. 96)

Although such an analysis can account for the results of the studies reviewed above, it is inconsistent with research that fails to find that alcohol increases aggressive behavior in the context of highly salient threats. Thus, Lang et al. (1975) conducted an experiment in which half the subjects were directly provoked by the insults of a confederate. Aggression was measured by shock intensities set by the subject in a subsequent teacher–learner task. The results indicated that there was no main effect of alcohol or alcohol by provocation interaction. Three recent studies by Rohsenow and Bachorowski (1984) all involved provocative situations and failed to find alcohol effects. In their first two experiments, provocation involved a negative evaluation of the subject's personality written by another subject. The third experiment involved an insulting supervisor. Aggression was measured in terms of the subject's retaliatory evaluations. Since provocation was held constant across all experimental conditions, the attentional analysis would predict alcohol main effects. In none of the three experiments was there a significant main effect of alcohol. Finally, and most damaging for an attentional analysis, a study by Zeichner et al. (1982) manipulated the extent to which the retaliatory behavior of the opponent was in focal attention and did not find an attention by intoxication interaction.

If alcohol does not increase aggression by enhancing the effects of salient, threatening cues, then why does alcohol increase aggression in the studies by Shuntich and Taylor (1972) and Boyatzis (1974) and interact with threat in the study by Taylor et al. (1976)? It is our view that alcohol does not have effects on social behaviors by enhancing the effects of salient cues, but rather by inhibiting the propensity of an individual to use higher-order encoding and elaboration processes necessary for self-control. Thus, research by Birnbaum et al. (1980) has shown that alcohol decreases the tendency of intoxicated individuals to employ higher-order encoding strategies. On the other hand, these researchers found that intoxicated individuals can benefit from such strategies when they are explicitly provided by the experimenter. Such findings suggest that alcohol will alter behavior to the extent that the individual is required to furnish his or her own encoding and elaboration strategies in order to infer a behavioral standard. The more explicit the cues to behave in a certain manner, the more likely sober and intoxicated individuals will act similarly. The more implicit the cues to behave in a certain manner, the more likely sober and intoxicated individuals will act differently.

Applied to the literature on aggression, such an analysis makes the following predictions: Situations that involve explicit cues to aggress (e.g., are explicitly threatening) and involve minimal cues not to aggress should result in equally high levels of aggression among sober and intoxicated individuals. Situations that involve explicit cues not to aggress (e.g., are explicitly nonthreatening) and involve minimal cues to aggress should result in equally low levels of aggression among sober and intoxicated individuals. Situations that involve explicit cues to aggress (e..g, have some explicitly threatening characteristics) and implicit cues not to aggress (i.e., require higher-order encoding and elaboration processes in order to recognize their nonthreatening characteristics) should result in greater aggression among intoxicated than sober individuals.

To illustrate this logic, let us consider the threat manipulation used in the study by Taylor et al. (1976). Subjects competed against a confederate in a reaction time–competition paradigm. In the "no-threat" condition, the confederate actively made a declaration of his pacifistic intentions toward the subject ("I have some strong convictions about hurting people and I'd feel more comfortable about this thing if I just set the one button [the lowest response allowed] all the time"; p. 940). According to the present analysis, such an explicit statement of pacifism should result in equally low levels of aggression among both sober and intoxicated subjects, and this was the case. In the "threat" condition, the confederate simply stated "no questions." Over the course of the experiment, the confederate then used low levels of retaliatory shock that were equivalent to the shock levels set by the explicitly pacifistic confederate in the nonthreat condition. The "threatening" confederate was thus threatening in the sense of occasionally shocking the subject, but also as pacifistic as possible within the constraints of the situation. Nevertheless, perception of this pacifism would require an inference based on the pattern of his shock settings. According to the present analysis, such potentially threatening but implicitly nonthreatening behavior by the confederate should result in greater aggression among intoxicated than sober subjects, and this was the case.

Such an analysis would appear to be able to integrate a large amount of the alcohol and aggression literature. The study by Shuntich and Taylor (1972) involved experimental conditions similar to the threat condition in the study by Taylor et al. (1976) and found similar results. They study by Bennett et al. (1969) involved no possibility for retaliation by the victim. Therefore, it can be classified as explicitly nonthreatening. Consistent with the present analysis, alcohol did not increase aggression. Boyatzis (1974) studied experimental parties involving competitive games. Most games involve explicitly competitive and implicitly nonthreatening ("it's only a game") characteristics. Consistent with the present analysis, alcohol increased aggressivelike behaviors. The studies by Lang et al. (1975) and the three studies by Rohsenow and Bachorowski (1984) involved explicit insults coupled with minimal cues discouraging retaliation. Consistent with the present analysis, intoxicated and sober subjects responded with equally high levels of aggressive retaliation. Finally, a study by Taylor and Gammon (1976) manipulated the degree to which subjects received social pressure by peers to reduce aggressiveness. Consistent with the present analysis, mild pressure was minimally effective in reducing the aggressive behavior of intoxicated subjects, whereas strong, sustained pressure resulted in a significant decrease in intoxicated aggressiveness.

While there have been no direct tests of the proposed analysis, several studies have provided indirect evidence that alcohol has effects on aggression by virtue of interfering with the processing of environmental information. In a study by Zeichner and Pihl (1979), subjects participated in what was described as a reaction time–pain perception experiment. Each subject was told that they would receive a tone of varying loudness from their partner. After a 5-sec exposure to the relatively unpleasant tone, subjects were to press a button that would both terminate the tone and send a shock to their partner. Shocks could be delivered for various durations and at various intensities. The subject was led to believe that after receiving the shock the partner would communicate to the subject the degree of pain he felt by means of the level of his subsequent tone choice.

For half the subjects in this experiment, the degree of tone loudness was fully correlated with the shock intensities chosen by the subjects. For the remaining subjects, the tone loudness varied randomly. The manipulation of random or correlated tone contingencies constituted one of the primary experimental variables. Given that the tones are correlated with previous shock intensity, one might reasonably expect that subjects would moderate their delivery of shocks in order to avoid loud tones. Given that the tones are random, one might reasonably expect that subjects would deliver more intense shocks in an effort to punish the irresponsible behavior of the other. In fact, this pattern was characteristic of subjects who had not consumed a beverage or had consumed a placebo they thought contained alcohol. Thus, these sober subjects behaved less aggressively when the loudness of the tones was correlated with the level of shocks than when these two variables were uncorrelated. On the other hand, subjects who consumed a beverage that actually contained alcohol were unaffected by the tone contingencies. Instead, these subjects apparently based their responses on the strength of the preceding tone regardless of whether it occurred in a random or correlated pattern. The authors ". . . speculate that although the intoxicated subjects attend preferentially to the arousing stimulus and respond to it, the other subjects attended to the consequences of their behavior. . . . The findings of the present study suggest that the occurrence of aggressive behavior following the ingestion of alcohol may in part be due to the individual's inability to process information pertinent to the consequences of his behavior." (p. 159)

A subsequent experiment by Zeichner and Pihl (1980) extended these findings to suggest that one of the crucial cognitive variables in the regulation of sober but not intoxicated behavior involves the attribution of intent. Using an experimental design similar to that employed by Zeichner and Pihl (1979), these authors varied what subjects were told about their partner's use of the tones. One-half of the subjects were told that the partner had no control over the tones. Therefore, by implication such tones were delivered with neutral intent. The remaining subjects were told that the partner had full control of the tones. High decibel levels were therefore indicative of malicious intent. The results of this experiment indicated that alcohol reduces the extent to which behavior is based on the intent manipulation. Thus, for measures of both shock intensity and duration there was an intent by drinking interaction such that the intoxicated group was less affected by the manipulation of intent than either a placebo or nondrinking control. Once again, the behavior of the intoxicated group was mainly a function of the decibel level of the preceding tone regardless of intent. Most strikingly, under neutral intent conditions, the correlation of tone level and shock intensity was highly significant in alcohol groups and nonsignificant in both placebo and nondrinking groups.

As Zeichner and Pihl (1980) point out, the exact mechanism responsible for such effects is unclear. In discussing their earlier studies, Zeichner et al. (1982) state ". . . that the alcohol may facilitate aggression through interference with related attentional processes" (p. 715). These authors therefore examined the issue of whether forcing an intoxicated subject to attend to the stimulus tone would affect aggressive responding. In a design similar to those used in the studies by Zeichner and Pihl (1979, 1980), subjects were given the task of recording the level of the stimulus tones. The pattern of the tones was prearranged by the experimenters to correspond to the level of shocks previously administered by the subject. The results indicated that subjects who were forced to attend to the intensity of the stimulus tone did not decrease subsequent aggressive behavior. Thus, despite focusing subjects' attention on the stimulus series, they remained less responsive to the contingency pattern (Zeichner and Pihl, 1979) and more responsive to the immediate cue of the stimulus tone. The authors recognize that these results contradict their hypothesis that alcohol affects aggression in their studies by virtue of interfering with attentional processes.

We interpret the results of these studies in terms of the encoding and elaboration model outlined earlier. From this perspective, the intoxicated individual does not moderate his aggressive behavior as a consequence of the manipulations in these three studies because in each case the meaning of the manipulation vis-a-vis the threatening character of the situation is implicit and requires higher-order encoding and elaboration in order to constitute a behavioral standard. Thus, intoxicated subjects are not responsive to the correlated–random pattern manipulation in the study by Zeichner and Pihl (1979) because it is an implicit manipulation of the degree of threat: Presentation of such a pattern requires an inference on the part of the intoxicated subject that given a particular pattern of play, the opponent is adopting a particular strategy, and therefore has a particularly threatening or nonthreatening intent which should form the basis for one's own standards of behavior. Similarly, the study by Zeichner et al. (1982) demonstrates that simply focusing subjects' attention on the intensity of the stimulus tone is not equivalent to rendering explicit the meaning of the tone pattern as an indication of intent. Finally, in the study by Zeichner and Pihl (1980), information about the extent to which the partner controls the level of the tone is one step removed from an explicit statement about the partner's intent. Again, it is not surprising that such a manipulation had little effect on intoxicated subjects' behavior.

Manipulations such as those employed by Zeichner and Pihl are in direct contrast to manipulations that have been shown to be effective in moderating the behavior of intoxicated subjects. Thus, in the study by Taylor et al. (1976), an explicit statement of pacifistic intentions by the partner ("I have some strong convictions about hurting people and I'd feel more comfortable about this thing if I just set the one button [the lowest response allowed] all the time"; p. 940) resulted in a significant reduction in aggression among intoxicated subjects. In a study by Taylor and Gammon (1976), highly explicit social pressure to decrease aggressiveness resulted in a significant reduction of aggression among intoxicated subjects. On the basis of this review, we feel that an encoding analysis of the effects of alcohol on social behaviors is more viable than an attentional analysis in accounting for the observed variation in results across studies.

The encoding analysis presented above is a variant of a model proposed by Hull (1981). From our current perspective, alcohol affects social behavior by decreasing the propensity of the individual to use higher-order encoding and elaboration strategies to render implicit environmental cues explicit with regard to self-regulatory standards of behavior. In a similar fashion, Hull (1981) argued that alcohol has effects on behavior by virtue of inhibiting the encoding of information according to its self-relevance, thus rendering the individual less self-aware. Since self-awareness has been associated with increased behavior in accord with internal and external standards of conduct (cf. Carver and Scheier, 1981), alcohol consumption is associated with decreases in self-regulation with respect to behavioral standards.

The model proposed by Hull (1981) built on two lines of research on cognitive processes. According to the literature on the cognitive effects of alcohol consumption, alcohol interferes with higher-order encoding processes (e.g, Birnbaum et al., 1980; Hartley et al., 1978; Weingartner et al., 1976). According to the recent literature on cognitive prototypes, self acts as a schema for encoding information (e.g., Markus, 1977; Rogers et al., 1977). Hull (1981) integrated these two literatures and proposed that alcohol affects processing of information about self by decreasing use of self-relevant encoding schemes. Three studies by Hull et al. (1983) demonstrated that (1) alcohol consumption resulted in decreases on traditional measures of self-awareness, (2) these effects were independent of subjects' alcohol expectancies, and (3) alcohol had these effects by inhibiting the encoding of information according to its self-relevance.

Given that alcohol consumption decreases self-awareness, it should have the opposite affective and behavioral consequences of variables that increase self-awareness. It should therefore decrease the correspondence of behavior with standards of appropriate conduct. In contrast to the effects of alcohol, manipulations that increase self-awareness have been shown to decrease aggressiveness in situations involving implicit standards against aggression (Scheier et al., 1974). Consistent with such reasoning, a recent study by Bailey et al. (1983) found that alcohol consumption increased and a self-awareness manipulation decreased aggressiveness. In addition, these authors found that an explicit cue to become self-aware appeared to overcome the effects of alcohol on aggression. Consistent with the analysis presented by Hull (1981), these authors suggest that alcohol may have effects on social behavior by decreasing the propensity (as opposed to the ability) of individuals to become self-aware.

The current analysis primarily differs from the earlier analysis by Hull (1981) in its breadth. While Hull was concerned with identifying self-relevant encoding as a key explanatory concept in alcohol effects, we are endorsing the broader view that alcohol affects behavior by inhibiting higher-order encoding and elaboration processes generally. Thus, we are arguing that alcohol affects social behavior by impairing the individual's understanding of the meaning of a set of environmental cues, one aspect of which concerns their self-relevance.

In summary, alcohol would appear to have pharmacological effects of increasing aggression in situations that contain both explicit cues to aggress (e.g., cues indicative of threat) and implicit cues not to aggress (e.g., cues that imply nonmalicious intentions). Situations that involve minimal cues to aggress and highly explicit cues not to aggress appear to result in equally low levels of aggression among intoxicated and sober subjects (e.g., the nonthreatening condition in Taylor et al., 1976). Situations that involve highly explicit cues to aggress and minimal cues not to aggress appear to result in equally high levels of aggression among sober and intoxicated subjects (e.g., Lang et al., 1975; Rohsenow and Bachorowski, 1984).

We have suggested that such a pattern of results is consistent with an analysis that alcohol affects social behavior by decreasing the propensity of the individual to use higher-order encoding and elaboration strategies in the self-regulation of behavior. Furthermore, we have argued that the literature on alcohol and aggression is not consistent with the notion that alcohol affects social behavior by focusing attention on the most salient cues (Pernanen, 1976; Taylor and Leonard, 1983; Zeichner et al., 1982).

Our analysis shares several features in common with a recent analysis by Steele and Southwick (1985). According to these authors, every social action involves both instigating and inhibitory cues. Under sober conditions, individuals weigh these factors in making a decision to act. However, alcohol dampens the effects of inhibitory cues, thereby allowing the instigating pressures to dominate the response. According to this analysis, alcohol should have its greatest effects on social behavior when both instigating and inhibitory cues are present in the same situations (a condition the authors refer to as involving "high inhibitory conflict"). Alcohol should have less of an effect when a situation involves either instigating or inhibiting cues (a condition of "low inhibitory conflict"). Such an analysis has its roots in early studies by Masserman and Yum (1946) and Conger (1951) on the effects of alcohol on approach–avoidance conflict. A principal difference between the analysis of Steele and Southwick (1985) and those of earlier researchers is their postulation of cognitive–perceptual as opposed to drive–motivational mechanisms of alcohol's effects.

While we are in essential agreement with many of the arguments advanced by Steele and Southwick (1985), it is unclear to what extent our conception of implicit and explicit cues for a behavioral standard overlaps with their conception of situations involving high inhibitory response conflict. From our perspective, a situation that involves both explicit cues to aggress and explicit cues not to aggress should result in less aggression by an intoxicated individual than a situation that involves explicit cues to aggress and implicit cues not to aggress (e.g., Taylor and Gammon, 1976). On the other hand, it would seem that one could derive the opposite prediction from the Steele and Southwick analysis insofar as the former condition would appear to involve more response conflict than the latter condition. Similarly, we would predict that intoxicated subjects may actually be less aggressive than sober subjects in a situation involving an implicit cue to aggress and minimal cues not to aggress (since the intoxicated individual may not infer the threat). It is unclear whether the Steele and Southwick analysis would make any predictions in such a situation.

The Effects of Alcohol Expectancies on Aggression. The effects of alcohol expectancies have also been investigated with respect to aggressive behaviors. In order to investigate the effects of expectancy on aggression, researchers have employed a balanced placebo design in which subjects consume either an alcoholic or placebo beverage and are told they are consuming either an alcoholic or placebo beverage. In general, it is predicted that alcohol expectancy increases aggressiveness independent of the effects of alcohol consumption by virtue of providing the individual with an excuse to engage in a desired but prohibited behavior.

In the first study to investigate the effects of alcohol expectancies on aggression, Lang et al. (1975) employed the balanced placebo design. As in much of the aggression literature, the experimental task involved a modified "aggression machine" (Buss, 1961). In contrast to previous findings, the cognition that one had consumed alcohol resulted in increased aggression (increased shock intensity and duration), whereas actual consumption had very little effect. Neither variable interacted with the provocativeness of the confederate's behavior. The lack of alcohol consumption effects has been discussed in the preceding section of the present chapter. With respect to the effects of expectancy, the authors argued that the results support the notion that the cognition that one has consumed alcohol provides the subject with a valid excuse to engage in an otherwise prohibited behavior: ". . . aggression may be attributed to the effects of alcohol, thus reducing the individual's own responsibilities for his actions" (Lang et al., 1975, p. 517).

Recent studies that have employed the balanced placebo design in order to assess the effects of alcohol expectancy on aggression have found mixed results.

An experiment by Kreutzer et al. (1984) found that both alcohol and expectancy independently increased self-reported hostility. In addition, alcohol increased subjects' use of profanity. In neither case did alcohol and expectancy interact. An experiment by Pihl et al. (1981) employed a reaction time–pain perception paradigm. These authors also found that both alcohol consumption and expectancy had effects of increasing different forms of aggressive behavior. In addition, alcohol and expectancy were found to interact such that the greatest aggression occurred when subjects' expectancies were inconsistent with their beverage content. Unfortunately, the implications of these results are ambiguous given that subjects' estimates of their degree of intoxication were unaffected by the expectancy manipulation. Finally, a study by Korytnyk and Perkins (1983) assessed the effects of alcohol consumption and expectancy on subjects' spontaneous tendency to deface property with graffiti. These authors found that alcohol consumption increased such vandalism. On the other hand, in contradiction to the findings of Lang et al. (1975), alcohol expectancy had a near significant effect of decreasing such vandalism.

Most troubling for the expectancy analysis are three recent studies by Rohsenow and Bachorowski (1984). In one of these experiments, subjects were negatively evaluated by a confederate and given an opportunity to retaliate in kind. Expectancy resulted in less retaliation. In another experiment, subjects were insulted by an experimenter. Once again, they were less likely to negatively evaluate the experimenter if they thought they had consumed alcohol. A third experiment found similar effects, but only for male subjects. The authors attempted to explain this discrepancy from previous research by claiming that the majority of their subjects expected alcohol to induce a pleasant mood. Furthermore, subjects who were in a pleasant mood were less aggressive. Therefore, expectancy caused less aggression by inducing a positive mood. This explanation is inadequate in several respects. First, the authors do not report the partial correlation analyses necessary to make such a case. Second, and more importantly, if such mood shifts are responsible for the expectancies in these experiments, why were they not operative in previous experiments?

The research on the effects of alcohol expectancies on aggression is thus less coherent than the research on the effects of alcohol consumption. One variable that may be related to the variation in expectancy effects across experiments is the extent to which subjects feel the need for a proper excuse to behave aggressively. A logical step in the development of the expectancy and aggression literature would be to assess the effects of expectancy under conditions that vary in the strength of the prohibited desire and need for a valid excuse. One possibility would involve selecting different types of individuals: people who feel inhibited about expressing anger and aggression may be particularly likely to express that aggression when they perceive that they have an appropriate excuse. A second possibility would involve manipulating the situational context of the aggressive response. In an interesting and socially relevant study, Rogers and Prentice-Dunn (1981) have suggested that some whites harbor ill feelings toward blacks that are usually restrained by public norms but are expressed in the form of heightened aggression when a proper excuse for such behavior is available. Given current reasoning for alcohol expectancy effects, it follows that such individuals should show greater aggressiveness toward a black than white target when they think they have consumed alcohol as opposed to a placebo. Future research might address this and related issues.

Alcohol and Aggression: Other Factors. Most of the experiments considered in the present review used male subjects. The results of the few studies that have employed females are ambiguous. Two studies by Rohsenow and Bachorowski (1984) used both male and female subjects. In one experiment, women were more aggressive as a function of alcohol consumption than were men, whereas men were less aggressive as a function of alcohol expectancy than were women. In a second experiment, there were no effects of sex. Obviously, additional research is necessary that investigates aggression among women as a function of alcohol consumption and expectancy before any conclusions can be drawn about the effects of this variable.

A second factor that has been considered as a potential moderator of the effects of alcohol on aggression is the level of frustration inherent in the task. Using a reaction time paradigm, Taylor et al. (1977) found no effects as a consequence of frustration. They concluded that frustration is a weak determinant of physical aggression and unrelated to the relationship of alcohol and aggression.

## Alcohol and Other Social Behaviors

Alcohol Consumption and Dyadic Interaction. A small collection of experiments have investigated social behaviors other than aggression. Most generally, these experiments support the hypothesis that alcohol consumption increases the inappropriateness of social behavior. Three studies have investigated dyadic interaction of sober or intoxicated individuals. A study by Smith et al. (1975) found that at low doses alcohol increased amount of and overlap in communications, and tended to decrease subjects' acknowledgment of their partner's statements. Higher doses further increased the degree of overlap in the conversations, although there was no additional increase in the amount of conversation. Similarly, Babor et al. (1983) found that alcohol consumption increased amount of verbal output and self-reports of cognitive confusion. In addition, alcohol consumption increased the amount of verbal hostility subjects expressed during a discussion of appropriate lifestyles. These authors endorse a cognitive model of the effects of alcohol on such social behaviors. Finally, in a study of dyadic self-disclosure, Rohrberg and Sousa-Poza (1976) found that alcohol consumption increased the depth, although not the amount of disclosure.

If it is assumed that such intoxicated self-disclosures were inappropriately deep relative to sober self-disclosures, each of these studies suggests that alcohol consumption leads to inappropriate interaction styles.

Alcohol Consumption and Risk Taking. Additional research has suggested that alcohol consumption increases the extent to which individuals adopt inappropriately risky strategies. Cohen (1960) found that intoxicated subjects tended to make larger estimates of their likelihood of success at driving a bus through a narrow opening than sober subjects and were more willing to take the risk. Teger et al. (1969) found that even when intoxicated and sober subjects perceived equal degrees of risk, intoxicated subjects were more likely than sober subjects to accept high levels of risk on a choice–dilemma questionnaire.

Alcohol Consumption and Helping. Consistent with the earlier analysis, two helping studies by Steele et al. (1985) suggest that the effects of alcohol on social behavior are a function of its effects on self-control. These behavioral effects are proposed to follow from the fact that alcohol interferes with processing of information related to response inhibition. Alcohol is proposed to have its maximum effect in situations that both instigate the individual to act and at the same time require response inhibition; situations that Steele et al. label high in inhibitory conflict.

In their basic experimental design, Steele et al. (1985) had intoxicated and sober subjects work on a task designed to be boring. The subjects were then approached with a request to help another individual by continuing to work on the boring task. When this request was low key (thus in the authors' language inducing a state of low inhibitory conflict), sober and intoxicated subjects volunteered to help at equally low levels. When the strength of the helping request was bolstered (thus increasing the conflict between inhibitory and instigatory cues), intoxicated subjects volunteered to help more than sober subjects. On the basis of these results, Steele et al. argue that intoxicated subjects have difficulty in situations that require response inhibition. Presumably this is because of the detrimental effect of alcohol on the processing of information required for inhibitory self-control.

In keeping with our earlier analysis, we would argue that the condition in which alcohol has effects involves explicit cues to act (the helping request) and implicit cues that require higher-order encoding and elaboration in order to provide a behavioral standard not to act (the boring nature of the task). By impairing encoding processes, alcohol results in behavior based on explicit cues: in this case, cues to help. Implicit cues, such as those involved in the nature of the task, are less likely to guide behavior because they are less likely to accurately register as cues in the intoxicated individual. In keeping with such an analysis, intoxicated subjects were less affected by the boring nature of the task: They found the task significantly more interesting and more fun than did sober subjects. Given these results, we see no reason to modify the earlier thesis that alcohol consumption results in decreased self-regulation and self-control. Once again, this effect would appear to be mediated by the effect of alcohol consumption on information processing.

Alcohol Consumption, Expectancy, and Sexual Behavior. Perhaps the clearest effects of alcohol expectancy can be found in the research on sexual arousal. In general, alcohol consumption has been shown to decrease sexual response to erotic stimuli in both males (decreased penile tumscence: Briddell and Wilson, 1976; Farkas and Rosen, 1976; Wilson et al., 1978) and females (decreased vaginal pressure pulse: Wilson and Lawson, 1976*a*, 1978). On the other hand, the cognition that one has consumed alcohol functions to increase sexual arousal in response to erotic stimuli (increased penile tumescence: Briddell et al., 1978; Wilson and Lawson, 1976*b*, increased reports of sexual arousal: Abrams and Wilson, 1983, Briddell, et al., 1978; Lang et al., 1980), although this effect appears to be restricted to males (Wilson and Lawson, 1978).

That alcohol has the pharmacological effect of decreasing sexual response has been known for some time (cf. Masters and Johnson, 1966). Why does the cognition that one has consumed alcohol function to increase sexual arousal in response to erotic stimuli? One possibility is that such responses are normally inhibited as a consequence of the subjects' perception that they are deviant and inappropriate. At the same time, what is deviant among sober individuals might actually be normative among intoxicated individuals. As a consequence, thinking that one has consumed alcohol may shift the standard for behavior in the direction of deviance. In this sense, expectancy that one has consumed alcohol leads to self-regulation with respect to a deviant standard of behavior.

Applied to the literature on sexual arousal, such an analysis predicts that individuals who are most concerned about the deviance of their response to erotica should show the greatest alcohol expectancy effects. This is precisely the case: Individuals who report the highest levels of sex guilt (Mosher, 1966) show the greatest effect of the expectancy manipulation (Lang et al., 1980; Lansky and Wilson, 1981). Apparently, thinking that one is intoxicated functions as a valid excuse for these individuals to become sexually aroused.

These results have implications for the issue of alcohol and self-regulation. Earlier it was argued that alcohol has the general *pharmacological* effect of inhibiting cognitive processes related to self-regulation. Now it would appear that *expectancy* that one has consumed alcohol leads to self-regulation with respect to a deviant standard of behavior. Recent studies on intoxication and sexual arousal provide support for both analyses. An experiment by Abrams and Wilson (1983) employed the balanced placebo design and measured subjects' delay of gratification to view erotic stimuli. In this case, longer delay was associated with seeing more film. The authors argued that long delay times were evidence of self-regulation with respect to the deviant standard of erotic desires. In keeping with this interpretation, alcohol expectancy was associated with longer delay times. At the same time, it has traditionally been assumed that short delay of gratification is evidence of the lack of self-control. Although the authors do not report the relevant statistics, a reanalysis of the data shows that alcohol consumption has a near significant effect ($p < .10$) of decreasing delay times. By implication, it would appear that while expectancy increases delay by altering the standard of behavior in a deviant direction, consumption decreases delay by inhibiting self-control.

A similar conclusion can be drawn from a study by Wilson and Niaura (1984). While previous research has found that alcohol consumption decreases sexual arousal, these authors found that consumption was associated with greater arousal when subjects were explicitly told to control their sexual response (there were no effects of expectancy). Once again, the implication is that alcohol consumption has the pharmacological effect of inhibiting self-control. As argued earlier, this may be a consequence of the effect of alcohol on cognitive processes required for self-regulation with respect to a behavioral standard.

Alcohol and Internal States. Alcohol consumption has generally been considered to result in certain physical sensations, to induce positive emotions, and to reduce negative emotions. Initial research on this topic has found inconsistent results: Sometimes alcohol has been found to reduce anxiety (e.g., Williams, 1966), sometimes it has been found to increase anxiety (e.g., Steffen et al, 1974), and sometimes it has been found to induce rather neutral physical states (e.g., Pliner and Cappell, 1974). For purposes of presentation, relevant studies will be considered in two separate categories. First, we will consider the effects of alcohol and expectancy of self-reported physical sensations. We will then consider their effects on mood states.

Alcohol, Expectancy, and Physical Sensations. There is little doubt that alcohol consumption induces specific physical sensations. At the same time, alcohol expectancies would appear to have little effect on such sensations. Thus, a study by Conners and Maisto (1979) used a six subscale measure of physical sensations. Alcohol consumption but not expectancy significantly increased subjects' self-reports of central stimulation, feelings of warmth-glow, dynamic-peripheral sensations, anesthetic sensations, and feelings of impaired function, but had no effect on gastrointestinal sensations. The results of a study by McCollam et al. (1980) are impressive in that these authors used the same scale and found precisely the same pattern of effects and noneffects. Using a combined scale that measured similar sensations, Vuchinich et al. (1979) also found increased sensation reports following alcohol consumption and no effects or interactions involving expectancy.

Alcohol, Expectancy, and Mood. Research on self-reported mood states shows a somewhat different pattern. In this case, alcohol and expectancy both appear to have independent, although inconsistent, effects. With respect to positive moods, subjects in studies by Conners and Maisto (1979), McCollam et al. (1980), and Vuchinich et al. (1979) all completed the Nowlis (1965) Mood Adjective Checklist (MACL). In general, both alcohol consumption and expectancy increased positive mood and the two variables did not interact. Thus, alcohol consumption was found to increase feelings of social affection (Conners and Maisto, 1979; McCollam et al., 1980; Vuchinich et al., 1979), surgency (Conners and Maisto, 1979; McCollam et al., 1980), and elation and egotism (McCollam et al, 1980).

In addition to alcohol consumption effects, Vuchinich et al. (1979) found that alcohol expectancy increased feelings of social affection, surgency, and elation, although Conners and Maisto (1979) and McCollam et al. (1980) found that expectancy only increased feelings of nonchalance. A possible reason for this discrepancy is that Vuchinich et al. (1979) actually manipulated positive mood by exposing subjects to humorous stimuli. Indeed, these authors found that subjects who thought they had consumed alcohol also showed increased amusement in response to the humorous material. It is possible, then, that the effects of expectancy depended in part on contextual cues that were absent in the studies by Conners and Maisto and McCollam et al. If expectancy has its effects by providing an excuse to engage in a desired but socially proscribed activity, it may be that expectancy increases expressions of positive emotion because it gives the person a proper excuse to express such emotion (e.g., laugh out loud) despite social conventions regarding decorum.

These results may shed light on the findings of an earlier study by Pliner and Cappell (1974). In contrast to the above findings regarding the effects of alcohol consumption, these authors found that although consumption increased positive mood, it did so only under social conditions involving a humorous cue. Under neutral conditions, alcohol simply increased reports of neutral physical symptoms. These authors did not attempt to control expectancies and indeed found a significant difference between alcohol and placebo conditions on ratings of perceived intoxication. It is possible, then, that their results reflect an expectancy effect similar to that found by Vuchinich et al. (1980) in response to a humorous cue.

The Nowlis MACL inventory used by McCollam et al. (1980) and Conners and Maisto (1979) also includes scales of negative emotions. (Vuchinich et al. do not report results for these subscales.) The main effects of alcohol, expectancy, and the alcohol–expectancy interaction were all nonsignificant across both studies for measures of anxiety, aggression, fatigue, sadness, and skepticism. The only effect on any of these scales in either study involved an interaction of expectancy and consumption rate reported by Conners and Maisto (1979): Subjects who were told they were drinking alcohol reported greater anxiety when they had to drink at a rapid rate. In addition to the Nowlis scale, McCollam et al. (1980) and Conners and Maisto (1979) also report effects on the Zuckerman et al. (1964) Multiple Affect Adjective Checklist. This scale includes subscales of anxiety, depression, and hostility. Again, the main effects of alcohol, expectancy, and the alcohol–expectancy interaction were all nonsignificant across both studies for measures of anxiety and hostility, although McCollam et al. (1980) did find that alcohol consumption decreased self-reported depression.

According to the studies by Conners and Maisto (1979) and McCollam et al. (1980), then, neither alcohol nor expectancy appears to have strong effects on negative mood. At the same time, neither of the studies attempted to manipulate subjects' mood. It is possible that alcohol and expectancies primarily have effects on negative mood in the context of a strong mood induction. Indeed, studies that have included anxiety inductions have been somewhat more likely to find a significant effect of alcohol consumption such that intoxicated subjects report feeling less anxious. Similar effects have not been found for manipulated expectancy. Polivy et al. (1976) manipulated anxiety by threatening subjects with electric shock for incorrect responses. They found that alcohol consumption decreased self-reported anxiety, whereas the cognition that one had consumed alcohol actually increased anxiety. Levenson et al. (1980) manipulated anxiety both through threat of shock and by having subjects make a self-disclosing speech. Consistent with Polivy et al., these authors found that alcohol reduced self-reported anxiety, although they found no effects of expectancy. Williams (1966) investigated alcohol effects in an experimental "party" setting and found that low levels of alcohol (4–6 oz 86 proof alcohol) reduced anxiety and depression, but that this improvement in mood disappeared at higher levels of intoxication (8 oz).

The findings of these studies are in contrast to findings by several researchers that alcoholics feel greater subjective distress after consuming alcohol (McNamee et al. 1968; Mendelson et al., 1964; Nathan and O'Brien, 1971; Steffen et al., 1974) and that intoxicated individuals are more likely to remain anxious than sober individuals during tasks involving the probability of being shocked (Dengerink and Fagan, 1978). In addition, five studies have investigated the effects of alcohol and/or expectancy on social anxiety in subjects anticipating an interaction with a partner of the opposite sex (Abrams and Wilson, 1979; Keane and Lisman, 1980, Exp. 1 and 2; Wilson and Abrams, 1977; Wilson et al., 1980). None of these studies found significant effects or interactions involving alcohol consumption or expectancy on measures of self-reported anxiety. Given the inconsistent nature of alcohol and expectancy effects, we hesitate to draw any general conclusions.

## Conclusions

Several conclusions can be drawn from this brief review. First, the cognition that one has consumed alcohol does appear to have important behavioral consequences. These effects would seem primarily to involve sexual arousal. Furthermore, these effects would seem to be due to the fact that the cognition one has consumed alcohol provides an attributional excuse to engage in a desired but prohibited action. With some exceptions, expectations would appear to have weak and conflicting effects on aggression, emotions, and physical sensations.

One possible reason for the inconsistency of expectancy effects is that even when individuals are equated with respect to their expectancies regarding beverage content, they may vary with respect to their expectancies regarding the way in which alcohol will affect their behavior (e.g., some subjects may believe that alcohol will facilitate performance, while others believe it will debilitate performance). Previous research has attempted to address this issue by manipulating subjects' expectancies regarding the effect of alcohol on their behavior. These studies have been relatively unsuccessful in manipulating such beliefs (Briddell and Wilson, 1976; Keane and Lisman, 1980). Given that such expectancies have proven relatively difficult to manipulate, a possible direction for future research is to classify subjects according to pre-existing beliefs about the effects of alcohol prior to their participation in the balanced placebo design.

In contrast to expectancy effects, alcohol consumption would appear specifically to increase aggression in contexts that involve explicit cues to aggress and implicit cues not to aggress. These effects would appear to be mediated by the pharmacological effect of alcohol of decreasing self-control. Similar decreased self-control has been observed with respect to other social behaviors. In addition alcohol consumption would appear to decrease physical sexual arousal, result in a particular pattern of physical sensations, and increase positive moods, although its effects on negative moods are inconsistent.

## 5.   THE MOTIVE TO CONSUME ALCOHOL

Much as in the literature on the social consequences of alcohol consumption, experimental analyses of causes of alcohol consumption have generally assumed the importance of either the pharmacological properties of the drug or individuals' expectancies regarding the drug. Models that assume the importance of the pharmacological properties of alcohol propose that people drink in order to attain a state of intoxication. Such consumption may be motivated out of a desire to reduce anxiety or tension directly or to disrupt cognitive processes involved in the awareness of anxiety-inducing cues (cf. Hull, 1981). Models that assume the importance of cognitive expectancies propose that people drink as a function of the cue that the beverage contains alcohol. This may be because of an expectancy that alcohol consumption disinhibits further alcohol consumption, or because of the self-presentational advantages of being labeled intoxicated (Jones and Berglas, 1978). We will consider evidence for each of these positions.

## The Motive to Achieve Intoxication

*The Tension Reduction Motive.* It has long been assumed that one motive to consume alcohol involves the individual's desire to reduce feelings of tension or anxiety (Yale University Center for Alcohol Studies, 1945). Conger (1956) provides an explicit formulation of this idea in learning theory terms. According to this approach (1) increased internal tension constitutes a heightened drive state, (2) alcohol consumption has the reinforcing property of lowering drive level by reducing tension, and (3) such drive-reducing reinforcement strengthens the alcohol consumption response. The cycle eventually leads to habitual drinking when alcohol consumption becomes a primary response to heightened internal tension. Cappell and Herman (1972) have termed this analysis the tension reduction hypothesis.

As formulated, the tension reduction hypothesis assumes both that alcohol is consumed for its tension reduction properties and that it has the effect of reducing tension. Evidence for these propositions is mixed. Comprehensive reviews by Cappell and Herman (1972) and Brown and Crowell (1974) found little support for the hypothesis that alcohol reduces tension. Similarly, recent reviews have also rejected traditional assumptions of the tension reduction hypothesis while leaving open the possibility that the model may have some validity in certain specifiable circumstances (cf. Pohorecky and Brick, 1983). In keeping with such conclusions, studies reviewed in the present chapter also reveal no consistent evidence that alcohol reduces anxiety, particularly in neutral settings.

Studies that have tested the hypothesis that tension motivates alcohol consumption have also found equally mixed results. Higgins and Marlatt (1973) found that threat of shock did not increase alcohol consumption. Holroyd (1978) found that subjects previously identified as high in social anxiety actually consumed less alcohol following negative as opposed to positive social evaluation. On the other hand, Miller et al. (1974) reported that anticipation of being personally evaluated by others increased alcohol consumption among alcoholics and Higgins and Marlatt (1975) reported similar findings using social drinkers. Finally, Marlatt et al. (1971) found that subjects who had been angered by a confederate consumed less alcohol if they were provided an opportunity to retaliate than if they were not offered such an opportunity. Evidence for the hypothesis that alcohol reduces tension and is consumed for its tension reduction properties would thus appear to be mixed as best.

*The Cognitive Impairment Motive.* Hull (1981) provided an alternative explanation of the results of these studies based on the notion that alcohol reduces self-awareness. Thus, while alcohol would appear to have social and behavioral consequences as a function of its effects on cognition, it is also possible that individuals consume alcohol in part in order to achieve a cognitive deficit. Extrapolating from the effects of alcohol on self-awareness, Hull (1981) proposed that alcohol consumption has the effect of disrupting processes necessary for self-evaluation. Thus, increased self-awareness has been associated with increased negative self-evaluation following failure and increased positive self-evaluation following success. Furthermore, it has been found that individuals try to avoid self-awareness following failure and increase self-awareness following success. Given that alcohol consumption decreases self-awareness (Hull et al. 1983), it follows that self-aware individuals should show increased alcohol consumption following failure and decreased consumption following success.

Hull and Young (1983) conducted an experiment to test this analysis. Subjects completed a questionnaire booklet that included the Self-Consciousness Inventory (Fenigstein et al., 1975). They then worked on intelligence-type tasks and were randomly assigned success or failure feedback. Immediate overt signs of pleasure–displeasure in response to this feedback were recorded as an indication of reactivity to the manipulation. Subjects were then released to participate in a second, ostensibly unrelated experiment during which they completed a mood measure and then tasted and rated a series of wines. This "wine-tasting" procedure has been used in several previous experiments to assess alcohol consumption under *ad lib.* conditions (e.g., Higgins and Marlatt, 1973, 1975).

For purposes of analysis, individuals were divided into high and low private self-consciousness groups. Three different measures were used to assess processes hypothesized by the self-awareness model of alcohol consumption. First, it was hypothesized that high self-conscious individuals would be more reactive than low self-conscious individuals in terms of overt displays of pleasure and displeasure in response to the self-relevant success–failure manipulation. Next, it was hypothesized that high self-conscious individuals would report more positive moods following success and more negative moods following failure than low self-conscious individuals. Finally, it was predicted that high self-conscious individuals would consume more wine following failure and less wine following success than low self-conscious individuals.

In accord with predictions, high self-conscious individuals showed significantly more overt pleasure–displeasure in response to the success–failure feedback than did low self-conscious individuals. In addition, analyses of the mood measures revealed the predicted interaction of self-consciousness and success–failure such that high self-conscious subjects who received failure feedback reported significantly more negative moods than high self-conscious subjects receiving success feedback. Low self-conscious individuals fell between these extremes and did not vary according to success–failure feedback. Finally, as predicted, there was a significant interaction of self-consciousness and success–failure on the measure of wine consumption. High self-conscious subjects given failure feedback drank significantly more wine than high self-conscious subjects given success feedback. Again, low self-conscious individuals fell between these extremes and did not vary according to success–failure feedback.

Additional tests further specified the mechanisms responsible for the effects of self-consciousness on alcohol consumption. Thus, there are two possible accounts of the above effects: (1) self-consciousness affects mood and mood motivates alcohol consumption because alcohol indirectly improves mood by reducing self-consciousness, or (2) self-consciousness affects mood and mood motivates alcohol consumption because alcohol directly improves mood. Although both accounts are consistent with the results of the analyses reported above, they make different predictions about the relationship between mood and alcohol consumption across high and low self-conscious individuals. If the first account is true, mood should only be correlated with alcohol consumption among high self-conscious individuals since only these individuals will show an improvement in mood as a function of self-awareness reduction. On the other hand, if the second account is true, mood should be correlated with alcohol consumption among both high and low self-conscious individuals since both groups show variations in mood and both will show improvements in mood as a function of the direct effects of alcohol on mood. Subsequent analyses clearly favored the first account. Mood was significantly correlated with alcohol consumption among high self-conscious ($r = .47$), but not low self-conscious ($r = .10$) individuals and these two correlations were significantly different from each other. It would thus appear that self-consciousness primarily affects alcohol consumption because alcohol indirectly affects mood by reducing self-awareness.

While these results support the self-awareness model of alcohol consumption, it is also necessary to determine if they can generalize to conditions outside the laboratory. Hull et al. (1986) report two studies that apply predictions from the model to alcohol consumption in the "real world." In both studies, the basic prediction was that life events indicative of success and failure should be more strongly related to alcohol consumption among individuals high as opposed to low in dispositional private self-consciousness. In the first study, this analysis was successfully used to predict relapse rates among detoxified alcoholics. Following release from the detoxification center, high self-conscious alcoholics who had experienced predominantly negative life events were particularly likely to relapse to pretreatment levels of drinking. High self-conscious alcoholics who had experienced predominantly positive life events were particularly likely not to relapse. The quality of life events had no effect on relapse rate of low self-conscious alcoholics. In the second study, the model was successfully used to predict alcohol consumption rates among high school adolescents. Again, high self-conscious individuals who were experiencing negative self-relevant events (poor grades in school) were especially likely to consume alcohol. High self-conscious individuals who were experiencing positive self-relevant events consumed significantly less alcohol. The quality of self-relevant events had a significantly smaller effect on the drinking of low self-conscious individuals. In addition, this interaction of self-consciousness and the quality of self-relevant events remained a significant predictor of consumption even after several other known predictors of alcohol use were statistically taken into account.

Together with the experimental study by Hull and Young (1983), these experiments indicate the generality of the self-awareness model. Alcohol use has been shown to be a joint function of an individual's level of private self-consciousness and personal success–failure. This interaction has been demonstrated across different populations (adolescents, college students, and middle-age adults), different patterns of alcohol use (social drinkers and alcoholics), different indicators of self-relevant life experiences (experimentally manipulated success–failure, reported academic performance, and reported performance of a significant role), and different measures of alcohol use (situationally measured wine consumption, self-reported alcohol consumption, and measures of alcoholic relapse). Furthermore, the result of the second study by Hull et al. (1986) suggest that this relationship is relatively independent of environmental and behavioral factors previously shown to be directly related to alcohol use.

## Expectancies and Alcohol Consumption

In contrast to an analysis of alcohol consumption as motivated by the pharmacological properties of the drug, recent researchers have also proposed that people drink out of a cognitive expectancy regarding alcohol's effects. As seen in a preceding section, alcohol expectancies are thought to affect social behaviors by providing an excuse to engage in a desired but prohibited action. Two recent analyses trade on the idea that a person may consume alcohol as a function of the attribution that he or she is intoxicated. In one case, it is proposed that for alcoholics the attribution that one is intoxicated functions as a cue to disinhibit the further consumption of alcohol. In the second case, it is proposed that individuals sometimes drink in order to achieve the attribution that their behavior is alcohol-impaired, thus absolving themselves of blame for subsequent failures while at the same time increasing their credit for subsequent successes.

Expectancies and Loss-of-Control Drinking. According to the disease model of alcoholism, one characteristic feature of the alcoholic condition is loss-of-control drinking. As described by Jellinek (1960), this state is reached "when the ingestion of one alcoholic drink sets up a chain reaction so that [individuals] are unable to adhere to their intention to 'have one or two drinks only' but continue to ingest more and more—often with quite some difficulty and disgust—contrary to their volition" (p. 41). Viewed in this way, loss-of-control drinking in alcoholics is a consequence of the pharmacological effects of the drug such that one drink functions to trigger subsequent consumption.

In opposition to this view, Marlatt et al. (1973) proposed that loss-of-control drinking may simply be a function of the cognition that one is consuming alcohol. This beverage expectancy may function as a cue to consume more alcohol.

Alternatively, in the words of the preceding section, alcohol expectancy may provide alcoholics with a viable excuse to lose control of their behavior and engage in the desired but prohibited action of consuming more alcohol. In either case, loss-of-control drinking is a function of cognitive expectancy and not the pharmacological effects of the drug.

Marlatt et al. (1973) contrasted the pharmacological and expectancy models of loss-of-control drinking in one of the first experiments to employ the balanced placebo design. Alcoholic and social drinker subjects tasted and rated a series of beverages that either did or did not contain alcohol and that they were told did or did not contain alcohol. The actual alcoholic content of the beverage had no effect on amount of alcohol subjects consumed. On the other hand, subjects who were told they were drinking alcohol drank significantly more beverage than subjects who were told they were drinking a placebo.

These results paralleled findings from an earlier study by Engle and Williams (1972). These authors also had alcoholics consume either an alcoholic or placebo beverage while expecting either an alcoholic or placebo beverage. Subjects then rate the degree of craving they experienced for alcohol. Although not originally analyzed in such a way as to clearly answer the question as to the locus of the "craving" effect, a reanalysis by Maisto et al. (1977) found results similar to those reported by Marlatt et al. (1973). Alcoholics who were told they were consuming alcohol reported greater craving for alcohol than those who were told they were drinking a placebo. On the other hand, actual beverage content had no effect on reported craving.

Taken together, the studies by Marlatt at al. (1973) and Engle and Williams (1972) support the notion that cognitions can affect alcohol consumption independent of the pharmacological characteristics of the drug. In this case, individuals drink as a function of alcohol expectancy either because of the cue value of the cognition or because the cognition provides an excuse to engage in a desired but prohibited behavior. Another recent model of alcohol use also emphasizes cognitive expectancies. In this case, however, people are proposed to drink as a means of minimizing their responsibility for failure and maximizing their responsibility for success.

The Self-Handicapping Model of Alcohol Use. Jones and Berglas (1978; Berglas and Jones, 1978) have proposed a self-handicapping model of alcohol use. According to this model, alcohol is sometimes consumed in an explicit attempt to render the causes of performance ambiguous. The model is based on the idea that alcohol debilitates performance at a wide variety of tasks. If one is intoxicated while performing these tasks, the actual cause of success or failure is ambiguous. Failure might be due to one's own lack of ability, but it might also be due to the debilitating effects of alcohol. On the other hand, the cause of success is less ambiguous and can only be attributed to abilities on the part of the individual sufficient to overcome the debilitating effects of the drug. The result is a no-lose situation: By adopting a self-handicap, one receives credit for successes while avoiding blame for failures.

Jones and Berglas (1978) proposed that one is particularly likely to resort to self-handicapping strategies under certain circumstances. These situations mainly involve the possibility of receiving information that is threatening to one's current level of self-esteem. For example, if a person is unsure of the actual causes of past successes, he or she may wish to give themselves the "benefit of the doubt" by avoiding highly diagnostic situations. In this way, one is able to maintain a high, if fragile, sense of self-esteem. Jones and Berglas (1978) theorize that situations in which past success feedback is perceived not to be contingent on actual performance are especially likely to promote fears of being "found out" and, hence, lead to the avoidance of diagnostic feedback. In a similar fashion, being propelled into new and more demanding situations as a consequence of past successes (the "Peter Principle") may result in a fear of diagnostic feedback indicating that one can no longer maintain past levels of success. The use of the self-handicapping strategy is one way in which to render any situation less diagnostic.

Jones and Berglas (1978) propose that self-handicappers may use a variety of external impediments (e.g., alcohol intoxication, lack of sleep before an exam, avoidance of lessons or practice at a sport, adoption of the "sick role") or may simply withdraw effort in an attempt to render performance ambiguous with regard to ability. They present two studies that investigate the hypothesis that drugs are sometimes used as a self-handicap. In both studies, college students received either contingent or noncontingent success feedback about their performance on a series of analogies. Following this manipulation and before retaking a similar analogy test, subjects were instructed to choose between drugs described by the experimenter either to facilitate or debilitate performance. The drugs were fictitious and the experiment was terminated after subjects reported the type and dosage level of the drug they wanted to receive. Drug type and amount served as the principal dependent measures. It was predicted that subjects who had received noncontingent success feedback would choose the performance-debilitating drug in order to self-handicap future performances.

Both studies provided basic support for the hypothesis that when one is uncertain that past successes are due to one's own ability (i.e, success is noncontingent on performance), drugs known to debilitate performance are preferred in order to reduce the diagnosticity of future performances. Furthermore, experiment 1 controlled for the possibility that this strategy is adopted for purely self-presentational reasons. Consistent with the Jones and Berglas (1978) analysis, it would appear that self-handicapping occurs in the service of one's private image of self. Experiment 2 controlled for the possibility that the effect is due simply to one's exposure to noncontingent events. Consistent with predictions, noncontingency in the absence of explicit feedback regarding one's success did not promote self-handicapping.

One problem with these results is that they only held for male subjects. Female subjects did not self-handicap in either experiment. The reasons for this sex difference remains unclear. Given that others (Smith et al., 1982, 1983) have reported evidence of the use of self-handicapping strategies (e.g., complaining of performance-debilitating anxiety or physical symptoms) by female subjects on very similar tasks, it would appear that the phenomenon of self-handicapping is not restricted to males nor specific to the tasks used by Berglas and Jones (1978). The latter studies differ from those of Berglas and Jones principally in that the self-handicapping strategies did not involve drug use, but rather involved self-reports of test anxiety and hypochondriacal symptoms. Subsequent research on self-handicapping through the use of drugs has only investigated male subjects. As a consequence, the reasons for the sex differences reported by Berglas and Jones remain unexplored and may be related to the use of *drugs* as means of self-handicapping.

Recent studies have further delimited the self-handicapping effect. Kolditz and Arkin (1982) argue that Berglas and Jones did not provide an adequate test of the hypothesis that self-handicapping is used in the service of a private rather than a public image of self. Specifically, they claim that the private condition in the Berglas and Jones experiment was insufficiently private. They attempted a precise replication of the Berglas and Jones experimental task and contingency manipulation with a more adequate manipulation of privacy. The results indicated that subjects were more likely to self-handicap following noncontingent than contingent success feedback, but *only* when the drug choice was made in public. Contrary to the theorizing of Jones and Berglas (1978), Kolditz and Arkin (1982) conclude that the self-handicapping use of drugs is principally in the service of a public as opposed to a private image of self.

Tucker et al. (1981) offer the only studies specifically devoted to the hypothesis that *alcohol* is sometimes consumed for its self-handicapping functions. They present two studies in which subjects anticipated working on a test of intellectual abilities, followed by an opportunity to consume a self-selected amount of alcohol, followed by a second test of intellectual abilities. As in Berglas and Jones (1978), the initial test of intellectual abilities involved either solvable or insolvable problems and hence constituted the manipulation of contingency. The principal dependent measures were amount of alcohol consumed and level of blood alcohol concentration. It was predicted that subjects would consume a greater amount of alcohol and have higher blood alcohol levels following noncontingent as opposed to contingent success feedback in an attempt to self-handicap performance on the second test. In addition to these variables, both studies manipulated the anticipated difficulty of the second test. The second test was described either as equal to or greater than the first test in difficulty. This manipulation was included in an attempt to test the "Peter Principle" hypothesis advanced by Jones and Berglas (1978) that individuals who have experienced contingent success are also likely to self-handicap when faced with increases in performance demands. Experiment 2 by Tucker et al. (1981) involved only these variables, whereas experiment 1 involved a more elaborate design.

Given the less elaborate nature of experiment 2, its results will be described first. In support of the notion of self-handicapping, subjects who experienced noncontingent success consumed more alcohol and had higher blood alcohol levels than subjects who experienced contingent success. However, this was true regardless of the anticipated difficulty of the subsequent test. The authors concluded that although the main self-handicapping hypothesis was supported, the additional Peter Principle hypothesis was not supported.

Experiment 1 differed from experiment 2 in two important respects. First, as in Berglas and Jones (1978, Exp. 1) no-feedback controls were included in addition to the success-feedback conditions. Second, subjects were provided study materials in the intertest period which they were told might or might not be helpful in preparing for the second test. In essence, this is a variation on the performance facilitating–debilitating choice offered in previous studies. However, unlike the drug choice measures used by Jones and Berglas (1978) and Kolditz and Arkin (1982), drinking alcohol and studying are not mutually exclusive. Amount of time spent studying during the intertest interval was surreptitiously recorded.

Once again, subjects who experienced noncontingent success had significantly higher blood alcohol levels and consumed marginally more alcohol than subjects who experienced contingent success. As in Berglas and Jones (1978, Exp. 1), there were no differences between subjects who experienced noncontingent and contingent tasks in the absence of performance feedback. While these results support the notion of self-handicapping, analyses of study time do not support the predictions. Thus, the facilitative behavior of study time was not the mirror image of the debilitative behavior of alcohol consumption. Subjects who anticipated a harder retest studied more (not less) than subjects who anticipated an identical retest.

Although Tucker et al. concede that the results of these studies support the principal predictions involving alcohol consumption as a self-handicapping strategy, they see two fundamental inconsistencies between their results and the model proposed by Jones and Berglas (1978). The first involves the failure to support the prediction of self-handicapping as a function of the Peter Principle of escalating performance demands. Second, and more troubling for the model, is the failure to find a reciprocal relation between alcohol consumption and study behavior. We feel that a possible explanation for these results is suggested by the findings of Kolditz and Arkin (1982). These authors found that the public nature of the self-handicap attempt was a key element in its strategic adoption.

Indeed, in the studies by Tucker et al., amount of alcohol consumed was public knowledge as was performance on the intellectual task. On the other hand, study time had to be surreptitiously observed. It is conceivable, then, that subjects self-handicapped with regard to alcohol consumption but not study time because the former behavior was public knowledge, whereas the latter behavior was not. While *post hoc,* such an hypothesis has the advantage of integrating a number of seemingly unrelated self-handicapping effects.

To briefly summarize these studies: It would appear that individuals try to avoid diagnostic feedback when they anticipate that it will reflect poorly on their self-image. One situation capable of eliciting such fears involves the anticipation of performance on a task on which one has succeeded in the past, but for ambiguous reasons. Under such conditions, adoption of an impediment is useful in maintaining one's fragile sense of high self-esteem: Subsequent failure can be attributed to the impediment, whereas subsequent success can only be attributed to abilities strong enough to overcome the impediment. This self-handicapping strategy appears to be a viable explanation of some forms of drug use including alcohol consumption. Whether it is used principally in the service of a public or private sense of self is a matter of current debate.

## 6.  CONCLUSIONS

On the basis of the experimental literature, it would appear that both alcohol consumption and expectancy independently affect subsequent alcohol consumption and a variety of social behaviors. Alcohol consumption would appear to have effects on these social behaviors by virtue of inhibiting cognitive processes related to self-regulation with respect to internal and external standards of conduct. At the same time, alcohol expectancy would appear to affect behavior by altering the standards of conduct in a deviant direction. Similarly, alcohol consumption would appear both to be motivated to achieve states induced by the pharmacological properties of the drug and out of expectancies regarding the effects of the drug on behavior. We are committed to the belief that the experimental tradition adopted in the present literature has resulted in real progress in understanding the causes and effects of alcohol consumption. At the same time, we hope that the current chapter will serve to draw attention to some of the conflicts in the existing literature and the lack of research on many aspects of social behavior.

## REFERENCES

Abrams, D. B., and Wilson G. T., 1979, Effects of alcohol on social anxiety in women: Cognitive versus physiological processes, *J. Abnorm. Psychol.* **88:**161–173.

Abrams, D. B., and Wilson G. T., 1983, Alcohol, sexual arousal, and self-control, *J. Pers. Soc. Psychol.* **45**:188–198.

Babor, T. F., Berglas, S., Mendelson, J. H., Ellingboe, J., and Miller, K., 1983, Alcohol, affect, and the disinhibition of verbal behavior, *Psychopharmacology* **80**:53–60.

Bailey, D. S., Leonard, K. E., Cranston, J. W., and Taylor, S. P., 1983, Effects of alcohol and self-awareness on human physical aggression, *Pers. Soc. Psychol. Bull.* **9**:289–295.

Bennett, R. M., Buss, A. H., and Carpenter, J. A., 1969, Alcohol and human physical aggression, *Q. J. Stud. Alcohol* **30**:807–876.

Berglas, S., and Jones, E. E., 1978, Drug choice as a self-handicapping strategy in response to non-contingent success, *J. Pers. Soc. Psychol.* **36**:405–417.

Birnbaum, I. M., Johnson, M. K., Hartley, J. T., and Taylor, T. H., 1980, Alcohol and elaborative schemas for sentences, *J. Exp. Psychol. Hum. Learn. Mem.* **6**:293–300.

Boyatzis, R. E., 1974, The effect of alcohol consumption on the aggressive behavior of men, *Q. J. Stud. Alcohol* **35**:959–972.

Briddell, D. W., and Wilson, G. T., 1976, The effects of alcohol and expectancy set on male sexual arousal, *J. Abnorm. Psychol.* **85**:225–234.

Briddell, D. W., Rimm, D. C., Caddy, G. R., Krawitz, G., Sholis, D., and Wonderlin, R. J., 1978, The effects of alcohol and cognitive set on sexual arousal to deviant stimuli, *J. Abnorm. Psychol.* **87**:418–430.

Brown, J. S., and Crowell, C. R., 1974, Alcohol and conflict resolution: A theoretical analysis, *Q. J. Stud. Alcohol* **35**:66–85.

Buss, A. H., 1961, *The Psychology of Aggression,* Wiley, New York.

Cappell, H., and Herman, C. P., 1972, Alcohol and tension reduction: A review, *Q. J. Stud. Alcohol* **33**:33–64.

Carver, C. S., and Scheier, M. F., 1981, *Attention and Self-Regulation: A Control Theory Approach to Human Behavior,* Springer-Verlag, New York.

Cohen, J., 1960, *Chance, Skill, and Luck, the Psychology of Guessing and Gambling,* Penguin Books, Baltimore.

Conger, J. J., 1956, Alcoholism: Theory, problem and challenge. II. Reinforcement theory and the dynamics of alcoholism, *Q. J. Stud. Alcohol* **17**:296–305.

Connors, G. J., and Maisto, S. A., 1979, Effects of alcohol, instructions, and consumption rate on affect and physiological sensation, *Psychopharmacology* **62**:261–266.

Dengerink, H. A., and Fagan, N. J., 1978, Effect of alcohol on emotional responses to stress, *J. Stud. Alcohol* **39**:525–539.

Engle, K. B., and Williams, T. K., 1972, Effect of an ounce of vodka on alcoholics' desire for alcohol, *Q. J. Stud. Alcohol* **33**:1099–1105.

Farkas, G., and Rosen, R. C., 1976, The effects of ethanol on male sexual arousal, *J. Stud. Alcohol* **37**:265–272.

Fenigstein, A., Scheier, M. F., and Buss, A. H., 1975, Public and private self-consciousness: Assessment and theory, *J. Consult. Clin. Psychol.* **43**:522–527.

Hartley, J. T., Birnbaum, I. M., and Parker, E. S., 1978, Alcohol and storage deficits: Kinds of processing, *J. Verb. Learn. Verb. Behav.* **17**:635–647.

Higgins, R. L, and Marlatt, G. A., 1973, Effects of anxiety arousal on the consumption of alcohol by alcoholics an social drinkers, *J. Consult. Clin. Psychol.* **41**:426–433.

Higgins, R. L, and Marlatt, G. A., 1975, Fear of interpersonal evaluation as a determinant of alcohol consumption in male social drinkers, *J. Abnorm. Psychol.* **84**:644–651.

Holroyd, K. A., 1978, Effects of social anxiety and social evaluation on beer consumption and social interaction, *J. Stud. Alcohol* **39**:737–744.

Hull, J. G., 1981, A self-awareness model of the causes and effects of alcohol consumption, *J. Abnorm. Psychol.* **90**:586–600.

Hull, J. G., and Bond, C. F., 1986, The social and behavioral consequences of alcohol consumption and expectancy: A meta-analysis, *Psych. Bull.* **99:**347–360.

Hull, J. G., and Young, R. D., 1983, Self-consciousness, self-esteem, and success-failure as determinants of alcohol consumption in malesocial drinkers, *J. Pers. Soc. Psychol.* **44:**1097–1109.

Hull, J. G., Levenson, R. W., Young, R. D., and Sher, K. J., 1983, Self-awareness reducing effects of alcohol consumption, *J. Pers. Soc. Psychol.* **44:**461–473.

Hull, J. G., Young, R. D., and Jouriles, E., 1986, Applications of the self-awareness model of alcohol consumption, *J. Pers. Soc. Psychol.* **51.**

Jellinek, E. M, 1960, *The Disease Concept of Alcoholism,* College and University Press, New Haven.

Jones, E. E., and Berglas, S., 1978, Control of attributions about the self through self-handicapping strategies: The appeal of alcohol and the role of underachievement, *Pers. Soc. Psychol. Bull.* **4:**200–206.

Keane, T. M., and Lisman, S. A., 1980, Alcohol and social anxiety in males: Behavioral cognitive, and physiological effects, *J. Abnorm. Psychol.* **89:**213–223.

Kolditz, T. A., and Arkin, R. M., 1982, An impression management interpretation of the self-handicapping strategy, *J. Pers. Soc. Psychol.* **43:**492–502.

Korytnyk, N. X., and Perkins, D. V., 1983, Effects of alcohol versus expectancy for alcohol on the incidence of graffiti following an experimental task, *J. Abnorm. Psychol.* **92:**382–385.

Kreutzer, J., Schneider, H. G., and Myatt, C. R., 1984, Alcohol, aggression, and assertiveness in men: Dosage and expectancy effects, *J. Stud. Alcohol* **45:**275–278.

Lang, A. R., Goeckner, D. T., Adesso, V. J., and Marlatt, G. A., 1975, Effects of alcohol on aggression in male social drinkers, *J. Abnorm. Psychol.* **84:**508–518.

Lang, A. R., Searles, J., Lauerman, R., and Adesso, V., 1980, Expectancy, alcohol and sex guilt and determinants of interest in and reaction to sexual stimuli, *J. Abnorm. Psychol.* **89:**644–653.

Lansky, D., and Wilson, G. T., 1981, Alcohol, expectations, and sexual arousal in males: An information processing analysis, *J. Abnorm. Psychol.* **90:**35–45.

Levenson, R. W., Sher, K. J., Grossman, L. M., Newman, J., and Newlin, D. B., 1980, Alcohol and stress response dampering: Pharmacological effects, expectancy, and tension reduction, *J. Abnorm. Psychol.* **89:**528–538.

Maisto, S. A., Lauerman, R., and Adesso, V. J., 1977, A comparison of two experimental studies of the role of cognitive factors in alcoholic's drinking, *J. Stud. Alcohol* **38:**145–149.

Markus, H., 1977, Self-schemata and processing information about the self, *J. Pers. Soc. Psychol.* **35:**63–78.

Marlatt, G. A., and Rohsenow, D. J., 1980, Cognitive processes in alcohol use: Expectancy and the balanced placebo design, in: *Advances in Substance Abuse: Behavioral and Biological Research, A Research Annual* (N. K. Mello, ed.), pp. 159–199, JAI Press, Greenwich, Connecticut.

Marlatt, G. A., Kosturn, C. F., and Lang, A. R., 1971, Provocation to anger and opportunity for retaliation as determinants of alcohol consumption in social drinkers, *J. Abnorm. Psychol.* **84:**652–659.

Marlatt, G. A., Demming, B., and Reid, J. B., 1973, Loss of control drinking in alcoholics: An experimental analogue, *J. Abnorm. Psychol.* **81:**233–241.

Masserman, J. H., and Yum, K. S., 1946, An analysis of the influence of alcohol on experimental neuroses in cats, *Psychosomatic Medicine* **8:**36–52.

Masters, W. H., and Johnson, V. E., 1966, *Human Sexual Response,* Little, Brown, Boston.

McCollam, J. B., Burish, T. G., Maisto, S. A., and Sobell, M. B., 1980, Alcohol's effects on physiological arousal, and self-reported effect and sensations, *J. Abnorm. Psychol.* **89:**224–233.

McNamee, H. B., Mello, N. K., and Mendelson, J. H., 1968, Experimental analysis of drinking patterns of alcoholics: Concurrent psychiatric observations, *Am. J. Psychiatry* **124:**1063–1069.

Mendelson, J. H., LaDou, J., and Solomon, P., 1964, Experimentally induced chronic intoxication and withdrawal in alcoholics: Part 3, Psychiatric findings, *Q. J. Stud. Alcohol* **2**:40–52.

Miller, P. M., Hersen, M., Eisler, R. M., and Hilsman, G., 1974, Effects of social stress on operant drinking of alcoholics and social drinkers, *Behav. Res. Ther.* **12**:67–72.

Mosher, D., 1966, The development and multi-trait multi-method matrix analysis of three measures of three aspects of guilt, *J. Consult. Psychol.* **30**:25–29.

Nathan, P. E., and O'Brien, J. S., 1971, An experimental analysis of the behavior of alcoholics and nonalcoholics during prolonged experimental drinking: A necessary precursor of behavior therapy? *Behav. Ther.* **2**:455–476.

Nowlis, V., 1965, Research on the mood adjective checklist, in: *Affect, Cognition, and Personality: Empirical Studies* (S. S. Tomkins and C. E. Izard, eds.), Springer, New York.

Pernanen, K., 1976, Alcohol and crimes of violence, in: *The Biology of Alcoholism: Social Aspects of Alcoholism,* Vol. 4 (B. Kissin and H. Begleiter, eds.), Plenum Press, New York.

Pihl, R. O., Zeichner, A., Niaura, R. , Nagy, K., and Zacchia, C., 1981, Attribution and alcohol mediated aggression, *J. Abnorm. Psychol.* **90**:468–475.

Pihl, R. O., Smith, M., and Farrell, B., 1984, Alcohol and aggression in men: A comparison of brewed and distilled beverages, *J. Stud. Alcohol* **45**:278–282.

Pliner, P., and Cappell, H., 1974, Modification of affective consequences of alcohol: A comparison of social and solitary drinking, *J. Abnorm. Psychol.* **83**:418–425.

Pohorecky, L. A., and Brick, J., 1983, *Stress and Alcohol Use,* Elsevier, New York.

Polivy, J., Schueneman, A. L., and Carlson, K., 1976, Alcohol and tension reduction: Cognitive and physiological effects, *J. Abnorm. Psychol.* **85**:595–600.

Rogers, R. W., and Prentice-Dunn, S., 1981, Deindividuation and anger mediated inter-racial aggression: Unmasking regression racism, *J. Pers. Soc. Psychol.* **41**:63–73.

Rogers, T. B., Kuiper, N. A., and Kirker, W. S., 1977, Self-reference and the encoding of personal information, *J. Pers. Soc. Psychol.* **35**:677–688.

Rohrberg, R. G., and Sousa-Poza, J. F., 1976, Alcohol, field dependence, and dyadic self-disclosure, *Psychol. Rep.* **39**:1151–1161.

Rohsenow, D. J., and Bachorowski, J. A., 1984, Effects of alcohol and expectancies on verbal aggression in men and women, *J. Abnorm. Psychol.* **93**:418–432.

Scheier, M. F., Fenigstein, A., and Buss, A. H., 1974, Self-awareness and physical aggression, *J. Exp. Soc. Psychol.* **10**:264–273.

Shuntich, R.J., and Taylor, S. P., 1972, The effects of alcohol on human physical aggression, *J. Exp. Res. Pers.* **6**:34–38.

Smith, R. C., Parker, E. S., and Noble, E. P., 1975, Alcohol's effect on some formal aspects of verbal social communication, *Arch. Gen. Psychiatry* **32**:1394–1398.

Smith, T. W., Snyder, C. R., and Handelsman, M. M., 1982, On the self-serving function of an academic wooden leg: Test anxiety as a self-handicapping strategy, *J. Pers. Soc. Psychol.* **42**:314–321.

Smith, T. W., Snyder, C. R., and Perkins, S. C., 1983, The self-serving function of hypochondriacal complaints: Physical symptoms as self-handicapping strategies, *J. Pers. Soc. Psychol.* **44**:787 –797.

Steele, C. M., and Southwick, L., 1985, Alcohol and social behavior I: The psychology of drunken excess, *J. Pers. Soc. Psychol.* **48**:18–34.

Steele, C. M., Critchlow, B., and Liu, T. J., 1985, Alcohol and social behavior II: The helpful drunkard, *J. Pers. Soc. Psychol.* **48**:35–46.

Steffen, J. J., Nathan, P. E., and Taylor, H. A., 1974, Tension-reducing effects of alcohol: Further evidence and some methodological considerations, *J. Abnorm. Psychol.* **83**:542–547.

Taylor, S. P., and Gammon, C. B., 1975, Effects of type and dose of alcohol on human physical aggression, *J. Pers. Soc. Psychol.* **32**:169–175.

Taylor, S. P., and Gammon, C. B., 1976, Aggressive behavior of intoxicated subjects: The effect of third-party intervention, *J. Stud. Alcohol* **37**:917–930.

Taylor, S. P., and Leonard, K. E., 1983, Alcohol and human physical aggression, in: *Aggression* (R. G. Geen and E. I. Donnerstein, eds.), pp. 77–101, Academic Press, New York.

Taylor, S. P., Gammon, C. B., and Capasso, D. R., 1976, Aggression as a function of alcohol and threat, *J. Pers. Soc. Psychol.* **34**:938–941.

Taylor, S. P., Schmutte, G. T., and Leonard, K. E., 1977, Physical aggression as a function of alcohol and frustration, *Bull. Psychonom. Soc.* **9**:217–218.

Teger, A. I., Katkin, E. S., and Pruitt, D. G., 1969, Effects of alcoholic beverages and their congener content on level and style of risk taking, *J. Pers. Soc. Psychol.* **11**:170–176.

Tucker, J. A., Vuchinich, R. E., and Sobell, M. B., 1981, Alcohol consumption as a self-handicapping strategy, *J. Abnorm. Psychol.* **90**:220–230.

Vuchinich, R. E., Tucker, J. A., and Sobell, M. B., 1979, Alcohol, expectancy, cognitive labeling, and mirth, *J. Abnorm. Psychol.* **88**:641–651.

Weingartner, H., Adefris, W., Eich, J. E., and Murphy, D. L., 1976, Encoding-imagery specificity in alcohol state dependent learning, *J. Exp. Psychol. Hum. Learn. Hum. Mem.* **2**:83–87.

Williams, A. F., 1966, Social drinking, anxiety, and depression, *J. Pers. Soc. Psychol.* **3**:689–693.

Wilson, G. T., and Abrams, D., 1977, Effects of alcohol on social anxiety and physiological arousal: Cognitive versus pharmacological processes, *Cognit. Ther. Res.* **1**:195–210.

Wilson, G. T., and Lawson, D. M., 1976a, Effects of alcohol on sexual arousal in women, *J. Abnorm. Psychol.* **85**:489–497.

Wilson, G. T., and Lawson, D. M., 1976b, Expectations, alcohol, and sexual arousal in male social drinkers, *J. Abnorm. Psychol.* **85**:587–594.

Wilson, G. T., and Lawson, D. M., 1978, Expectancies, alcohol, and sexual arousal in women, *J. Abnorm. Psychol.* **87**:358–367.

Wilson, G. T., and Niaura, R., 1984, Alcohol and the disinhibition of sexual responsiveness, *J. Stud. Alcohol* **45**:219–224.

Wilson, G. T., Lawson, D. M., and Abrams, D. B., 1978, Effects of alcohol on sexual arousal in male alcoholics, *J. Abnorm. Psychol.* **87**:609–616.

Wilson, G. T., Abrams, D. B., and Lipscomb, T. R., 1980, Effects of increasing levels of intoxication and drinking pattern on social anxiety, *J. Stud. Alcohol* **41**:250–264.

Yale University Center for Alcohol Studies, 1945, *Alcohol, Science, and Society: Twenty-nine Lectures with Discussions As Given at the Yale Summer School of Alcohol Studies*, Greenwood Press, Westport, Connecticut.

Zeichner, A., and Pihl, R. O., 1979, Effects of alcohol and behavior contingencies on human aggression, *J. Abnorm. Psychol.* **88**:133–160.

Zeichner, A., and Pihl, R. O., 1980, Effects of alcohol and instigator intent on human aggression, *J. Stud. Alcohol* **41**:265–276.

Zeichner, A., Pihl, R. O., Niaura, R., and Zacchia, C., 1982, Attentional processes in alcohol-mediated aggression, *J. Stud. Alcohol* **43**:714–724.

Zuckerman, M., Lubin, A., Vogel, L., and Valerius, E., 1964, Measurement of experimentally induced affects, *J. Consult. Psychol.* **28**:418–425.

7

# The Role of Naltrexone in the Treatment of Opioid Dependence

CHARLES P. O'BRIEN and GEORGE E. WOODY

## 1. TREATMENT FOR OPIOID DEPENDENCE

For many years clinicians considered the treatment of opioid dependence with pessimism. But, over the last 15 years, even as the problem of addiction has grown, we have come to know more about the physiological and psychological bases of drug dependence; and we have developed a number of new pharmacologic and psychosocial modalities of treatment. These have met with varying degrees of success and have given physicians and patients alike varying degrees of hope or encouragement.

The use of methadone in a clinical setting has been successful in many cases, but methadone itself has certain liabilities and does not suit all patients (see Chapter 8). Naltrexone, an opioid antagonist approved by the Food and Drug Administration in late 1984, is a different pharmacological approach. As long as it is taken by the patient on a regular basis, it will prevent readdiction to opioids. It is thus a viable and often rewarding alternative to methadone or to approaches that depend on total abstinence. For the physician, nurse, or other professional person who is dependent on opioids and would not be permitted to work while on a methadone maintenance program, naltrexone has been and can be remarkably effective. In short, naltrexone accompanied by psychotherapy, job counseling, urine testing, and, when necessary, psychoactive medication can be another very helpful means of treatment. After a brief look at the rationale for antagonist therapy and its relationship to the pharmacology of naltrexone,

CHARLES P. O'BRIEN and GEORGE E. WOODY • Psychiatry Service, Philadelphia Veterans Administration Medical Center and University of Pennsylvania, Philadelphia, Pennsylvania 19104.

we will consider the types of patients for whom naltrexone seems to be most appropriate and perhaps the most effective, some of the studies of its use in a clinical setting, and suggestions as to how treatment may be conducted in a typical protocol, comprising detoxification, induction, and stabilization.

## Opioid Antagonists

Opioid antagonists such as naltrexone are substances that bind to opioid receptors but do not produce opioid effects (Martin et al., 1973). By acting in this way they block or compete with both exogenous opioids and endorphins. When an antagonist is present in sufficient quantity to occupy all or most opioid receptors, agonist substances such as morphine cannot reach and bind to the receptors. An injection of heroin will have little or no agonist effect, and the pattern of addiction (or readdiction) may be interrupted.

It is generally agreed that naltrexone is pharmacologically effective. When an individual is protected by naltrexone, the effects of opioids are blocked or attenuated in a dose-related fashion (O'Brien et al., 1975). After a 150- or 200-mg dose of naltrexone, significant antagonism of injected opioids can persist for 72 hr. The duration of the pharmacological efficacy of naltrexone exceeds the time one would expect from its plasma level kinetics (Meyer et al., 1984). It is important to note, however, that naltrexone antagonizes competitively; it does not block absolutely. In the controlled conditions of the laboratory, some opioid effects can be perceived by subjects injecting opioids while receiving naltrexone, but the attenuation is such that even fairly large doses of heroin are relatively unrewarding.

## Safety of Naltrexone

Toxicity data from animal studies indicate that naltrexone has a wide margin of safety; the $LD_{50}$ in several species is quite high, and studies of long-term administration also indicate lack of toxicity (Rosenkrantz, 1984; Christian, 1984). In animal studies, a full range of testing has found naltrexone to be noncarcinogenic; one model even suggests some antitumor activity. In the more than 2000 opioid abusers who have taken naltrexone over the past 10 years, the drug has shown few clinically evident signs of toxicity. The endocrine changes associated with naltrexone and noted in normals have tended to occur mainly in acute studies. In normal subjects abruptly given full doses of naltrexone, dysphoria has been reported, but postaddicts gradually inducted have few complaints. In chronic dosing, no evidence has appeared so far of any clinically significant long-term endocrine changes, tumorigenic effects, or clinically relevant reproductive effects. Up to 800-mg doses per day have been given under short-term experimental conditions to subjects without apparent toxicity. However, obese subjects treated with 300 mg naltrexone per day for 4–6 weeks developed elevated transaminase levels.

This dose is six times the dose used to prevent readdiction, but this finding raises the possibility of liver toxicity in some subjects even at lower doses. Consequently, naltrexone should never be used in patients in liver failure. However, the mild liver problems seen in the majority of street heroin addicts are not a contraindication to naltrexone treatment. Our group and other clinical investigators prescribing 350 mg naltrexone per week in relapse prevention programs have not found clinically evident liver problems in our patients (Arndt et al., 1985; Judson et al., 1981). Still we recommend pretreatment laboratory testing and continued liver profiles especially serum glutamic-pyruvic transaminase (SGPT) monthly and then quarterly or biannually in long-term treatment cases.

Several cases of skin rash and one case of idiopathic thrombocytopenic purpura that cleared on cessation of the drug have been reported; these may have represented immune reactions. The side effect most commonly reported for naltrexone has been gastrointestinal distress; in some cases, this may represent a mild precipitated opioid withdrawal reaction rather than a side effect in the usual sense. In summary then, naltrexone has been remarkably free of side effects in the vast majority of cases.

## Pharmacological versus Clinical Efficacy

The pharmacological efficacy of naltrexone is unquestioned, but is the drug clinically efficacious for the treatment of opioid dependence? One of the most vexing aspects of that complex question involves the difficulty of devising a double-blind trial of an antagonist drug versus placebo. In the case of naltrexone, for instance, the structure of such a trial has to be based on the known pharmacological activity of naltrexone. Patients are told they will receive either an antagonist or a placebo. If they test by using opioids, they can break the blind. If they fail to test, then the placebo has, as it were, borrowed the pharmacological activity of naltrexone and there is no advantage to being on the antagonist drug and subject to its possible side effects. Results of the one large placebo-controlled trial of naltrexone (Hollister, 1978) have, in fact, followed this pattern. The overall study showed few drug–placebo differences, but among the small number of subjects who tested the blockade, those on naltrexone showed a significant advantage. The naltrexone patients tended to use heroin once or twice toward the beginning of the study and then stopped opioid drug use. Those on placebo continued to use heroin resulting in significantly more positive urines. The large number of drop-outs in both groups dominates the picture and we must conclude that the placebo-controlled approach does not verify overall clinical efficacy for naltrexone.

Antagonist therapy of opioid abuse is based, as we have seen, on the fact that a drug such as naltrexone effectively blocks or significantly attenuates the agonist effects of an opioid (Sideroff et al. 1978). If an individual experiences craving or withdrawal and takes an opioid while he is maintained on naltrexone, the hoped-for agonist effect is blocked or unrewarded. Protected in this fashion, the individual can expose himself to conditions that might encourage relapse without incurring the danger of readdiction. The addicted individual who wishes to "kick" his habit must struggle not only with the phenomenon of protracted abstinence (which may last for 3 to 6 months) (Martin and Jasinski, 1969), but with the conditioned withdrawal responses associated with the sensory phenomena and rituals experienced during his addiction (O'Brien et al., 1984), or with any other events that may stimulate craving or drug-taking behaviors. Any of these phenomena can increase the temptation to resume opioid use. If the addict under treatment with naltrexone does decide to test the blockade and does not get high, this unrewarding use of opioid may be considered *extinction* of a conditioned or otherwise rewarding response and potentially beneficial (Brahen et al., 1984).

## 2.  TREATMENT CANDIDATES

Of the many types of addicts, which might be described as suitable candidates for treatment with naltrexone? In general, at least five categories of patients have been found to be suitable candidates for naltrexone, each group being treated in an appropriate context in which naltrexone is an adjunct (Washton et al., 1984; Kleber and Kosten, 1984). These groups are the working-class or the "street" addict, who may come directly from illegal heroin use; the patient on methadone, who has not been on an illegal drug but has been taking a long-acting oral narcotic for a substantial period; the suburban or middle-class addict; the business executive; the health care professional with access to licit narcotics; and the currently "clean" individual coming out of a hospital, prison, or therapeutic community.

Those who have the most to lose are the ones who have the most to gain. As Washton and colleagues noted, middle- and upper-class patients typically enter treatment with good jobs to protect, career and family supports, and histories of adequate or above-average functioning (Washton et al., 1984):

> In general, patients with social and economic stability have a better prognosis for remaining in naltrexone treatment. Involvement in work and other productive activities, as well as living with a supportive mate who is not a drug abuser, tends to facilitate treatment outcome.

Addicted business executives, physicians, and other health care professionals have proved in general to be excellent candidates for naltrexone because their jobs are in jeopardy and they are very strongly motivated. The external factors such as fear of losing a job or medical license can provide important motivation early in treatment. Treatment with naltrexone may also often be useful in helping released prison inmates make a transition to a productive drug-free life (Brahen et al., 1984).

Another team of investigators has studied what patient characteristics are predictive of a longer duration of treatment and found that patients who were employed and/or married at the start of naltrexone therapy were more likely to stay in treatment longer. They also found that employment at the beginning of naltrexone therapy and length of therapy (at least 30 days) were significantly related to better outcome at one-month follow-up (Greenstein et al., 1983).

## Working-Class and Street Addicts

In Philadelphia, the Veterans Administration/University of Pennsylvania antagonist program between 1974 and 1983 treated 327 patients with naltrexone (Greenstein et al., 1984). Most of the patients were from working-class backgrounds and had become addicted by purchasing drugs "on the street," with all of the illegal activity associated with this practice. Just under half were employed and they ranged in age from 19 to 47 years (mean, 28). They had been addicted for 9 years on average, with a range of 20 years or more to less than 2 years. Over 80% were veterans, 60% were black. Most patients had multiple treatment experiences with previous detoxifications, residence in therapeutic communities, and/or varying periods of methadone maintenance. Most patients, in fact, had first learned of naltrexone while on methadone.

One-fifth of the patients came to treatment after incarceration or detoxification and were drug-free, but most required detoxification from methadone or street opioids. Before the first dose of naltrexone and 2 to 14 days after discontinuation of methadone or heroin, a naloxone challenge was given to determine the presence of residual dependence and to avoid precipitating more prolonged and severe withdrawal with naltrexone. When the opioid-free interval was less than 5 days, many patients experienced withdrawal after receiving naloxone, particularly if they had been on high-dose methadone. To avoid problems of precipitated withdrawal, all patients should wait at least 5 to 7 days before the naloxone challenge and naltrexone should not be administered until the naloxone challenge is clearly negative. In some patients, clonidine may be useful in speeding up the induction process. Over the 6-day induction period, naltrexone dosage was increased from 25 to 50 mg/day, and patients were then scheduled to receive 100-mg and 150-mg doses of naltrexone on three successive clinic visits each week. Benzodiazepines, such as oxazepam or flurazepam, were prescribed throughout treatment when indicated. Patients saw a staff nurse, counselor, or psychiatrist at each visit and were offered individual and family therapy.

Treatment was primarily supportive and emphasized developing alternatives to drug-oriented behavior and the improvement of interpersonal relationships and level of functioning. If patients missed naltrexone doses for 5 or more days, they had to pass a naloxone challenge before they were restarted. Drug-screening urines were collected once or twice per week.

Of the 327 patients who entered the naltrexone program, over one-third failed to complete induction; for those who continued (60%), treatment time ranged from 1 week to 12 months (mean, 2 months). One-fourth of the subjects had more than one treatment episode. One-fifth of the patients experienced clinically significant anxiety or depression, even after stabilizing on naltrexone. The anxiety and depression seen may represent disorders uncovered during opioid detoxification rather than side effects of naltrexone. Those who began the naltrexone program as outpatients had a much better chance of remaining in treatment, perhaps because they had undergone a slow tapering of methadone or because, if they had been on heroin or low-dose methadone, detoxification was less difficult for them. Certainly, when beginning naltrexone, outpatients had a lower incidence and severity of precipitated withdrawal.

Almost half the patients (44%) tested the opioid-blocking effect of naltrexone at least once. No overdoses were reported. Of the drug-screening urines collected during the study, 94% were negative for opioids. A few patients increased their alcohol consumption during the early part of treatment. Of the 38 patients who continued naltrexone treatment for more than 30 days, 22 (58%) were opioid-free at 1-month follow-up; of 28 patients who continued naltrexone for 7 to 30 days and were evaluable at 6-month follow-up, 9 (32%) were opioid-free; an equal percentage was still opioid-free at 6-month follow-up among 37 evaluable patients who had received naltrexone for more than 30 days.

The motivations of many of the patients were ambivalent. The early high drop-out rate may reflect the discomfort produced by methadone withdrawal as well as the fact that all the patients were volunteers with no significant external motivation such as the threat of losing a job. Some patients discontinued naltrexone after finding out that it did indeed block opioid effects but produced no agonist or "high" effect itself. Some patients later admitted that they volunteered for the naloxone challenge and for naltrexone treatment only in order to "clean the methadone out" of their systems.

The concurrent use of minor tranquilizers may have aided the retention of patients, a large number of whom because restless during induction and anxious even after stabilization. Some patients may have dropped naltrexone prematurely because it produced no euphoria and was therefore nonreinforcing. Minor tranquilizers with mild reinforcing properties may enable patients to remain in treatment until they make enough changes in lifestyle to ensure abstinence.

## The Methadone-Treated Patient

Some methadone-maintained patients can benefit from naltrexone as an adjunct to drug-free counseling following a slow detoxification. Methadone patients who successfully detoxify are usually those with social and vocational supports, who are relatively free of psychiatric, medical, or legal problems, and who have not used illicit drugs for an extended period of time. These patients often find naltrexone to be an extra insurance against relapse. The studies described above contain mixtures of street heroin addicts and methadone-maintained patients.

## The Suburban or Middle-Class Addict

The experience of a group of 13 suburban California clinics, treating patients from a variety of socioeconomic levels (middle-upper white to low-poor minority), has been that clinicians trained to utilize an appropriate induction protocol can effectively use naltrexone to treat volunteer opioid addicts (Tennant et al., 1984). The study group of 160 patients who received at least one dose of naltrexone was made up of experienced addicts (mean opioid use, 10.5 years) most of whom ($n = 114$, 71.3%) were active heroin addicts. Another subset of the group ($n = 42$, 26.3%) were former heroin addicts who had been maintained with methadone, propoxyphene napsylate, or Levo-alpha-acetyl methadol (LAAM) prior to naltrexone treatment. Most individuals were employed ($n = 102$, 63.8%).

Of the 114 heroin addicts, 59 were detoxified with clonidine, guanabenz acetate, or symptomatic medication over a 20-day period before receiving naltrexone. The remaining 55 patients were heroin addicts who had been detoxified elsewhere and had been heroin-abstinent for more than 10 days. Before the first dose of naltrexone, patients had to self-report opioid abstinence and show no withdrawal after a naloxone challenge. The first dose of naltrexone was 50 mg, with the dosage increased to 100 mg on Monday and Wednesday and to 150 mg on Friday. Patients then attended the clinic on those days to receive their doses of naltrexone and paid a fee of $20 per week unless they had insurance coverage. Each week patients had a urine screen for opioids and other substances of abuse, with an alcohol breath test and examination for fresh needle marks. Each patient met weekly with a counselor.

Patients remained in treatment a mean of 50.7 days (range, 1 to 635), a total of 44 (27.5%) dropping out in the first week and 27 (16.9%) remaining for over 90 days. On the basis of their experiences with drop-outs from the naltrexone program, Tennant and colleagues (1984) recommend several procedures for induction in outpatients. Detoxification should be carried out with clonidine, guanabenz acetate, or symptomatic medication; after heroin dependence, opioid abstinence prior to beginning naltrexone should be maintained for 5 or more days; and after medical maintenance with methadone, propoxyphene napsylate, or LAAM for 10 or more days. They recommend an intramuscular challenge of naloxone (0.8 mg), which can be repeated if it is positive every 24 to 48 hr until it is negative. They concluded that the first dose of naltrexone should be 50 mg although others have recommended one-half tablet (25 mg).

Very little drug or alcohol abuse (1% to 3%) occurred during naltrexone treatment. Retention averaged only a few weeks but was clearly helped by counseling. The investigators (Tennant et al., 1984) cautioned that "use of naltrexone appears to promote long periods of opioid abstinence, but it does not prevent relapse following treatment."

## The Addicted Business Executives

Data from other studies based on specialized intensive programs at two hospitals (Regent Hospital, New York, and Fair Oaks Hospital, Summit, New Jersey) demonstrate even more vividly the efficacy and usefulness of naltrexone in treating addicted business executives as well as physicians. In the first part of the study, subjects consisted of 114 business executives addicted to heroin, methadone, or prescription opioids. All had been addicted for at least 2 years. Average age was 30 years, 92% were white, 77% males, 43% married; average income was $42,000 per year. All subjects began treatment under serious threat of losing their job, career, or family; almost half entered under stipulated and unequivocal job jeopardy, with legal consequences in some cases.

The major goals of treatment, as elaborated by the investigators (Washton et al., 1984), were "to help patients (1) recognize their addiction problem, the potential for relapse, and the need for continued treatment; (2) reduce sources of stress and change their lifestyle to minimize chances of relapse; (3) develop ways of coping with stress without resorting to drug use; and (4) return to work and family with a continued commitment to maintaining abstinence."

Treatment with an intensive, highly structured abstinence-oriented program on a specialized unit staffed by a multidisciplinary treatment team began with initial detoxification using clonidine, in addition to group therapy, peer group meetings, family meetings, individual therapy, educational sessions, and physical exercise. This inpatient program lasted 4 to 10 weeks and was followed by outpatient naltrexone aftercare for a minimum of 6 months. It was emphasized to patients that the inpatient program was only the initial phase of a more comprehensive recovery plan, including outpatient naltrexone aftercare (100 mg on Monday and Wednesday, 150 mg on Friday) for at least 6 months in combination with group and/or individual therapy, self-help groups, and urine monitoring. Before hospital discharge, all patients signed a contract for aftercare treatment specifying these details and stipulating that participation in the outpatient program required abstinence from opioids and all other mood-altering substances, including alcohol.

With naltrexone administered as part of a multidisciplinary treatment program in this group of highly motivated patients, the results of treatment were impressive. All 114 patients were effectively detoxified with clonidine, underwent naltrexone induction, and continued naltrexone during the transition from hospital to home. Of the group, 70 (61%) completed 6 months of outpatient treatment with no missed visits and no drug-positive urines. An additional 23 patients (20%) discontinued naltrexone before 6 months, but remained in the program with drug-free urines; the remaining patients (19%) were lost to follow-up or known to have relapsed. At 12- to 18-month follow-up, 64% were still opioid-free. Patients who completed at least 6 months of treatment were more likely to be opioid-free, employed, and continuing therapy or a self-help group (68%) than were those who discontinued treatment shortly after hospital discharge. Patients who began treatment under strong job jeopardy remained in treatment significantly longer (mean, 185 days) than did those who did not have the risk of job loss (mean, 101 days).

In a separate part of the same study, the investigators reported treating 15 opioid-dependent physicians in the same type of inpatient–outpatient programs utilizing naltrexone. The physicians had been addicted for 2 to 15 years, were all white males, aged 25 to 62 years (mean, 42), and in large part (76%) were married. All 15 were successfully detoxified with clonidine; 11 (74%) completed 6 months of naltrexone in the outpatient program and returned to medical practice. Three others discontinued naltrexone; two of these remained opioid-free and active in the treatment program. All patients who completed 6 months of treatment were still opioid-free and had returned to practice at 1-year follow-up.

These retention rates in executive and physician addicts were several times greater than those reported in previous studies of other populations. Several factors contributed to this. Certainly all the patients were highly motivated and had a great deal to lose by not succeeding in treatment. But the multidisciplinary treatment team was also highly motivated; and the highly structured, intensive, and vigilant nature of the inpatient and outpatient programs contributed largely to the successful outcomes.

## The Health Care Professional

In a collaborative study by private practitioners in California, 60 health care professionals (32 physicians, 16 registered nurses, 8 dentists, 2 pharmacists, and 2 veterinarians) were treated with naltrexone (Ling and Wesson, 1984). Of the physicians, 16 were in medical specialties, 6 were surgeons, and 10 anesthesiologists; most were participants in the diversion program of the California

Medical Association. Ages of the subjects ranged from 25 to 61 years, most being between 30 and 39 years old; 13 were 40 and older. Two thirds of the subjects were male.

The five drugs found to be most abused were meperidine, cocaine, morphine, fentanyl, and propoxyphene; 31 patients reported abuse of at least 2 drugs, and 32 reported drinking regularly. The length of addiction ranged from 2 months to 18 years, with a median of just over 2 years; one third of the patients had been addicted 4 or more years.

The length of naltrexone treatment ranged from 1 month to over 2 years, with an average of 8 months; 32 patients received more than 6 months of treatment; 52 patients had a single treatment episode, 5 had two, and 3 had three treatment episodes. Of the 60 patients, 29 were much improved on naltrexone treatment; 18, moderately improved; and 13, slightly improved. Compared with all professional groups combined, many more physicians showed "much" improvement ($p < .05$). Compared with those who drank, more nondrinkers also showed "much" improvement ($p < .01$). Those receiving naltrexone for longer than 6 months were judged "much improved" significantly more often than those receiving the drug for 6 months or less ($p < .01$). Length of addiction or use of a second drug did not seem to affect treatment outcome. Patients receiving naltrexone were remarkably free of side effects. Several patients complained of gastrointestinal disturbances, but at least two later admitted feigning symptoms to get off naltrexone so as to resume drug use. Several reports of headache attributed to naltrexone were counter-balanced by the three patients' reports of marked improvement in their migraine headaches while they were receiving the drug. There were no significant changes in routine laboratory values during the course of the study. The few side effects reported were mild, except for one patient who discontinued treatment because of intolerable dysphoria and difficulty with concentration for 48 hr, even after a dose of only 10 mg.

On the basis of these findings, Ling and Wesson (1984) concluded that "naltrexone can be easily, safely, and effectively used in physicians' offices for adjunctive treatment of narcotic addiction in selected groups of patients." Their study group was highly selective so that outcomes cannot be extrapolated securely to other patient populations. All participants were highly motivated; all had a great deal to lose by relapsing to narcotic abuse; all participated voluntarily. Naltrexone was recommended but not required. Each investigator treated each patient privately according to his own clinical style. A very important feature of this study is that all patients participated in peer-support group activities: Alcoholics Anonymous or Narcotics Anonymous group meetings, 90-minute sessions together two evenings a week. The groups seemed to offer valuable psychosocial support for the participants in conjunction with naltrexone maintenance. In these meetings, in fact, relapse to drug use and deterioration in other aspects of the recovery program usually became evident even before illicit drug use was detected in urine testing. The length of retention in these patients was quite remarkable compared with that in previously studied groups.

The use of naltrexone may offer a partial solution to the difficult problem of how to treat addiction in health care professionals. In the cases of such patients, restriction of access to drugs by external means is difficult, because if they continue in their professions, they have access to narcotics in their work environment. Methadone maintenance is not a practical alternative for them because many state laws prohibit health professionals from practicing while taking opioid drugs. Complete abstinence is difficult to ensure when opioids are readily available. The use of naltrexone, which effectively blocks the agonist effects of opioids and thus makes them unavailable, is a particularly attractive approach to ensuring that the health professional is not under opioid influence while practicing. If the treatment program involves observed ingestion of naltrexone, the patient has the added benefit of being "certified" as not abusing opioids. Thus the suspicion of relapse that often occurs when the health professional returns to work can be eliminated. If there are interpersonal problems on the job, these can be worked on without the shadow of readdiction interfering.

## The Currently "Clean" Addict Coming Out of a Hospital, Prison, or Therapeutic Community

Naltrexone also provides a valuable, nonaddictive alternative to methadone for the opioid-addicted prison inmate who is motivated toward rehabilitation, but could not survive in the traditional work-release setting. In 1972, a work-release program incorporating naltrexone was begun at the Nassau County (New York) Jail (Brahen et al., 1984). Addicted inmates, who had once been excluded automatically from such a program because of their general unreliability, could request admission if they received naltrexone regularly and had routine urine checks and an alcohol breath test when indicated. Vocational, educational, and counseling services including weekly psychotherapy sessions were provided. During the early transitional period on naltrexone, counseling was found to be of the utmost importance since many inmates had very few social supports. In addition to the work-release program, Brahen and colleagues also used naltrexone in the postprison treatment of inmates with addiction histories. They were encouraged to enroll in a hospital-based outpatient clinic where they could continue to take naltrexone or at least continue to receive counseling and regular urine monitoring. The outpatient program has provided an alternative to methadone maintenance for former inmates likely to become readdicted.

Over a 10-year period, the program including naltrexone has been utilized to treat 691 patients (636 males, 55 females). Of these, 279 have been treated at the work-release facility and 412 at the hospital outpatient clinic. After a full medical history, complete medical examination, and psychiatric evaluation, the patient signs an informed consent and treatment contract and is inducted into the naltrexone treatment program. Inmates can begin a job or training program after only one 150-mg dose of naltrexone; the most efficacious maintenance regimens have been those in which patients are medicated two or three times each week. Inmates in the program use naltrexone for approximately 4 months. Monitoring the inmates's drug use (with random urine samples) as well as his general health and behavior is mandatory during the first weeks of treatment. Each inmate sees an individual counselor at least one night per week, an important aim being to guard against the substitution of a nonopioid substance for an opioid since some of the patients continue to have a liability to addiction. A vocational rehabilitation counselor works also with each patient to determine the potential usefulness of various vocational training programs. (Many work-release participants in the naltrexone program have never worked before and believed they could never hold a job!) Finally, a social worker makes appropriate referrals and community contacts for patient inmates before their reentry to the community. About 20% of the released inmates have continued to receive naltrexone at the hospital outpatient clinic; and even after discontinuing naltrexone, many have remained in the outpatient program for urine monitoring and counseling.

The Narcotic Antagonist Work-Release Program has been operating successfully since 1972. Success in the program for addicts on naltrexone was the same as that for inmates with no addiction history. When preincarceration drug arrests have been compared with arrests after release, patients treated with naltrexone have had significantly fewer drug arrests after release than did subjects not receiving naltrexone. Naltrexone seems particularly well-suited to treatment of selected opioid-addicted inmates in a structured rehabilitation program.

## 3.  NALTREXONE INDUCTION AND STABILIZATION

Thus, naltrexone has been useful with a very wide spectrum of patients. Because of various difficulties sometimes encountered during the induction period and a high drop-out rate during the first month of treatment in many studies, the actual usefulness of naltrexone has not matched its pharmacological effectiveness (Kleber and Kosten, 1984). The reasons for this are many and are not equally valid for middle-class addicts, street addicts, and methadone patients. Different types of patients often seem to respond best to different pharmacological and psychological strategies.

### Inpatient versus Outpatient Induction

Inpatient induction is more expensive both for the facility and for the patient; it may be somewhat more successful. Certainly such induction makes it easier for the patient to receive psychological and licit pharmacological help. Individuals who have been maintained on methadone are usually less willing to become inpatients than are street addicts. The latter often seek help and treatment at a time of crisis and they usually need a variety of interventions, including inpatient treatment. These patients may present with mixed substance abuse and require detoxification from sedatives as well as from narcotics. Methadone-maintained subjects often come to naltrexone programs with a stable relationship to a methadone counselor already established so that it is difficult for them to change to the naltrexone treatment staff. Of course it should be possible for the same counselor or therapist to continue to treat the former methadone patient while he is on naltrexone and later drug-free. Also, their use of a long-acting opioid (methadone) regularly over a long period increases the risk of a longer duration of withdrawal symptoms and the risk of unpleasant reactions to the early doses of naltrexone.

The type of withdrawal method used also differs according to the type of addict. For individuals addicted to heroin or short-acting narcotics, the usual method is methadone substitution and withdrawal. If a clean urine is obtained by the seventh day off methadone, the addict is offered one or two naloxone challenge doses. If no symptoms follow, the subject can be given 25 mg naltrexone and the next day 50 mg. It is possible to treat any withdrawal symptoms with clonidine (usually 0.1 to 0.3 mg) or other symptomatic measures. As Kleber and Kosten (1984) emphasize, "it is better to err on the side of longer rather than shorter time intervals between the last dose of narcotic and the first dose of naltrexone, especially if clonidine is not available." If naltrexone is started too soon and the patient experiences withdrawal, he is usually not willing to try the drug again. Clonidine, however, can significantly shorten the length of time needed to begin naltrexone and thus reduce the problems of that difficult, drug-free interval. For individuals maintained on methadone, the dose is gradually reduced to zero, to be followed by 5 or more days of abstinence prior to the naloxone challenge. Clonidine is sometimes used when the methadone dose is in the range of 25 mg or below to allow a speedier methadone dose reduction.

## Clonidine-Assisted Detoxification

In administering clonidine as part of a naltrexone induction, certain precautions should be observed. This is not an FDA-approved use for this antihypertensive agent, but it has been found to be helpful over the past 8 years (Gold et al., 1978). Clonidine has been used to withdraw patients from doses as high as 70 mg of methadone per day. Usually, however, the patient is brought down gradually to 25 mg methadone per day or its equivalent. Details of the clonidine method in conjunction with naltrexone induction have been described by Kleber and Kosten (1984). The most common side effects of clonidine are sedation and hypotension. Patients should be discouraged from operating machinery or driving during the first few days. Even though it may cause sedation during the day, clonidine may worsen insomnia associated with detoxification. Patients to be treated with clonidine should not have had tricyclic antidepressants for at least 2 weeks and these drugs should not be used in combination with clonidine. With the vigorous use of clonidine, naltrexone can be started 3 to 5 days after drugs such as heroin and 5 to 7 days after the last dose of methadone in well-motivated patients who are willing to tolerate temporary withdrawal discomfort (Kleber and Kosten, 1984).

## Psychosocial Support Systems

During naltrexone stabilization, the major problems to be dealt with may be protracted abstinence symptoms (Martin and Jasinski, 1969), with craving for opioids, and the patient's need for social support systems. Some of the withdrawal symptoms may be conditioned responses related to drug-associated stimuli and not side effects of naltrexone. Drug craving generally has been reported to diminish or disappear during the first month on naltrexone. Protracted abstinence symptoms may be manifested as anxiety or insomnia. Other psychiatric symptoms previously suppressed by opiates may appear. It is critical to get the patient involved in psychotherapy or counseling at an early stage of treatment. Individual, group, or family therapy may be indicated. Family therapy has been found to be particularly useful in naltrexone programs. During the first month of naltrexone therapy, addicts in family therapy have been reported to have a significantly higher rate of retention than addicts not in family therapy.

## 4.  COMPLIANCE ISSUES

The problems of compliance in the patient taking naltrexone must be addressed. Unlike methadone or LAAM, naltrexone produces no physical dependence and has no immediate positive reinforcing properties, no inherent characteristics that would make an unmotivated addict seek it out and want to continue treatment (Grabowski et al., 1979). With most patients, compliance must be ensured and built into the treatment program. Observed ingestion in front of a trusted staff member, colleague, wife, or partner is convenient and has much to recommend it. One other dose each week might be taken at the clinic or in the office of the therapist. Of course, the patient must also give a random urine sample for complete drug screenings. Successful programs usually have a structured system for ensuring compliance. This may involve treatment contracts and reports to a medical board or to an employer. The structure should be applied to all patients including physicians and business executives as well as street addicts.

## 5.  LENGTH OF TREATMENT ON NALTREXONE

The question of length of treatment cannot be answered specifically. The data indicate that most treatment is short term, particularly in street addicts where the mean duration of therapy in most studies is a matter of weeks. However, there is evidence that even as little as 1 month of outpatient treatment with naltrexone improves outcome at 6-month follow-up (Greenstein et al., 1983). Also, patients may have several episodes of short-term treatment before a long period of abstinence is induced (Judson and Goldstein 1984). Our present practice is to ask that each new patient for antagonist treatment make a commitment of 6–12 months of outpatient protection with naltrexone. This allows time for recovery from protracted abstinence symptoms, the establishment of new behavior patterns in the opioid-free state, and progress in family, group, or individual therapy. Of course we recognize that much of the therapy will be short term, but 6–12 months should be a goal.

Another side of the question concerning duration may occur in patients doing well: What length of naltrexone treatment is enough? This has come up in our experience when dealing with physicians who must continually work in the presence of drugs. It has also arisen with impulsive drug users who have relapsed many times in the past. They find naltrexone to be free of side effects and only a minor inconvenience. They feel "protected" when taking naltrexone and their families, employers, or colleagues also feel more secure. In these cases we have had patients treated for 2–3 years of continuous naltrexone. Thus far we are aware of no chronic toxicity problems and no evidence of tolerance to the antagonist effects. From a psychotherapy standpoint, one should think of eventual independence and termination from therapy, but for those with a history of relapse who must work in the presence of opioid drugs, very long-term protection with naltrexone may be practical. Even after naltrexone is discontinued, *psychotherapy*—whether formal psychoanalytically oriented psychotherapy, cognitive-behavioral therapy, family therapy, or extinction therapy—*must be continued*. Psychotherapy, in fact, must be given a role in the treatment program at least equal to that of naltrexone itself. Complemented by psychotherapy and perhaps by job counseling, and often supported by observed ingestion of the antagonist, naltrexone as part of a comprehensive treatment program can then provide the clinical efficacy that its pharmacological efficacy seems to promise.

## REFERENCES

Arndt, I. O., in press, 1986, Cacciola, J. S., McLellan, A. T., O'Brien, C. P., A re-evaluation of naltrexone in recovering opiate addicts, in: *Problems of Drug Dependence, 1985*, NIDA Research Monograph Series, Rockville, Maryland.

Brahen, L. S., Henderson, R. K., Copone, T., and Kordal, N., 1984, Naltrexone treatment in a jail work–release program, *J. Clin. Psychiatry* **45**(2):49.

Christian, M. S., 1984, Reproductive toxicity and teratology evaluations of naltrexone, *J. Clin. Psychiatry* **45**:7.

Gold, M. S., Redmond, D. E., and Kleber, H. D., 1978, Clonidine blocks acute opiate withdrawal symptoms, *Lancet* **2**:599.

Grabowski, J., O'Brien, C. P., and Greenstein, R. A., 1979, Effects of contingent payment on compliance with a naltrexone regimen, *Am. J. Drug Alcohol Abuse* **6**:355.

Greenstein, R. A., Evans, B. D., and McLellan, A. T., 1983, Predictors of favorable outcome following naltrexone treatment, *Drug Alcohol Depend.* **12**:173.

Greenstein, R. A., Arndt, I. O., McLellan, T. A., O'Brien, C. P., and Evans, B. D., 1984, Naltrexone: A clinical perspective, *J. Clin. Psychiatry* **2**:25.

Hollister, L., 1978, Report of the National Research Council Committee on Clinical Evaluation of Narcotic Antagonists: Clinical evaluation of naltrexone treatment of dependent individuals, *Arch. Gen. Psychiatry* **35**:335.

Judson, B. A., and Goldstein, A., 1984, Naltrexone treatment of heroin addiction: One year follow-up, *Drug Alcohol Depend.* **13**:357.

Judson, B. A., Carney, T. M., and Goldstein, A., 1981, Naltrexone treatment of heroin addiction: Efficacy and safety in a double–blind dosage comparison, *Drug Alcohol Depend.* **7**:325.

Kleber, H. D., and Kosten, T. R., 1984, Naltrexone induction: Psychologic and pharmacologic strategies, *J. Clin. Psychiatry* **2**:29.

Ling, W., and Wesson, D. R., 1984, Naltrexone treatment for addicted health-care professionals: A collaborative private practice experience, *J. Clin. Psychiatry* **2**:46.

Martin, W. R., and Jasinski, D. R., 1969, Physiological parameters of morphine in protracted abstinence, *J. Psychiatr. Res.* **7**:9.

Martin, W. R., Jasinski, D. R., and Mansky, P. S., 1973, Naltrexone, an antagonist for the treatment of heroin dependence, *Arch. Gen. Psychiatry* **28**:784.

Meyer, C. M., Straughn, B. A., Lo, M., Schary, L. W., and Whitney, C. C., 1984, Bioequivalence, dose-proportionality and pharmacokinetic of naltrexone after oral administration, *J. Clin. Psychiatry* **45**(2):15.

O'Brien, C. P., Greenstein, R., Mintz, J., and Woody, G. E., 1975, Clinical experience with naltrexone, *Am. J. Drug Alcohol Abuse* **2**:365.

O'Brien, C. P., Childress, A. R., McLellan, A. T., and Ternes, J. W., 1984, Use of naltrexone to extinguish opioid-conditioned responses, *J. Clin. Psychiatry* **45**(9):53.

Rosenkrantz, H., 1984, Physiologic and morphologic changes and incidence of neoplasms in mice and rats fed naltrexone HC1 for 24 months, *J. Clin. Psychiatry* **45**(2):11.

Sideroff, S. I., Charuvastra, V. C., and Jarvik, M. E., 1978, Craving in heroin addicts maintained on opiate antagonist naltrexone, *Am. J. Drug Alcohol Abuse* **5**(4):415.

Tennant, F. S., Rawson, R. A., and Cohen, A. J., 1984, Clinical experience with naltrexone in suburban opioid addicts, *J. Clin. Psychiatry* **45**(2):42.

Washton, A. M., Pottash, A. C., and Gold, M. S., 1984, Naltrexone in addicted business executives and physicians, *J. Clin. Psychiatry* **45**:9.

8

# Update on Methadone Maintenance

GEORGE E. WOODY and CHARLES P. O'BRIEN

## 1. INTRODUCTION

Narcotic substitution therapy, of which methadone maintenance is the most prominent example, was first introduced by Dole and Nyswander (1965). Methadone very quickly came into wide use for the treatment of opioid dependence and it has continued to be one of the most commonly used treatments. Much additional investigative work has been done since methadone maintenance began and this work has been conducted in diverse areas. These include evaluation of treatment outcome; descriptive studies of the behavioral and psychiatric disorders seen in methadone-maintained addicts; evaluation of ancillary treatments that may improve outcome, particularly those treatments that affect the psychiatric disorders of methadone-maintained patients; development of instruments to measure the spectrum of problems seen in opioid addicts; studies of the value of matching specific types of patients to specific ancillary treatments; testing the efficacy of maintenance drugs other than methadone; reviewing the outcome of patients treated with high versus low methadone doses; studying various methods of methadone detoxification; and evaluating the effects of opioid dependence and methadone treatment on neonates.

Most of this work has been aimed at evaluation and refinement of the original concept and technique of narcotic substitution therapy. This chapter will briefly describe work that has been done in each of these areas.

## 2. EVALUATION OF OUTCOME

A number of large-scale studies have been conducted to re-examine the original positive findings regarding the efficacy of methadone maintenance.

GEORGE E. WOODY and CHARLES P. O'BRIEN ● Psychiatry Service, Philadelphia Veterans Administration Medical Center and University of Pennsylvania, Philadelphia, Pennsylvania 19104.

Maintenance treatment for opioid dependence has always been controversial, with both proponents and critics. Part of this ambivalence may be inherent within the maintenance concept itself, which substitutes a legally prescribed narcotic for an illicitly obtained and self-administered one. By doing so, one chooses to deliver a treatment that maintains, rather than discontinues, the problematic narcotic use. This approach differs from that seen in many other areas of medicine where immediate attempts are made to do something to reverse the underlying symptoms of the disease, such as the surgical removal of an inflamed appendix. The rationale for the maintenance approach has been that if narcotic substitution is delivered with proper controls and appropriate supportive services, the addict/patient will do better than he or she would have done if maintenance treatment had not been used. It is important to note that methadone treatment programs have developed in environments where illicit opiates or opioids are easily available. Opioid addicts who are not on methadone generally have little difficulty obtaining narcotics, provided they can pay for them. The methadone programs claim to provide a positive alternative for addicts within this type of environment. One of the major accomplishments over the last several years has been to re-examine this basic concept.

The most systematic and comprehensive re-examination is found in a review that was carried out recently by the National Institute of Drug Abuse (NIDA). This agency sponsored a series of meetings that summarized and critiqued many aspects of methadone treatment, including outcome (Cooper et al., 1983). Several large-scale studies were available for inclusion in this review, particularly those of Sells (1979) and McLellan et al. (1982). These studies were supportive of the initial positive findings of Dole and Nyswander (1965), which were that methadone-treated patients showed significant gains when compared with addicts not in treatment. These gains were seen in reductions of illicit drug use, reductions in crime, and increased rates of employment. Substantial reductions in criminal activity by maintenance patients was further documented by the findings of Ball et al. (1983), who studied treated and untreated addicts in Baltimore and Philadelphia. Addicts who were in methadone treatment reported significant reductions in criminal activity compared with those reported during comparable time periods prior to treatment. Comprehensive arrest, penal, hospital, and other institutional data were also available to verify the interview reports.

All these studies are somewhat "soft" in that many rely on historical controls or on patient self-reports. Independent measures such as police records, urine test results, or documentation of employment are available in some studies (Ball et al., 1983) but not in others. We know of only two studies that were prospective and used random assignment. The first (Dole et al., 1969) was weakened somewhat by the use of a relatively narrow (but important) range of outcome measures (relapse, employment, and crime) and by the lack of description of psychosocial treatments that were (or were not) available to the control (nonmethadone) group.

However, the results of this prospective study were overwhelmingly in favor of the methadone group (Dole et al., 1969). A second controlled prospective study took place in Hong Kong (Newman and Whitehill, 1979). One hundred heroin addict volunteers were randomly assigned to either methadone or placebo. All received a broad range of supportive services. After 32 weeks, only 10% of the controls remained in treatment, while 76% of the methadone group continued. The rate of convictions for criminal activity was more than twice as great in the control patients. The impressive thing about these studies is that they *all* found that methadone treatment was effective. Thus, we cannot escape concluding that a substantial amount of data are now available that indicate that methadone maintenance is an effective treatment when administered to patients who are chronic narcotic addicts, who are actively addicted, and who are living in an environment where narcotics are readily available. No evidence is available that maintenance is a steppingstone toward permanent discontinuation of narcotic addiction, however. Rather then curing addiction, methadone maintenance appears to provide control over its adverse personal and social consequences. No work is available that measures the level of physiological dependence on opioids prior to methadone treatment and then relates this to dose and outcome. Methadone treatment outcome data derive from approved programs that are periodically inspected for compliance with Food and Drug Administration requirements for maintenance treatment. The FDA mandates current physiological dependence plus other characteristics that signify long-term addiction, but the regulations do not define criteria that quantify degrees of dependence. The only study even remotely evaluating outcome as it relates to degrees of physiological dependence is that of McLellan et al. (1983), who showed that the amount of drugs used and the length of the drug dependence did not relate to treatment outcome. Thus, information on this point is very sketchy, but the limited data available point to features other than levels of physiological dependence as major determinants of outcome.

## 3.   BEHAVIORAL AND PSYCHIATRIC DISORDERS OF ADDICTS

Another area that was studied and reviewed recently is that concerning the psychiatric and behavioral problems seen in opiate addicts, both while in and out of treatment. Both medical and lay personnel have noted that opioid addicts often display serious behavioral and psychiatric problems. The successful management of these problems is of great importance to both the addict and the community. Failure on the part of a treatment program to successfully control their patients' behavioral disorders has sometimes resulted in community pressure to close the program, and in some cases this pressure has achieved its intended result. Inability to treat the psychiatric disorders of the addict/patient has contributed to poor treatment response and sometimes premature termination from therapy.

## Behavioral Disorders

A second NIDA review (Grabowski et al., 1984) described some of the behavioral problems demonstrated by addicts and ways to manage them. Most participants in this review felt that many of these problems originate from drug-seeking behavior that is, of course, a characteristic of the addiction syndrome. Such problems include fabricating stories to obtain controlled substances, buying or selling illicit drugs, attempting to falsify urine samples to avoid loss of take-home methadone doses, or attempting to divert methadone at the pharmacy window.

A serious problem reported by many programs was loitering. The most common motive for loitering appeared to be social contact, but a considerable amount of drug dealing was also observed to take place (Hunt et al., 1984). Persistent loitering was felt to serve as a nidus for the development of other behavioral problems such as arguments or fights. Loitering was also noted to be frightening to people who happened to be in the vicinity of the program and who were not familiar with the personalities and lifestyles of addicts. Threats, disruptive behavior, fighting, or even carrying weapons were reported occasionally. These problems were thought to be related to the addiction itself, to personality disorders or other psychiatric problems, or to socioenvironmental circumstances. Participants in the NIDA conference (Grabowski et al., 1984), which reviewed these problems, were in reasonably good agreement in feeling that a combination of support and structure involving specified rules that include suspension of those who display serious behavioral problems was essential for effective control of these disorders.

## Psychiatric Disorders

A number of studies were reviewed that described the types of psychiatric problems seen in opiate addicts. These studies found that almost every psychiatric illness that can occur in nonaddicts also occurs in addicts. The most thorough evaluation of psychiatric disorders in addicts was that done by Rounsaville et al. (1982*b*). A sample of 533 opiate addicts was given a thorough and careful psychiatric evaluation as part of a comprehensive diagnostic study. Depression was the most commonly diagnosed illness among addicts, with about 60% of the sample having had some form of depression at least once. The next most common problem was alcoholism, followed by antisocial personality disorder and anxiety disorders. Occurring with a much lower frequency were schizophrenia, other types of personality disorders, mania, and hypomania. Eighty-five percent of the patient sample were found to have had a psychiatric disorder in addition to opiate dependence at some time in their lives. Not included in this study but also seen regularly are acute situational reactions that involve intense but transient feelings of anger, anxiety, or depression; psychiatric disorders complicated by medical conditions such as hepatitis; and illnesses or injuries that produce chronic pain such as pancreatitis, sickle cell anemia, or trauma resulting in nerve root irritation. Rounsaville et al. (1982*b*) also studied untreated addicts and found them to have types of psychiatric illnesses similar to patients who were in treatment and in relatively similar proportions. The major difference between treated and nontreated addicts was that the latter group was less likely to have a current psychiatric illness. This finding can be interpreted as indicating that concurrent psychiatric illness serves as a motivator to seek treatment.

## Depression

Special attention was given to depression, which was found to be the most commonly diagnosed psychiatric disorder in opiate addicts. The course of the depression seen in methadone-maintained patients was also studied by Rounsaville et al. (1982*a*). It was found that a substantial reduction in symptoms of depression occurred within the first month of methadone treatment. However, the amount of depression that remained, even after this initial improvement, was substantially greater than that observed in nonaddicted controls. These researchers also found that the depressions seen, although usually diagnosed as major depressive disorder, were of mild to moderate intensity and were usually not of the severely retarded or psychotic levels that are commonly seen on psychiatric inpatient units. Another finding that emerged from this study was that those addicts who remained depressed beyond the initial 1-month period usually improved within 6 months, even without specific antidepressant treatment. It was also found that some patients who were not initially depressed at program intake became depressed at some later point.

These findings, when taken together, painted the following picture of the depressions experienced by addicts: (1) they usually fit the criteria for major depressive disorder, although other types of depression are also seen; (2) they are very common at program intake but often improve within the first month of treatment; (3) they are of mild to moderate intensity; (4) they usually improve without specific antidepressant treatment; however, recovery may take 4–6 months or longer; (5) they tend to recur; and (6) they are often precipitated by situational factors such as the loss of important relationships, arrest, or job difficulties.

## 4. TREATMENTS FOR ADDICTS WITH PSYCHIATRIC DISORDERS

Studies were also conducted that examined the efficacy of treatments for methadone-maintained addicts who have specific psychiatric disorders, such as depression. These studies included examinations of pharmacotherapy and psychotherapy.

### Tricyclic Antidepressants

Four controlled studies have provided evidence that doxepin is an effective drug for methadone-maintained addicts who are clinically depressed (Woody et al., 1975; McBride et al., 1982; Titievsky et al., 1982; Woody et al., 1982). Doxepin was found to reduce the signs and symptoms of depression more quickly than occurred when the depression was not treated by specific antidepressant medication. The benefits seen appeared similar to those obtained with doxepin in depressed patients who are not addicts in that it accelerated the usual course of recovery.

Few other antidepressants have been studied; however, the less-sedating tricyclics desipramine and imipramine have been tried and they do not appear to be generally useful (Kleber et al., 1983; Woody et al., 1982). Thus, the mechanism for the improvement seen with doxepin may relate to its antianxiety properties as much as its antidepressant effects. However, practically speaking, doxepin has been found to be a useful adjunctive therapy for the more refractory depressions seen in methadone-maintained addicts. These depressions are those that have endogenous signs and symptoms such as anorexia, insomnia, psychomotor retardation or agitation, and suicidal thoughts. Beck Depression Inventory (Beck, 1972) scores would be in the range of 15 to 20 or more and patients tried on a tricyclic would be those who have not responded to counseling or other psychological treatments.

The treatment recommendation that emerges from the studies of depressive illness and of doxepin as an antidepressant treatment is as follows: many of the depressions seen in addicts are transitory and are best left to improve without additional antidepressant therapy. However, those depressions that last longer than 2–4 weeks may be improved upon by adding doxepin to the ongoing treatment. No antidepressants other than doxepin, imipramine, or desipramine have been studied in these patients. Tricyclics other than doxepin may work. Amitriptyline has been observed to be abused by methadone-maintained addicts, in the same manner as diazepam, and should be used with great caution or not at all (Cohen et al., 1978). We are not aware of reports describing abuse of doxepin, and we have not seen it clinically.

## Psychotherapy

Other work studied the efficacy of professional psychotherapy with nonpsychotic addicts when added to routine counseling services in a methadone program. Two major studies were completed in this area and their results varied. One showed no differences in outcome when interpersonal psychotherapy was added to drug counseling in a full-service methadone treatment program (Rounsaville et al., 1983). This study, although well-designed, did not succeed in engaging many patients in the psychotherapy, and it also had a high drop-out rate. The recruitment methods used suggested that some patients who did participate were probably unusually resistant to treatment, since they were referred to the research project as a last resort prior to being discharged from the program for showing little improvement.

A second study of psychotherapy had a higher recruitment rate, fewer drop-outs, and a larger total number of subjects (Woody et al., 1983). This study provided good evidence that professional psychotherapists can provide additional benefits to those obtained with paraprofessional drug counselors. The treatments used are described in detail elsewhere (Woody et al., 1983). One of the most interesting findings from this work emerged when outcome was assessed for patients classified as low, mid, or high severity on the basis of the number and intensity of their psychiatric symptoms as rated by standard assessments such as the Beck Depression Inventory (Beck, 1972) or the SCL-90 (Derogatis, 1970). Low-severity patients made considerable and approximately equal progress with the additional psychotherapy or with counseling alone. Mid-severity patients had better outcomes with the additional psychotherapy than with counseling alone, but counseling was associated with numerous significant improvements. High-severity patients made little progress with counseling alone, but made considerable progress if they had the additional psychotherapy.

Patients with only antisocial personality disorder showed little response to the additional therapy, but others, especially those with depression, had a good response. It was also found that there was considerable variation in outcome among therapists, with some achieving consistently better outcomes than others (Luborsky et al., 1985). Those therapists who had the best results were noted as having the best doctor–patient relationships and also as being the most successful in conforming to the specifications of their particular school of therapy.

The overall results of these studies, and of other similar but less comprehensive ones (Abrams, 1979; Connett, 1980; LaRosa et al., 1974; Resnick et al., 1980; Stanton and Todd, 1982; Willett, 1973), indicate that professional psychotherapy can probably make a valuable contribution to ongoing treatment services, particularly if it is targeted to that segment of the nonpsychotic addict population with moderate to high levels of psychiatric symptoms. The therapy program is not always simple to apply to this population (Rounsaville et al.,

1983; Stanton and Todd, 1982; Woody et al., 1983) and must be well-integrated into the overall drug rehabilitation program. It must also employ skilled therapists who relate well to opioid addicts. Few programs are using professional psychotherapists at this time. Our impression is that available services could be improved, especially for the most disturbed patients, by the selective use of skilled psychotherapists who have an interest in working with opiate-addicted individuals.

## Other Psychotropic Medications

The last 10 years have also seen the use of psychotropic drugs other than tricyclic antidepressants for addicts with psychiatric disorders. Ciccone et al. (1980) and Kleber (1983) reviewed evidence for the efficacy of these psychotropic medications, including antianxiety agents, lithium, and disulfiram. These reviews indicated that each of these medicines may be helpful for addicts with the specific disorder for which the drugs have been found helpful in other, nonaddict populations. Kleber (1983) advised a conservative approach to using benzodiazepines. Many clinicians have noted that diazepam and also quite possibly alprazolam (R. Millman, personal communication, 1985) have a significant abuse potential with this population and should be prescribed sparingly or not at all.

Disulfiram (Antabuse) has been found to be useful in opiate addicts who are also alcoholics. Up to 25% of patients on methadone programs have been reported to have drinking problems. While these problems can frequently be controlled by behavioral means such as breathalizer testing, disulfiram may be indicated. Liebson et al. (1973) found the rewarding effects of methadone to compliment the use of disulfiram in a small successful study. A large Veterans Administration collaborative study (Ling et al., 1983), however, found that the placebo control method worked poorly for disulfiram comparison. The patients tended to test with alcohol to determine whether they were receiving placebo. While the VA investigators found no disulfiram–placebo differences in the methadone-maintained patients, the drinking problems in both groups showed improvement, probably due to the structure imposed by the disulfiram study. The investigators concluded that willingness to accept disulfiram was significant in the beneficial effects observed and that there was no evidence of adverse interaction between methadone and disulfiram.

## 5.  MEASURING THE PROBLEMS SEEN IN OPIOID ADDICTS

Other work focused on documenting and measuring the types of problems seen in methadone-maintained addicts and on matching patients to treatments. This work was made possible by the development of the Addiction Severity

Index (ASI) by McLellan et al. (1980*a*). The ASI is a structured, 40-min, clinical research interview designed to assess problem severity in seven areas commonly affected by addiction: medical, legal, drug, alcohol, employment, family, and psychiatric. In each of these areas, objective questions are asked that measure the number, extent, and duration of problem symptoms in the patient's lifetime and in the past 30 days. The patient also supplies a subjective report of the recent severity and importance of the symptoms within the problem areas. Severity is operationally defined on all ASI scales as the number, duration, frequency, and intensity of symptoms in that specific area. Severity is rated by a trained technician from a patient's answers to questions regarding the lifetime and recent (past 30 days) experience of problem symptoms. The interviewer assimilates the two types of information to produce seven global ratings reflecting problem severity in each area. The patient's answers form the baseline against which subsequent follow-up interviews can be used to assess improvement. These ten-point (0 to 9) ratings have been shown to provide reliable and valid general estimates of problem severity for both alcoholics and drug addicts.

## 6.   PATIENT–PROGRAM MATCHING

Patient–program matching is a concept that could produce considerable progress in methadone treatment. The heterogeneity of methadone patients, as described earlier, provides ample justification for studies of patient–program matching. In addition there appears to be tremendous variability among methadone programs themselves, occasionally in regard to dose policies (high dose vs. low dose), but particularly in regard to the ancillary services that are available, the quality of treatment facilities, and the levels of discipline imposed. Some programs have very little physician coverage whereas others have one or more physicians actively involved in the treatment program. Counseling services, psychotherapy, social work, vocational counseling, and other aspects of drug rehabilitation services are similarly unevenly distributed among programs. Some states require paraprofessional drug counselors to pass an examination and others appear to have no competency standards. Thus, the variability among programs adds another dimension to patient–program matching.

Very little work has been done in this important area. Some of the treatment-outcome studies mentioned in this chapter (antidepressant drugs, psychotherapy, disulfiram, contingency contracting) represent attempts to match patients to treatments. McLellan and his colleagues have completed several studies of patient–program matching. Follow-up studies done using the ASI showed that patients with different symptom profiles showed different responses to residential versus outpatient treatments. For example, patients with significant sedative abuse or those with a highly unstable living arrangement appeared to do better in a residential program than in an outpatient methadone program. Generally, patients with significant amounts of psychiatric symptoms and those with stimulant abuse did better in methadone maintenance than in residential drug-free programs (McLellan et al., 1980*b*, 1983). This probably occurred because these more disturbed patients need support and medication. Psychotropic drugs are usually not prescribed in a residential drug rehabilitation program and treatment may involve confrontation and stress. Methadone itself may have psychotropic effects and methadone programs tend to utilize supportive therapy and ancillary psychotropic medications in addiction to methadone. These studies also found that a global rating of psychiatric severity was a good predictor of outcome in both residential and outpatient programs (McLellan et al., 1983). Patients with low levels of psychiatric symptoms did reasonably well in either outpatient or inpatient programs. Patients with mid-levels of psychiatric symptoms did better than patients with high levels and the patients with high ratings of psychiatric symptoms did not do well in either outpatient or inpatient programs. Matching of patients to programs appeared to be most helpful for those with mid-level psychiatric symptoms and least necessary for low-severity patients.

Another earlier attempt in the area of patient–program matching was that formulated by Goldstein (1976) regarding a sequential drug treatment strategy in which first methadone, then Levo-alpha-acetyl methadol (LAAM), and finally naltrexone are used in a progression from daily street opiate use to successful abstinence. We have not seen this progression to occur with any regularity either in the literature or in our experience. The psychiatric, social, and learned aspects of opioid dependence appear to be much stronger determinants of outcome then the pharmacological ones that are implicit in this theory. However, the efficacy of such a pharmacological progression, perhaps combined with psychosocial treatments, remains to be studied.

## 7.   TESTING THE EFFICACY OF MAINTENANCE DRUGS OTHER THAN METHADONE

### LAAM

Other work was done to evaluate new drugs that may be used in place of methadone for narcotic substitution therapy. The most promising of these was LAAM. LAAM has pharmacological effects similar to methadone, but with LAAM their onset is slower and their duration is longer. One dose of LAAM can suppress opioid withdrawal symptoms for 48–72 hr. LAAM appears to have no serious toxicity when used properly; however, some patients report an increase in psychomotor activity while on LAAM, possibly due to its having greater stimulant effects than methadone (Ling et al., 1980). The major advantage of

LAAM is that patients need come to clinic only three times per week. This is especially beneficial for those who must travel long distances or who have long hours or irregular work schedules. It is also desirable from a public health standpoint because its use almost eliminates the problems caused when patients sell or otherwise improperly use methadone that is dispensed to consume outside the clinic. Studies done to date show that treatment results for those who remain on LAAM compare favorably with those obtained with methadone, although LAAM's slower onset probably contributes to the higher initial drop-out rates seen when LAAM is compared with methadone (Ling et al., 1980). LAAM is still classified as an investigational drug by the FDA largely because it has had no commercial sponsor to shepherd it through the approval procedures. If it is eventually approved for general use, treatment programs would be in a position to offer most patients either LAAM treatment three times per week or daily methadone. This policy could lead to a marked improvement in the safety of methadone by practically eliminating the diversion of take-home methadone to illicit street users.

## Propoxyphene

Propoxyphene napsylate, a drug with weak narcotic effects, was also tested for efficacy as a maintenance drug. Propoxyphene can suppress abstinence symptoms in patients who have low levels of physical dependence, and open clinical trials showed that some addicts could be maintained on propoxyphene (Tennant, 1974). However, double-blind studies comparing low doses of methadone (maximum 36 mg/day) with high doses of propoxyphene (maximum 1200 mg/day in divided doses) showed that propoxyphene was not nearly as effective as methadone (Woody et al., 1981). Drop-out rates and street drug use were considerably higher in patients maintained on propoxyphene than in those treated with methadone. The results of these studies were instrumental in propoxyphene being judged inappropriate for use as a maintenance treatment.

## 8.  BEHAVIORAL TREATMENTS TO IMPROVE OUTCOME

Attention was also focused on behavioral treatments, especially as they apply to patients who demonstrate a poor response to methadone therapy. One serious problem that occurs with some maintenance patients is persistent drug use in spite of methadone doses that are more than sufficient to suppress withdrawal symptoms and that also provide some degree of cross-tolerance to injected opioids thus reducing or eliminating the "reward" of using street opioids. Patients who continue to use opiates in spite of being treated with adequate doses of methadone are often not keeping regular counseling appointments and usually show little initiative to engage in constructive activities such as school or work. In these cases, some recommend raising the methadone dose to the highest allowable levels (80–100 mg/day). Sometimes this process is combined with a treatment contract that states that the patient must accomplish certain behaviors within a specified period of time or face suspension from the program (Bigelow et al., 1984). Goals to be achieved can include drug-free urine specimens, keeping all counseling appointments, and demonstrating proof that job-seeking behavior is occurring. Selectivity in choosing patients for contracts is important, as is choosing realistic goals. A recent study shows that 50–60% of carefully selected patients who were given a treatment contract succeeded by achieving the specified goals within the allotted time (Dolan et al., 1985). Those who fail a contract can be offered the option of transferring to another methadone program or of entering another treatment modality, such as a therapeutic community or narcotic antagonist (naltrexone) treatment.

## 9.  HIGH- VERSUS LOW-DOSE METHADONE TREATMENT

The last several years saw a partial resolution of the debate over whether patients need high methadone doses (more than 60 mg) or whether they will do just as well on lower doses (50 mg or less). Several studies explored this question and the conclusion was that patients can do well on any dose, that there are many variables that influence outcome, but that patients on high doses generally use fewer drugs and tend to do better than those on low doses (McGLothlin and Anglin, 1981). Thus, dose can be important, but for most patients it is only one of many variables that influence outcome.

## 10.  DETOXIFICATION FROM METHADONE

Another series of studies examined detoxification from methadone maintenance. Several areas were explored. One group of investigations examined the proportion of detoxified addicts who remained abstinent. Stimmel et al. (1977) followed up 335 persons who were successfully detoxified from methadone maintenance for as long as 6 years. Of the 269 persons located, 35% were narcotic-free, 58% had returned to narcotic use, and 8% were either jailed or deceased. This study concluded that while abstinence after narcotic dependence is possible, it is not a realistic goal for all. Premature detoxification was associated with a high relapse rate. Trained staff were able to identify those candidates who were most likely to succeed, although the exact criteria for identification were not specified. Kleber (1977) described similar findings and also found that successful withdrawal is possible, but only for a small proportion of patients.

Senay et al. (1977) studied rate of methadone withdrawal as a determinant of successful detoxification. One hundred twenty-seven successfully maintained addicts were randomly assigned to four treatment groups and it was found that patients who were detoxified at the rate of 3% per week were more likely to successfully detoxify then those who were reduced by 10% per week. Other studies demonstrated that clonidine, a nonopioid α-adrenergic agonist already approved for use as an antihypertensive drug, can suppress the autonomic symptoms of opiate withdrawal (Gold et al., 1978). Patients treated with clonidine can obtain significant relief from withdrawal symptoms; however, tolerance develops to the effect in 7–10 days. Many programs now use clonidine as an adjunct to routine services during detoxification. A comprehensive review on clonidine has been written by Ginzberg (1983).

The conclusions one draws from all the work on detoxification are as follows: some patients can be successfully detoxified, but candidates should be carefully selected by trained staff as those most likely and able to achieve success; premature detoxification can be harmful; a slow decrease (3% per week or less) will provide the best chance for success in an outpatient methadone detoxification program; and clonidine can be a useful short-term aid to detoxification.

## 11.   OPIOIDS AND PREGNANCY

Opioid-dependent women have been found to have increased rates of miscarriage and to give birth to children of reduced birth weight. Furthermore, children born to opioid-dependent women often, but not always, experience narcotic withdrawal symptoms shortly after birth. The neonatal mortality rate for the opioid-exposed fetus is between 3 and 4.5%, with the majority of complications being related to low birth weight. The causes for this relationship are not exactly clear, but they probably relate to poor nutrition, fetal anoxia produced by changes in placental blood flow when the mother experiences the opiate withdrawal syndrome, and a direct effect of the opioid drug on the developing fetus.

These problems appear to be reduced by methadone maintenance. However, if maintenance is used to maintain pregnant addicts, it should be given only to those who are unlikely to successfully detoxify and it should be given in the lowest possible effective dose. The incidence of withdrawal symptoms seen in newborns is probably reduced by lower methadone doses. Neonatal withdrawal, when it occurs, can be treated with paregoric or phenobarbital.

Detoxification is not advised prior to 14 weeks gestation because of the potential risk of inducing abortion. Detoxification also should not be performed after the 32nd week of pregnancy because of possible withdrawal-induced fetal stress.

Most patients who are maintained during pregnancy are controlled on 20–35mg methadone per day. Methadone-maintained women have often been found to complain of increasing withdrawal symptoms as pregnancy progresses. This is probable due to the pregnancy producing an increased extracellular fluid space with a consequent lowering of plasma methadone levels. These symptoms should be treated with elevation of the oral dose sufficient to suppress the withdrawal symptoms. These findings are discussed in detail by Kaltenbach and Finnegan (1984).

Wapner and Finnegan (1981) reviewed five longitudinal studies that evaluated methadone-exposed infants throughout the first two years of life. The results of these studies suggested that no long-term developmental sequelae are directly associated with methadone exposure *in utero*. Although differences were often found between methadone-exposed infants and comparison infants, developmental scores for the methadone-exposed infants were well within the normal range. Follow-up data at age 14 exposed to methadone *in utero* revealed no significant differences in cognitive performance between methadone-exposed and comparison children, although scores for both groups were in the low normal range (Wapner and Finnegan, 1981).

The question regarding whether there are permanent developmental sequelae associated with *in utero* methadone exposure has not been conclusively resolved at this time. The number of follow-up studies is small, and most encompass only the first two years of life. Although the findings are relatively consistent in finding no serious problems, a general consensus regarding their interpretation has not emerged due to the relatively small amount of data currently available. Work is currently being done in this area, thus future years should see more information on these important issues.

## REFERENCES


Abrams, J., 1979, A cognitive behavioral versus nondirective group treatment program for opioid addicted persons: An adjunct to methadone maintenance, *Int. J. Addict.* **14:**503–511.

Ball, J. C., Shaffer, J. W., and Nurco, D. N., 1983, The day to day criminality of heroin addicts in Baltimore—a study in the continuity of offense rates, *Drug Alcohol Depend.* **12:**119–142.

Beck, A. T., and Beck, A. W., 1972, Screening depressed patients in family practice, *Postgrad. Med.* **52:**81–85.

Bigelow, B. E., Stitzer, M. D., and Leibsan, I., 1984, The role of behavioral contingency management in drug abuse treatment, in: *Behavioral Intervention Techniques in Drug Abuse Treatment* (J. Grabowski, M. L. Stitzer, J. Henningfield, eds), pp. 36–52, NIDA Research Monograph Series No. 46, Rockville, Maryland.

Ciccone, P. E., O'Brien, C. P., and Khatami, M., 1980, Psychotropic agents in opiate addiction: A brief review, *Int. J. Addict.* **15**(4):449–513.

Cohen, M., Hanbury, R., and Stimmel, B., 1978, Abuse of amitriptyline, *J. Am. Med. Assoc.* **240:**1372–1373.

Connett, G., 1980, Comparison of progress of patients with professional and paraprofessional counselors in a methadone maintenance program, *Int. J. Addict.* **15:**585–589.

Cooper, J. R., Alterman, F., Brown, B. J., and Czechowicz, D. (eds.), 1983, *Research on the Treatment of Narcotic Addiction,* U.S. Government Printing Office, Rockville, Maryland.

Derogatis, L. R., 1970, Dimensions of outpatient neurotic pathology: Comparison of a clinical vs. an empirical assessment, *J. Consult. Clin. Psychol.* **34**(2):164–171.

Dolan, M. P., Black, J. L., Penk, W. E., Robinowitz, R., and Deford, H. A., 1985, Contracting for treatment termination to reduce drug use among methadone maintenance treatment failures, *J. Consult. Clin. Psychol.* **53**(4):549–551.

Dole, V. P., and Nyswander, M. E., 1965, A medical treatment for diacetylmorphine (heroin) addiction, *J. Am. Med. Assoc.* **193**(8):646–650.

Dole, V. P., Robinson, J. W., Orraca, J., Towns, E., Seargy, P., and Caine, E., 1969, Methadone treatment of randomly selected criminal addicts, *N. Engl. J. Med.* **280**:1372–1375.

Ginzberg, H., 1983, Use of clonidine or lofexidine to detoxify from methadone maintenance or other opioid dependencies, in: *Research on the Treatment of Narcotic Addiction* (J. R. Cooper, F. Altman, B. J. Brown, and D. Czechowicz, eds., pp. 174–224, U.S. Government Printing Office, Rockville, Maryland.

Gold, M. S., Redmond, D. E., and Kleber, H. D., 1978, Clonidine blocks acute opiate withdrawal symptoms, *Lancet* **2**:599–602.

Goldstein, A., 1976, Heroin addiction, sequential treatment employing pharmacological supports, *Arch. Gen. Psychiatry* **33**:353–358.

Grabowski, J., Stitzer, M., and Henningfield, J. E. (eds.), 1984, *Behavioral Intervention Techniques in Drug Abuse Treatment,* NIDA Research Monograph Series No. 46, U.S. Government Printing Office, Rockville, Maryland.

Hunt, D., Lipton, D. S., Goldsmith, D. S., and Strug, D. L., 1984, Problems in methadone treatment: The influence of reference groups, in: *Behavioral Intervention Techniques in Drug Abuse Treatment* (J. Grabowski, M. L. Stitzer, and J. E. Henningfield, eds.), pp. 8–22, NIDA Research Monograph Series No. 46, U.S. Government Printing Office, Rockville, Maryland.

Kaltenbach, K., and Finnegan, L. P., 1984, Developmental outcome of children born to methadone maintained women: A review of longitudinal studies, *Neurobehav. Toxicol. Teratol.* **6**:271–275.

Kleber, H. D., 1977, Detoxification from methadone maintenance: The state of the art, *Int. J. Addict.* **12**:807–820.

Kleber, H. D., 1983, Concomitant use of methadone with other psychoactive drugs in the treatment of opiate addicts with other *DSM-III* diagnoses, in: *Research on the Treatment of Narcotic Addiction* (J. R. Cooper, F. Altman, G. S. Brown, and D. Czechowicz, eds.), pp. 119–149, U.S. Government Printing Office, Rockville, Maryland.

Kleber, H. D., Weissman, M., Rounsaville, B., Wilber, C., and Prusoff, B., 1983, Imipramine as treatment for depression in methadone treated addicts, *Arch. Gen. Psychiatry* **40**:649–653.

LaRosa, J. C., Lipsius, J. H., and LaRosa, J. H., 1974, Experiences with a combination of group therapy and methadone maintenance in the treatment of heroin addiction, *Int. J. Addict.* **9**:605–617.

Liebson, I., Bigelow, G., and Flamer, R., 1973, Alcoholism among methadone patients: A specific treatment method, *Am. J. Psychiatry* **130**:483–485.

Ling, W., Klett, J., and Gillis, R., 1980, A cooperative clinical study of methadyl acetate, *Arch. Gen. Psychiatry* **37**:908–911.

Ling, W., Weiss, D. G., and Charuvastra, C. V., 1983, Use of disulfiram for alcoholics in methadone maintenance programs, *Arch. Gen. Psychiatry* **40**:851–854.

Luborsky, L., McLellan, A. T., Woody, G. E., O'Brien, C. P., and Auerbach, A., 1985, Therapist success and its determinants, *Arch. Gen. Psychiatry* **42**:602–611.

McBride, D. C., Westie, K. S., and Goldstein, B. J., 1982, The alleviation of depression in a population of narcotics users, in: *Final Report to the National Institute of Drug Abuse.*

McGlothlin, W. H., and Anglin, D., 1981, Long-term follow-up of clients of high- and low-dose methadone programs, *Arch. Gen. Psychiatry* **38**:1055–1063.

McLellan, A. T., Luborsky, L., Woody, G. E., and O'Brien, C. P., 1980a, Improved diagnostic instrument for substance abuse patient: The addiction severity index, *J. Nerv. Ment. Dis.* **168:**26–33.

McLellan, A. T., Druley, K. A., O'Brien, C. P., and Kron, R., 1980b, Matching substance abuse patients to appropriate treatments, *Drug Alcohol Depend.* **3:**189–195.

McLellan, A. T., Luborsky, L., O'Brien, C. P., Woody, G. E., and Druley, K. A., 1982, Is treatment for substance abuse effective? *J. Am. Med. Assoc.* **247:**1423–1428.

McLellan, A. T., Luborsky, L., Woody, G. E., Druley, K. A., and O'Brien, C. P., 1983, Predicting response to alcohol and drug abuse treatments: Role of psychiatric severity, *Arch. Gen. Psychiatry* **40:**620–625.

McLellan, A. T., Childress, A. R., Griffith, J., and Woody, G. E., 1984, The psychiatrically severe drug abuse patient: Methadone maintenance or therapeutic community? *Am. J. Drug Alcohol Abuse* **10:**77–95.

Newman, R. G., and Whitehill, W. B., 1979, Double-blind comparison of methadone and placebo maintenance treatments of narcotic addicts in Hong Kong, *Lancet* **8141:**485–488.

Resnick, R. B., Washton, A. M., Stone-Washton, N., *et al.,* Psychotherapy and naltrexone in opioid dependence, in: *Problems of Drug Dependence, 1980,* (L. S. Harris, ed.), pp. 109–115, NIDA research monograph 34, DHHS publication (ADM) 81–1058. U.S. Dept. of Health and Human Services, National Institute on Drug Abuse.

Rounsaville, B. J., and Kleber, H. D., 1985, Untreated opiate addicts: How do they differ from those seeking treatment? *Arch. Gen. Psychiatry* **42:**1072–1077.

Rounsaville, B. J., Weissman, M. M., Crits-Christoph, K., Wilber, C., and Kleber, H. D., 1982a, Diagnosis and symptoms of depression in opiate addicts, *Arch. Gen. Psychiatry* **39:**151–156.

Rounsaville, B. J., Weissman, M. M., Kleber, H. D., and Wilber, C. H., 1982b, The heterogeneity of psychiatric diagnosis in treated opiate addicts, *Arch. Gen. Psychiatry* **39:**161–166.

Rounsaville, B. J., Glazer, W., Wilber, C. H., Weissman, M. M., and Kleber, H. D., 1983, Short-term interpersonal psychotherapy in methadone-maintained opiate addicts, *Arch. Gen. Psychiatry* **40:**630–636.

Sells, S. B., 1979, Treatment effectiveness, in: *Handbook on Drug Abuse* (R. L. Dupont, A. Goldstein, and J. O'Donnell, eds.), pp. 105–118, National Institute on Drug Abuse.

Senay, S. C., Dorus, W., Goldberg, F., and Thornton, W., 1977, Withdrawal from methadone maintenance. Rate of withdrawal and expectation, *Arch. Gen Psychiatry* **34:**361–367.

Stanton, M. D., Todd, T., (eds.), 1982 *The Family Therapy of Drug Abuse and Addiction,* Guilford Press, New York.

Stimmel, B., Goldberg, J., Rotkopf, E., and Cohen, M., 1977, Ability to remain abstinent after methadone detoxification, *J. Am. Med. Assoc.* **237:**1216–1220.

Tennant, F. S., 1974, Propoxyphene napsylate (Darvon-N) treatment of heroin addicts, *J. Natl. Med. Assoc.* **66:**23–24.

Titievsky, J., Guillermo, S., and Barranco, M., 1982, Doxepin as adjunctive for depressed methadone maintenance patients: A double-blind study, *J. Clin. Psychiatry* **39:**151–156.

Wapner, R. J., and Finnegan, L. P., 1981, Perinatal aspects of psychotropic drug abuse, *Perinatal Med.* **20:**384–417.

Woody, G. E., O'Brien, C. P., and Rickels, K., 1975, Depression and anxiety in heroin addicts: A placebo-controlled study of doxepin in combination with methadone, *Am. J. Psychiatry* **132:**4,447–450.

Woody, G. E., Mintz, J., Tennant, F., O'Brien, C. P., McLellan, A. T., and Marcovici, M., 1981, Propoxyphene for maintenance treatment, *Arch. Gen. Psychiatry* **38:**898–900.

Woody, G. E., O'Brien, C. P., McLellan, A. T., Marcovici, M., and Evans, B., 1982, The use of anti-depressants with methadone in depressed maintenance patients, *Ann. N.Y. Acad. Sci.,* **1982:**120–127.

Woody, G. E., Luborsky, L., McLellan, A. T., O'Brien, C. P., Beck, A. T., Blaine, J., Hermer, I., and Hole, A., 1983, Psychotherapy for opiate addicts. Does it help? *Arch. Gen. Psychiatry* **40**:639–645.

Woody, G. E., McLellan, A. T., Luborsky, L., O'Brien, C. P., 1984, Severity of psychiatric symptoms of a predictor of benefits from psychotherapy: The Veterans Administration-Penn Study, *Am. J. Psychiatry* **141**:1172–1177.

# 9

# Alcohol and Opiate Dependence
## Re-evaluation of the Victorian Perspective

SHEPARD SIEGEL

## 1.  INTRODUCTION

> One of the great weaknesses of the discipline that concerns itself with problems of substance abuse is that there is a strong tendency to ignore both the cross-cultural experience and the historical data that are available (Bourne, 1982, p. xvii).

In terms of mortality and morbidity and social and financial costs, alcohol abuse is clearly a major medical problem in our society today (e.g., Popham et al., 1984). There are considerable methodological problems in studying alcohol treatment efficacy, but at least some observers have concluded that the overall efficacy of treatment for alcoholism is not great (see Smith, 1981).

In the last century, one approach to intractable alcoholism was to transform the alcohol dependency into an opiate dependency. Although such a strategy would seem outrageous today, recent developments suggest that we can benefit by a review of the Victorian perspective on alcoholism and its relationship to opiate use.

## 2.  ALCOHOL AND OPIATES IN THE 19th CENTURY

> The High Victorians, in fact, were often high (MacInnes, 1967, p. 67).

There was virtually no restrictive legislation concerning either opium or alcohol during most of the 19th century in Great Britain or America. Although

---

SHEPARD SIEGEL ● Department of Psychology, McMaster University, Hamilton, Ontario, L8S 4K1, Canada.

both substances were freely available, and widely used, concerns about their deleterious effects on health were expressed by many people throughout the century (see Berridge and Edwards, 1981; Courtright, 1982; Morgan, 1981). Most authorities agreed that although restraint in the use of both substances was desirable, excessive use of opium was a lesser evil than excessive use of alcohol. For example, when a paper was presented to the Royal Medical and Chirurgical Society in 1840 advocating that the Society pass a resolution discouraging excessive opium eating, there was little enthusiasm for the adoption of the resolution. The report of the Society's meeting indicates that Queen Victoria's own physician, Sir Benjamin Brodie, argued that the resolution might accomplish more harm than good, as opium eating was not as life-threatening as gin drinking: "if he had to choose between the two, he should prefer the use of opium" (Jeffreys, 1840, p. 383).

The alleged relative detrimental effects of alcohol and opium had important public policy implications. In the 19th century, there was concern in England about the effects of opiate use in colonial India. In 1893, a committee of the British Medical Assocation sent questionnaires concerning native opiate use to over 100 British physicians in India. The near unanimous response was that curtailment of native opium use would be followed by a compensatory increase in alcohol consumption, and that such a change would be for the worse (see review by Peters, 1981). The committee's report was influential in Britain's continuing policy of support of India's opium cultivation and her continued enforcement (by way of the Opium Wars) of the profitable commerce in India's opium with China. Indeed, to some extent the British justified the smoking of opium in China on similar grounds: "Should the Chinese abandon opium they will infallibly fall back on their old, but now happily vanquished enemy, alcohol, when their last state will be worse than their first" (Davenport, 1904, p. 28). As recently summarized by Morgan (1981), "late nineteenth-century experts generally doubted that opium damaged the body as much as alcohol did. Nor did it seem to cause organic changes that might be passed on to offspring. Opium seemed more respectable than alcohol in its observable effects on individuals" (p. 89). Among the reasons for opium eating revealed in a 1873 survey of all physicians in Massachusetts was "the superior 'gentility' of opium to alcohol as a stimulant" (Anonymous, 1873, p. 330). Indeed, "the hypodermic kit was preferable to the hip flask or decanter in many respectable homes" (Morgan, 1981, p. 21): The 1897 Sears Roebuck catalogue featured many hypodermic kits, but none of the paraphernalia of alcohol use, such as hip flasks, wine glasses, or decanters (Israel, 1968). Throughout the 19th century, the temperance movement was much stronger than the antiopium lobby (see Kissin, 1982).

The perceived safety of opiate abuse, relative to alcohol abuse, was further illustrated by the actions of life insurance companies. Actuarial science—differential life insurance premiums for populations with different health risks—

started in the beginning of the 19th century. With respect to intemperance, concern focused on the drinker rather than the opium user (Berridge, 1977).


## 3. RECIPROCITY BETWEEN ALCOHOL AND OPIUM IN THE 19th CENTURY

> In Sears Roebuck mail order catalogues at the turn of the century two pages were devoted to drug therapies for morphine addiction and alcoholism respectively. The drug being sold for morphine addiction consisted mainly of alcohol; a good part of the drug for alcoholism consisted of tincture of opium, a relative of morphine (Goodwin, 1981, p. 97).

As illustrated by the previously presented comparisons between opium and alcohol, the 19th century physician perceived some reciprocity between the use of the two substances. Use of opium was associated with abstinence (or at least moderation) with respect to alcohol, and vice versa. A correspondent to the *Lancet* in 1851 asserted that opium addiction and alcoholism constitute two classes of the disease of intoxication, "narcosis and alcoholsis," with the two intoxicants being "antagonistic or curative of each other" (Pidduck, 1851, p. 251). In fact, there was considerable interest in the use of each chemical as a cure for dependence on the other.


## The Alcohol Treatment of Opiate Addiction

> . . . for habitual narcotism, nothing in the way of remedy is so efficacious as alcohol (Pidduck, 1851, p. 251).

Alcohol was occasionally recommended as a treatment for opium addiction by some Victorian physicians (e.g., Pidduck, 1851; Wilson, 1889). However, since alcohol was considered to be more dangerous than opium, this treatment had a potential disadvantage; the opiate addict might simply shift his or her dependence to the more pernicious alcohol. This iatrogenic complication of the alcohol treatment was recognized. A note in the *Lancet* in 1876 described such cases: "The (morphine-addicted) wife of a medical man who saw in a book on Materia Medica that alcohol was considered an antidote for morphia became a confirmed drunkard. The author observed four similar cases" (Anonymous, 1876, p. 330). The alcohol treatment must be "judicious" (Wilson, 1889, p. 569) and "used sparingly" (Anonymous, 1876, p. 331): "The danger of replacing a vicious habit of morphine taking by the more brutalizing vice of chronic alcoholism must be ever held in view" (Wilson, 1889, p. 568). Mattison cautioned that "when the nervous system is losing the loved morphia impression, it will take kindly to alcohol . . . let the physician beware lest, in the effort to aid his patient in escaping one peril, he but involves him in another yet greater" (Mattison,

1885, pp. 37–38) . . . "The shores of the post-poppy land is strewn with the wrecks of those who, after escape from narcotic peril, have taken to rum" (Mattison, 1902, p. 29).

## The Opiate Treatment of Alcoholism

> . . . if it be desirable, as it is, that a loved one may be preserved as long as possible to his family circle, with all his faults—and who is there without them—the preference on the score of health and length of days should by all means be given to morphine in lieu of alcohol (Black, 1889, p. 539).

More common than the prescription of alcohol for excessive opiate use was the prescription of opiates for excessive alcohol use—especially for the treatment of alcohol withdrawal symptoms (Crothers, 1902; French, 1903; Parrish, 1867; Pereira, 1842). For example, Cobbe (1895) noted:

> It is now a fad among physicians to inject morphine; so that when a business or professional man awakens in the morning overcome of the [alcohol] excesses of the previous night and finds himself incapable of reaching his office, the physician who is sent for knows that a morphine tablet will restore the equilibrium for a little while and the syringe is put in requisition (p. 173).

It was reported that, in such cases, sufferers often found that the cure not only alleviated the symptoms of alcohol withdrawal, but also the tendency to drink too much. The patient often switched his chemical dependence: "When these [alcoholic spirits] are taken in excess, morphin [*sic*] is given to lessen the violence of the symptoms. The spirits are then abandoned and morphin taken up" (Crothers, 1902, p. 94–95). Habitual opiate use initiated in this manner was mentioned in many early turn-of-the-century medical texts and journal articles (see review by Terry and Pellens, 1928).

It should be emphasized that although opiates may have been preferred to alcohol, and recommended in cases of excessive alcohol use, it is not the case that opium was perceived as innocuous. There were many people in the 19th century who warned of the dangers of excessive opiate use, but it *is* true that the near-unanimous opinion was that if a choice of abused substances had to be made, opiates were preferred (see Morgan, 1981, pp. 89–90). For example, Macnish (1859) noted, "in attempting to cure the habit of drunkeness opium may sometimes be used with advantage" (p. 154). Although the opium treatment might transform the alcoholic into an opium user, this would be an improvement: "Of two evils, however, we should always choose the least: and it is certain, that however perniciously opium may act on the system, its moral effects and its power of injuring reputation are decidedly less formidable than those of ordinary intoxicating agents" (Macnish, 1859, p. 154). A similar point of view prevailed in America: the Proceedings of the 1878 meeting of the Richmond (Virginia) Academy of Medicine (Brittan, 1878) indicate that although there was concern about the opium habit, there was an appreciation of its value if it supplanted alcohol abuse. Dr. J. B. McCaw described to the meeting a case in which "he would recommend this substitution of one vice for another":

> For instance, in the case of a gentlemen acquaintance who was so addicted to his wines, etc., that he was fast becoming a common sot, disgracing self and family during his alcoholic intoxications, Dr. McCaw advised that if the man could not resist the desire for a stimulant, he should take to the opium habit as preferable to the alcoholic. The result was, that his patient did leave off alcoholic stimulants, become an opium-eater, resumed his social position, and his family became content to let him take the opium as long as he would not touch wine or whiskey, etc. This satisfaction with opium continued for sixteen years, and he restored himself to the confidence and affections of his acquaintances" (Brittan, 1878, pp. 882–883).

Support for opiate maintenance as a treatment for alcoholism accumulated. Eleven years after Dr. McCaw's observations in Virginia, Dr. J. R. Black of Ohio presented a summary (Black, 1889). Black prefaced his arguments for the morphine-substitution treatment by noting that alcoholism is very refractory to all usual treatments: "The writer may be allowed to be justified in his skepticism of them all when he states that after forty years of superior means for observation he has never met with a single instance of permanent reformation among confirmed drinkers" (Black, 1889, p. 537). Although instances of treatment success may be offered from time to time, "it is found that the reformation amounts only to this, that they have simply substituted opium stimulation for the alcoholic" (Black, 1889, p. 537). Moreover, this substitution has beneficial consequences. Black echoed others of the era in indicating that although opiate dependence may not be healthy, it is to be much preferred to alcohol dependence: "as a whole, the use of morphine in place of alcohol is but a choice of evils, and by far the lesser" (Black, 1889, p. 538). Black went on to discuss the reasons why morphine was to be preferred to alcohol as a drug of abuse. Morphine, unlike alcohol, does not lead to disreputable appearance and behavior: "There is no strong odor of the breath, no swollen red and turgid face, no staggering gait, no violent language and conduct in the case of the morphine eater" (Black, 1889, p. 538). In addition, alcohol "rarely fails to arouse the combative, destructive and unfeeling brutality in a man's nature, while morphine seldom shows any effects of the kind" (Blac, 1889, p. 538). Black noted that alcohol damages the health more than morphine—delerium tremens, liver and kidney cirrhosis, and vascular problems result from alcohol, but not morphine. He summarized the case for morphine substitution of alcohol as follows:

> On the score, then, of a saving to the individual and his family in immediate outlay and of incurred disability, on the great diminution of peace disturbers and of crime, whereby an immense outlay will be saved by the State; on the score of decency in behavior instead of perverse deviltry, of bland courtesy instead of vicious combativeness, of quiet geniality instead of cruel destructiveness; on the score of a lessened liability to fearful diseases and the lessened propagation of pathologically inclined

blood, I would urge the substitution of morphine instead of alcohol for all to whom such a craving is an incurable propensity. In this way I have been able to bring peacefulness and quiet to many disturbed and distracted homes, to keep the head of the family out of the gutter and out of the lock-up, to keep him from scandalous misbehavior and neglect of his affairs, to keep him from the verges and actualities of delirium tremens horrors, and, above all, to save him from committing, as I veritably believe, some terrible crime that would cast a lasting and deep shadow upon an innocent and worthy family circle for generation after generation (Black, 1889, p. 540).

Black noted that his prescription would benefit society, as well as the individual addict: "The mayors and the police courts would almost languish for lack of business; the criminal dockets, with their attendant legal functionaries, would also have much less to do than they have now—to the profit and well-being of the community" (p. 541).

The *Cincinnati Lancet-Clinic,* where Black presented his recommendation for the morphine treatment of alcoholism, published comments from readers concerning articles appearing in the journal. Examination of every issue of the *Cincinnati Lancet-Clinic* for the two years following the publication of Black's paper reveals no correspondence concerning the paper. An implication of this is that the recommendation that some alcoholics would profit by opiate maintenance was, less than 100 years ago, simply not contentious.


## 4.   TEMPERANCE AND ANTIOPIATE LEGISLATION

. . . the habit of chewing opium is alleged to have become very prevalent in the British Islands and in the United States, especially since the use of alcoholic drinks has been to so great an extent abandoned by certain classes of people under the influence of the fashion introduced by total abstinence societies (Stillé, 1868, p. 704)

The perception of the relative health-damaging properties of the two chemicals reversed by the early 20th century. Our current cultural and legislative treatments of the recreational use of opiates and alcohol have resulted from a variety of interrelated social and political forces (see Berridge and Edwards, 1981; Courtright, 1982; King, 1972; Morgan, 1981; Musto, 1973). One factor was the growth of the temperance movement throughout the 19th century. As would be expected from an alcohol–opiate reciprocity perspective, an increase in opium use concomitant with alcohol rejection was perceived. A prevalent view was that many who had "taken the pledge" merely shifted their chemical dependence from alcohol to opium. For example, Pereira (1842) claimed that individuals who no longer drink wine "in some fit of repentance," or because "the preservation of their character forbits them the use of wine" (p. 1746) frequently use opium instead. A correspondent to the *Lancet* in 1851 noted, "since the crusade of the teetotalers against spirit-drinking there is great reason to believe that the practice of taking opium is on the increase" ("Medicus," 1851, p. 169). Some years later, the opioidogenic effect of alcohol prohibition was more succintly noted: "the poor drunkards seek a substitute for the alcohol from which they have been forcibly separated, and they think they find it in the extract of poppy" (Cobbe, 1895, p. 173). To counter the growing political strength of American temperance organizations during the early part of this century, anti-prohibition groups argued that alcohol prohibition would result in a great increase in opiate use (see Bellis, 1981, p. 8; Musto, 1973). In fact, when the Eighteenth Amendment to the United States constitution was adopted in 1920, this drug-substitution argument was used to justify increasingly stringent antinarcotic legislation and enforcement—a legacy that persists to this day: "In this sense prohibition, which divided the nation, unified the public in a condemnation of narcotic abuse and maintenance" (Musto, 1973, p. 157; for a similar view from a Canadian perspective, see Murphy, 1922, p. 61).


## 5.  RECIPROCITY BETWEEN ALCOHOL AND OPIATES IN THE 20th CENTURY

> Searching the contemporary drug literature offers rather the same pleasures and excitements as browsing in an antique shop where there is the chance of picking up a good piece of Victoriana (Berridge and Edwards, 1981, p. 248).

Victorian beliefs about drug effects were based on unsystematic observation and introspection, rather than rigorous clinical and experimental research. Although some of the earlier observers of the drug scene displayed a seeming prescience with respect to later developments in the science of pharmacology (Siegel, 1982a, 1983), others merely affirmed unscientific Victorian prejudices (Berridge and Edwards, 1981). It is of interest to evaluate recent epidemiological and experimental research relevant to the 19th century belief of reciprocity between alcohol and morphine. In fact, the results of the 20th century research often substantiate the more casual observations of the previous century. For example, the Victorian belief that a unitary "disease of intoxication" (e.g., Pidduck, 1851) was expressed by excessive use of either alcohol or opiates was supported by Cohen's (1976) analyses of responses to questions concerning alcohol and drug use by alcoholics and opiate addicts, respectively. The responses by abusers of both chemicals were very similar: "that is, addicts use drugs for the same reasons that alcoholics use alcohol" (Cohen, 1976, p. 610). Alcoholics and opiate addicts also display very similar profiles in a variety of psychometric tests, such as the Minnesota Multiphasic Personality Inventory (see reviews by Carroll, 1982; Gottheil et al., 1982).

As discussed previously, the Victorian practitioner's belief concerning the interchangeability of alcohol and opiates had implications for the way in which individuals dependent on one of the substances were treated. The logic of these recommendations may be assessed with the benefit of hindsight provided by the extensive, 20th century literature concerning alcohol–opiate interactions.

## Alcohol in the Treatment of Opiate Addiction

> On two occasions she successfully withdrew herself from opium using a Lao folk remedy consisting of herbal medications and rice whiskey . . . She found that drinking four glasses of rice whiskey a day helped, even without the herbs (Westermeyer, 1983, p. 435).

The occasional 19th century prescription of alcohol for the treatment of opiate addiction has been reiterated more recently. In the middle of this century, Child et al. (1950) reported a "spectacular" (p. 23) success in an opiate addict treated with intravenous alcohol—a finding subsequently replicated (Child, 1951). Contemporary opium addicts in Laos who try to stop their use of the drug, sometimes find alcohol a useful substitute (Westermeyer, 1982, 1983). Many North American opiate addicts also discover the value of alcohol as a substitute drug (Gold et al., 1984).

Recent research with animals has indicated that morphine withdrawal symptoms in rats and mice are ameliorated by alcohol (Blum et al., 1976; Ho et al., 1979; Ho, 1980*b;* Pack and Ferko, 1977); that rats undergoing morphine withdrawal will preferentially consume alcohol (Ho, 1980*b;* Ho et al., 1976), and that rats that self-administer morphine intravenously will continue to self-administer if the chemical reinforcement is changed to alcohol (Smith et al., 1975). (However, no evidence of alcohol attenuation of morphine withdrawal [Jones and Spratto, 1977] or oral consumption of a morphine solution [Gelfand and Amit, 1976] has also been reported.)

In the 19th century, as discussed previously, the danger of transforming an opiate addict into an alcoholic was recognized. More recently, some investigators have reiterated this concern; opiate addicts undergoing treatment may discover the value of alcohol as a substitute, and thus simply transfer their dependence from opiates to alcohol (A. Cohen, 1976; S. Cohen, 1982; Green and Jaffe, 1977; Waldorf and Biernack, 1979).

## Opiate Use Initiated to Relieve Alcohol Withdrawal Distress

> I noticed then that if I took a hypo [of morphine] the craving for liquor would leave and with it that terrible nervous feeling. I thought I had discovered a good way to sober up and to stop the craving for alcohol (Van Slyke, 1945, p. 29).

The frequent 19th century observation that habitual opiate use often started when the drug was used to ameliorate the distress of alcohol withdrawal was substantiated in a study by Lawrence Kolb, published in 1925, of 230 drug addicts (in morphine maintenance clinics, prisons, hospitals, and "in good social standing in various parts of the United States" [Kolb, 1925, p. 300]): "Early in the study it was discovered that some patients had fallen victims to opium as a result of using it or having it prescribed for them in the course of treatment for [alcohol] sprees. The usual history was that the physician prescribed morphine during several sprees, until the patient found out about it and thereafter treated himself by the same remedy until addicted" (Kolb, 1925, p. 304). A subsequent study by Kolb of a different population of addicts, all of whom first received opiates for medical reasons (Kolb, 1928), similarly indicated that almost half the male addicts were heavy drinkers prior to opiate use. This route of opium addiction was further confirmed in Pescor's (1943) analysis of the clinical records of 1036 patients admitted to the narcotic treatment facility at Lexington Kentucky during a 1-year period shortly after it opened (July 1, 1936 to June 30, 1937). These patients frequently indicated that they initially self-administered opiates to relieve alcohol withdrawal symptoms: "These 'shots' are taken closer and closer together until finally the drug displaces the alcohol and the patient becomes a drug addict" (Pescor, 1943, p. 3). Similar results were reported by Westling (1958) in a study of Finnish drug addicts treated between 1930 and 1954, and by O'Donnell (1969) in a study of a sample of patients admitted to the Lexington hospital during the first 25 years of its existence (1935–1959). In addition, several autobiographical accounts of repentant, early 20th century drug addicts describe their initiation to opiate use when they suffered severe alcohol withdrawal symptoms, with the treatment recommended by a knowledgeable addict (e.g., MacMartin, 1921, pp. 32–38) or physician (e.g., Van Slyke, 1945, pp. 28–29). As recently as 1955, it was noted that physicians in Finland frequently prescribed morphine for "controlling states of mental unrest during inebriation, particularly during its late stages" (Eerola et al., 1955, p. 253).

## Opiate-Induced Decrease in Alcohol Use—Studies with Humans

Opium and alcohol are mortal enemies (Cocteau, 1958, p. 26).

Victorian physicians frequently noted that the alcoholic, once introduced to opiates as a treatment for alcohol withdrawal, would often abandon excessive alcohol use in favor of the drug. Many more recent retrospective reports of alcohol use by drug addicts have substantiated this observation. In Kolb's (1928) study of opiate addicts introduced to the drug by medical prescription or self-medication for disease, he noted an inverse relationship between the use of alcohol and opiates in men. Thirty-three of the 45 male drug addicts had been heavy drinkers before they became narcotic addicts, but only one of these continued to drink (albeit moderately) when regularly using morphine. When these men abstained from opiates, they resumed drinking. Many, many subsequent clinical and epidemiological studies have confirmed Kolb's (1928) findings concerning the high degree of reciprocity between alcohol and opiate use (in chronological order: Haggard and Jellinek, 1942, p. 160; Pescor, 1943; Isbell et al., 1955; Stevenson et al., 1956; Westling, 1958; O'Donnell, 1969; Willis, 1969; Weppner and Agar, 1971; Brown et al., 1973; Schut et al., 1973; Waldorf, 1973; Goodwin et al., 1975; McGlothlin et al., 1977; Barr and Cohen, 1979). Some of these reports are especially compelling. In an intensive study of 100 consecutive admissions of heroin addicts to a British Columbia prison farm, the use of alcohol was evaluated:

> Perhaps the most remarkable relationship between alcohol and heroin is the addict's nearly complete rejection of alcohol after he has accepted heroin addiction. Of 100 consecutively convicted addicts, 63 stated their use of alcohol was decreased after they started to use heroin. Alcohol ceased to have any real attraction for them, and to some of them, the idea of an alcoholic beverage was repulsive. The majority completely discontinued the use of alcohol while using heroin (Stevenson et al., 1956, p. 83).

The respondents' comments characterize their attitude to alcohol while using heroin: "I don't want alcohol when on drugs"; "I just don't drink any more. No use for drinking"; "Alcohol has no attraction for me"; "Alcohol makes me sick, I don't want it" (Stevenson et al., 1956, p. 84).

As part of O'Donnell's follow-up study of admissions to the Lexington, Kentucky drug addiction treatment facility during the years 1935–1959, he made a detailed analysis of the relationship between narcotic and alcohol use (O'-Donnell, 1969, pp. 135–142): "Case material makes it clear that almost all subjects preferred the effects of narcotics to those of alcohol" (p. 137); "The usual attitude expressed was not that the addict had little or no desire for alcohol when narcotics were available, but that he positively avoided using alcohol" (p. 138).

Brown et al. (1973) questioned matched groups of heroin addicts and non-addict alcoholics about their histories of alcohol use. They reported that the heroin addicts, prior to their use of heroin, drank *more* alcohol than the alcoholic respondents; however, once the addicts used heroin, their consumption of alcohol decreased markedly.

Goodwin et al. (1975) assessed drinking and drug use by a sample of United States Army Vietnam veterans. Alcoholism was a serious problem with these men both before and after service in Vietnam. While in Vietnam, however, where both heroin and alcohol were freely available, these same men tended to choose the opiate and refrain from excessive alcohol use.

In summary, the available data obtained by questioning opiate addicts suggest that the drug decreases alcohol consumption. Indeed, it has often been noted that the effects of alcohol are actually unpleasant while the individual is addicted to opiates.

## Opiate-Induced Decrease in Alcohol Use—Studies with Animals

> Together these studies show that morphine, given by any of several routes, strongly suppresses alcohol consumption for over a week after its administration in at least two strains of rats, under a variety of experimental conditions (Sinclair et al., 1973a, p. 426).

Results of recent retrospective reports are congenial with the frequent 19th century assertion that opiate use serves to decrease alcohol use. Of course, being correlational, the 20th century epidemiological findings do not prove that assertion. As might be expected, the experimental data necessary to constitute such proof are not available with humans. Such experimentation has, however, been performed with animals. There are now several demonstrations that voluntary alcohol consumption, in a variety of species, is substantially attenuated by opiate administration.

The first such demonstration was that of Sinclair et al. (1973a). In a series of experiments, rats were given free access to food, water, and alcohol solution (10%, v/v) for a prolonged period of time (ranging from about 140–200 days). In these circumstances the rats eventually drank approximately 16 ml of alcohol solution per day. Following injection of a large dose of morphine (60 mg/kg), ethanol consumption fell by about 80%, and remained suppressed from preinjection levels for about 1 week. This suppression of alcohol intake was not merely because the opiate so narcotized the subjects that they were generally behaviorally inactive; concomitant intake of water was not reduced, and food intake was depressed to a lesser degree and for a shorter period than alcohol. In one experiment, the specificity of the morphine-induced suppression of alcohol ingestion was assessed. Rats were given access to a saccharin solution 10 days prior to morphine administration (alcohol, water, and food were present throughout the experiment). Although morphine again profoundly affected alcohol consumption, it did not affect consumption of the distinctively flavored saccharin solution. Finally, Sinclair et al. (1973a) demonstrated that morphine administered to alcohol-experienced rats during a period of alcohol abstinence eliminated the increase in consumption typically observed when alcohol is again made available (the "alcohol-deprivation effect," Sinclair and Senter, 1968; Sinclair et al., 1973b). Results of a subsequent experiment from the same laboratory (Sinclair, 1974) extended the original findings. A single morphine injection (30 mg/kg) effectively suppressed alcohol intake in rats without very extensive prior experience with alcohol ingestion (as little as one day).

In another evaluation of the effect of morphine on alcohol consumption, Ross et al. (1976) used hamsters as subjects; hamsters, in contrast with many other animals, normally prefer alcohol to water (Arvola and Forsander, 1961). Ross et al. (1976) found that this preference was reversed for 24 hr by a single injection of 7.5 mg/kg morphine, attenuated by the same dose of the analgesically active $(-)$ isomer, levorphanol (and unaffected by the inactive $(+)$ isomer, dextrorphan). Ross et al. (1976) also reported that hamsters' ingestion of alcohol is *enhanced* if they are injected with the opiate antagonist, naltrexone (having never been injected with an agonist).

The ability of opiates to suppress voluntary ethanol consumption has been demonstrated in mice, as well as confirmed in rats (Ho, 1980*a;* Ho et al., 1976). Moreover, the finding of Sinclair et al. (1973*a*) that the effects of morphine are specific to ethanol consumption has been confirmed by Ho (1980*b*). He reported that in mice, although morphine suppressed alcohol consumption, it does not suppress sucrose or quinine consumption. Recently, Ho (1982) reported that an endogenous opiate (met-enkaphalin), as well as morphine, decreases voluntary ethanol consumption in rats (with the decrease induced by both met-enkephalin and morphine being reversible by naltrexone).

In summary, the experimental data concerning the ability of opiates to suppress voluntary alcohol consumption in rats, hamsters, and mice are remarkably parallel to the epidemiological data suggesting that opiates suppress alcohol drinking in humans. The animal data indicate that opiates do *not* merely reduce the consumption of all fluids; rather, the drug's effect appears to be specific to alcohol. The results of recent animal experiments, then, further attest to the validity of the Victorian practitioner's belief that opiates reduce alcohol ingestion.


## 6.  PRIMING AND ALCOHOL–OPIATE INTERACTIONS

> . . . once he takes any alcohol into his system, something happens, both in the bodily and mental sense, which makes it virtually impossible for him to stop. The experience of any alcoholic will confirm that . . . We are without defense against the first drink (Anonymous, 1939, pp. 34–35).

The use of opiates to reduce avidity for alcohol is relevant to the alcoholic's use of alcohol to satisfy his or her craving for alcohol; following sufficient alcohol consumption, this craving is satiated. In this sense, the alcohol–opiate reciprocity view maintains that opiates substitute for alcohol in satisfying the alcohol craving.

With respect to alcohol-induced alterations in alcohol consumption, there appears to be a paradoxical effect of dose. A sufficiently large amount of alcohol will be followed by termination of alcohol consumption. A small amount, then, might be expected to have a small effect in reducing the desire for alcohol. However, it has frequently been reported that a small dose of alcohol will have a "priming" effect—it will augment the craving for additional alcohol and enhance subsequent alcohol consumption. This "loss of control" initiated by a priming dose is incorporated in the dogma of Alcoholics Anonymous (Anonymous, 1939). It has some empirical support (Hodgson et al., 1979) and has been subject to various theoretical analyses (Hinson and Siegel, 1980; Stewart et al., 1984). A view stressing parallels between alcohol and opiates suggests that the dose-dependent, bidirectional effect of alcohol on alcohol consumption should be paralleled by similar effects of opiate dose on alcohol consumption. That is, although opiates generally decrease alcohol consumption, a small dose might actually enhance alcohol consumption (see Reid and Hunter, 1984). It does not appear that there have been any studies, either with humans or animals, that have systematically evaluated the effect of a wide range of opiate doses on the modulation of alcohol intake. Recently, however, studies by Reid and Hunter and colleagues (Beaman et al., 1984; Reid and Hunter, 1984) have evaluated the effect of morphine on ethanol consumption in rats, using a dose of morphine usually considered to be small for this species—smaller than that used in previous studies of morphine-induced alteration in ethanol intake: 2.5 mg/kg (with 10 mg/kg being additionally evaluated by Beaman et al., 1984). In contrast with all the earlier studies, which used higher opiate doses in rats, these investigators reported that their small morphine doses increased ethanol intake. They obtained similar findings with "remarkably small doses of diprenorphine" (Hunter et al., 1984, p. 46). In this study, the nonmonotonic effect of opioids on alcohol intake was supported by the observation that the smallest doses of diprenorphine (.005 and .025 mg/kg) enhanced intake more than the largest dose evaluated (.05 mg/kg). Such findings, when taken together with others in the literature, suggest that parallels between alcohol and opiates extend to the phenomenon of primed alcohol consumption; whereas a substantial opiate dose, like a substantial alcohol dose, will decrease subsequent alcohol intake, a small dose of either may serve to increase alcohol intake.

## 7.  POSSIBLE REASONS FOR ALCOHOL–OPIATE RECIPROCITY

One may also speculate that a common pathway shared by all drugs of abuse could involve endorphin and the opiate (endorphin) receptors (Goldstein, 1977, pp. 400–401).

Various considerations suggest that opiates and ethanol are two very different drugs: "it is difficult to envisage the biochemical, physiological or metabolic pathways that complex phenanthrene-type alkaloids would have in common with a simple 2-carbon molecule" (Blum et al., 1980c, p. 371). Moreover, it seems clear that ethanol and opiates have discriminably different subjective effects. It is likely that people can tell whether they are intoxicated with alcohol or narcotized with opiates. Certainly rats can discriminate the internal state produced by parenterally administered opioids from that produced by parenterally administered ethanol; if trained such that a particular response leads to a reward only when they are under the effects of opioids, they do not make that response while under the effects of ethanol, and vice versa (Shearman and Herz, 1983; Winter, 1975).

Although the two drugs do have very different effects, there are similarities in their actions that have suggested to some investigators that they have a common mode of action: "alcohol and a wide variety of psychotropic drugs seem to share a final common pathway because both produce a feeling of euphoria and both can relieve symptoms of stress and discomfort" (Keeley and Solomon, 1981, p. 100). Such a common pathway for the two substances was proposed by Davis and Walsh in 1970. They suggested that products of alcohol metabolism, tetrahydroisoquinolines (TIQs), possess some opiatelike properties; thus some effects of alcohol, in common with opiate drugs, might be mediated by stimulation of brain opiate receptors. Although the hypothesis was initially criticized (e.g., Goldstein and Judson, 1971; Halushka and Hoffmann, 1970; Seevers, 1970), more recent evidence has provided impressive support for the contribution of TIQs and endogenous opioids to the effects of alcohol (see Blum et al., 1980c). The various hypotheses are still speculative, but, in simplified form, they suggest that both alcohol and opiates effect stimulation of central opiate receptors.

In support of an endorphinergic contribution to alcoholism, recent findings indicate that alcohol alters opiate receptor binding in living neural cells (Anokhina et al., 1983; Charness et al., 1983; Levine et al., 1983) and brain endorphin levels in rats (Schulz et al., 1980). Moreover, analyses of the cerebrospinal fluid endorphin levels in human alcoholics indicate altered endorphin homeostasis (Borg et al., 1982; Genazzani et al., 1982). Additionally, results of some studies have suggested that opiate antagonists can reverse the effects of alcohol in animals and people (for review, see Jeffcoate et al., 1981). In one dramatic clinical report (Lyon and Antony, 1982), two comatose patients, initially believed to suffer from opiate overdose, were administered naloxone. The opiate antagonist promptly aroused both patients. Toxicology tests subsequently indicated that the original diagnosis of opiate overdose was erroneous. Although there was no evidence of recent opiate use in either case, both individuals had high blood alcohol concentrations. It would appear that naloxone reversed the effects of ethanol-induced coma, much as this opiate antagonist reverses the effect of an opiate-induced coma. It should be noted that although such reports have usually been interpreted as supporting a common, endogenous opioid-mediated effect on both alcohol and opiates, other interpretations must be entertained (see Badawy and Evans, 1983; Dole et al., 1982). Moreover, some investigators have reported no evidence of antagonism of ethanol's effects by naloxone (e.g., Bird et al., 1982).

Some interpretations of the interchangeability of alcohol and morphine dependence suggest a common biogenic basis (Nichols, 1972). In 1903, French asserted that "the surest victims" of morphinism are "those who inherit an alcoholic tendency" (French, 1903, p. 608). More recently, evidence for common endorphinergic pathways for both substances has suggested to several investigators that genetic differences in the endogenous opiate system are responsible for the reinforcing effect of both morphine and alcohol (e.g., Blum et al., 1980*a*, 1982*b;* Ho and Allen, 1981; Terenius, 1982; Verebey and Blum, 1979). Recent research has evaluated the genetic bases of the commonality between alcoholism and opiate dependence. Results of studies of a genetic link between the two chemical dependencies in humans have been equivocal; data both supporting (Tennant, 1976) and refuting (Hill et al., 1977) such a link have been presented. Results from animal studies have been clearer. Strains of rats and mice that display a high preference for alcohol generally display a high preference for morphine (see Satinder, 1982). Although a subline of morphine-preferring mice does not display an alcohol preference (Whitney and Horowitz, 1978), and a study with inbred strains of Tryon rats has provided evidence of independent genetic transmission of alcohol and morphine consumption (Hill, 1978), most of the animal studies support the existence of a possible genetic link between alcohol and morphine preference (recent reviews are provided by Kakihana and Butte, 1980, and Satinder, 1982). Furthermore, Blum et al. (1980*b*) recently reported that alcohol-preferring strains of mice display a greater sensitivity to opiate receptor actions (as measured by ethanol-induced inhibition of electricially stimulated contractions of the vas deferens) than alcohol-avoiding strains. This finding is congenial with a common, genetic, endorphinergic interpretation of the substitutability of alcohol and opiates.

Recently, it has been suggested that both alcoholism and opiate dependence are manifestations of a common nutritional deficiency, with people dependent on either chemical having a high (and unfulfilled) niacin requirement (Cleary, 1985). According to this analysis, the vitamin deficiency facilitates the effects of TIQs and exogenous opiates on central opiate receptors.

In summary, results of recent studies of apparent commonalities in action between ethanol and opiates are suggestive of an endorphinergic mechanism, but the available data are conflicting and controversial. The area is one of intense research activity, and undoubtedly a clearer picture of the similarities and differences in the mechanisms of action of these drugs will soon emerge. It should be noted, however, that the issue of the reciprocity between ethanol and opiates can be addressed independently of the mechanisms of such reciprocity.


## 8.  20th CENTURY VIEWS TOWARD SUBSTITUTION OF ALCOHOL WITH OPIATES

> Intoxication is not what it used to be (Straus, 1983, p. 3).

The Victorian practitioner who recommended opiate dependence as an alternative to alcoholism did so in the belief that the cure, although not without complications, was much preferable to the disease. Although we currently treat the two substances as if opiates were more pernicious, this recent reversal is not readily attributable to post-Victorian developments in pharmacological knowledge. Rather, as indicated previously, complex social and political factors caused recreational use of opiates to be illicit, and recreational use of alcohol to be licit. Several 20th century scientists have lamented this turn of events. Among the most vocal was Lawrence Kolb, who was Assistant Surgeon General of the United States Public Health Service and first superintendent of the service's addiction treatment hospital in Lexington, Kentucky. In 1925, Kolb noted that "some drunkards are improved socially by abandoning alcohol for an opiate," and acknowledged that this is "substitution of a lesser for a greater evil" (Kolb, 1925, p. 312–313). Kolb reiterated this position in subsequent papers (Kolb, 1927, 1928), and, as late as 1958, noted that "all informed people know that alcoholics are more dangerous both to themselves and to society than are morphine and heroin addicts" (Kolb, 1958, p. 97). In Pescor's (1943) extensive survey of alumni of the Lexington Hospital, in which he provided data supporting the substantial trade-off between alcohol and opiate use, he concluded that "most of the alcoholics are better off on drugs than they are on alcohol," except for the "greater social disapproval" engendered by drug use (Pescor, 1943, p. 13). Other researchers have more recently noted the improved health and social functioning of the opiate addict while he is taking the narcotic, compared with his functioning while taking alcohol, and have indicated the mischievous consequences of medical and legislative hindrances to such an alternative to alcoholism (Brotman and Freedman, 1968; O'Donnell, 1969; Waldorf, 1973).

The Victorian view of the salutary effects of substituting opium dependence for alcohol dependence appears quaint. The more recent echos of this policy, however, appear ludicrous. Indeed, despite its long history, the "opiate treatment" is generally not considered worthy of mention in contemporary summaries of pharmacological treatments of alcoholism (e.g., Rada and Kellner, 1979; Sellers et al., 1981; but for an exception, see Jaffe, 1984). It should be noted, however, that the most recent edition of the authoritative pharmacological reference work, *Goodman and Gilman's, The Pharmacological Basis of Therapeutics*, acknowledges the importance of diaglogue on this topic: ". . . if the only alternative to the use of opioids is the compulsive use of alcohol, there are many who would take the view that opioid dependence is far less destructive to the individual and society, and that some provision should be made to permit that particular individual to use opioid drugs" (Jaffe, 1985, p. 533).


## 9.   WHY OPIATE SUBSTITUTION IS NOT NOW RECOMMENDED AS A TREATMENT FOR ALCOHOLISM

Trying to understand drug addiction by studying drugs makes about as much sense as trying to understand holy water by studying water (Szasz, 1974, p. xv).

We do not now permit opiates to be used as an alternative to compulsive alcohol use. Moreover, there does not seem to be much serious sentiment that this substitution be evaluated as a technique for ameliorating the alcohol problem. As recently pointed out by Jaffe (1984), "there have been no controlled studies of the effects of prescribed opiates on the affective state of alcoholics and of the course of alcoholism in alcoholics who have not been previously dependent on opiates" (p. 482). The reasons for our rejection of this approach (despite its long history and recent epidemiological and experimental data attesting to its validity) may be surmised. They reveal our current cultural and legislative biases toward the phenomenon of addiction.

## Abstinence as Goal of Treatment

> Merely replacing the alcohol with another drug means that we basically subscribe to the concept of chemical escape (Ewing, 1972, p. 93).

One reason why there is little enthusiasm for the evaluation of opiate maintenance as a substitute for excessive alcohol abuse is the pervasive view that the goal of drug-dependence treatment should be abstinence. Pharmacological aid in reaching this goal may be acceptable, but the mere trading of one dependence for another is not usually considered to be desirable.

The goal of a long-lasting, drug- and alcohol-free state is laudable, but success in attaining the goal has been, at best, problematic (see Smith, 1981). The epidemiological data, like the experimental data, offer little encouragment to therapeutic strategies that aim to achieve abstinence. Since the end of prohibition, there have been enormous gains in alcoholism treatment facilities, funding for treatment, and training of treatment professionals. Despite these measures, there has been no general trend for alcohol consumption to decrease in North America or Europe, nor for the various indices of alcoholism (e.g., liver cirrhosis, public drunkenness convictions, and hospital admissions with a primary diagnosis of alcoholism or alcohol psychosis) to decrease (see Lieber, 1982). Obviously, these data do not prove that our treatments are ineffective; the alcohol-related problems might be much more severe without the current interventions. The available experimental and epidemiological data, however, do suggest that the usual treatments for alcoholism, whether or not assisted by any of a variety of currently acceptable drug therapies, are of limited value (or, at best, are of value for a limited proportion of the alcoholic population).

In any event, even the most enthusiastic 19th century advocates of the opiate treatment of alcoholism did not recommend the substitution as an initial treatment. Rather, opiates were to be used when all other treatment methods had failed, and the alcoholic was posing an imminent danger to himself and/or others. That is, *care* for the alcoholic was a suitable alternative when *cure* was impossible.

## Potential Complications of Multiple Drug Abuse

> Polydrug abuse—the use of multiple drugs—is common, as are its complications
> (Jaffe et al., 1980, p. 9).

The 19th century practitioner believed that the alcoholic would abandon the dangerous alcohol for the less pernicious opiate. Recent evidence in support of this belief has been presented. An argument against such a substitution treatment may center on the possibility that the treatment has the potential of converting many alcoholics into polydrug abusers, dependent on both alcohol and opiates. Thus, according to this argument, the opiate treatment will exacerbate, rather than relieve, the alcoholic's chemical dependency problems.

As discussed previously, there is a substantial body of epidemiological and experimental evidence that opiates do substitute for alcohol and replace (rather than supplement) alcohol as an abused drug. Nevertheless, there is considerable concern about the issue of concomitant opiate and alcohol abuse by opiate-dependent persons. The bases for this concern rest on reports of combined alcohol–opiate abuse by both heroin addicts and methadone-maintenance patients.

Alcohol Use by Heroin Addicts. Evidence that some heroin abusers are also alcohol abusers has been summarized by Green and Jaffe (1977). For example, Rosen et al. (1973) questioned male heroin addicts admitted to an inpatient addiction treatment facility in Pennsylvania about alcohol use. As expected on the basis of a reciprocity view, the majority of the patients who abused alcohol prior to opiate use no longer did so when they started using heroin; however, a substantial minority (about one third) of the drug addicts continued to abuse alcohol while addicted to heroin. Such evidence of the use of both heroin and alcohol by the same individuals has suggested to some investigators that opiates would not be completely effective in abolishing a desire for alcohol (Green and Jaffe, 1977).

Instances in which illicit heroin increases alcohol consumption may be explicable on the basis of the priming effect discussed previously (Section 6). Although a substantial heroin dose would be expected (on the basis of reciprocity view) to decrease alcohol consumption, a very small heroin dose may, like a small dose of alcohol, increse such consumption. Indeed, illicit heroin packages, in contrast to licit opiate supplies, often contain only very tiny amounts of drug (Primm and Bath, 1973).

In fact, detailed analyses of alcohol use by heroin addicts suggests that often the mode of such use is explicible on the basis of an opiate–alcohol reciprocity perspective. In instances of polydrug abuse, the various chemicals may be used concomitantly (i.e., the addict chooses to experience the combined intoxicating effects of alcohol and opiates) or sequentially (i.e., the addict alternates between alcohol and opiate use) (see review by Belenko, 1979). In fact, most cases of alcohol abuse by heroin addicts are of the latter type. More-over, alcohol typically is used in a *substitutive* fashion; the addict who temporarily cannot obtain opiates uses alcohol (see Jackson and Richman, 1973; Maddux and Elliot, 1975; Mitcheson et al., 1971). Thus, the fact that a minority of heroin addicts are also classified as alcoholics does not persuasively argue against an opiate-maintenance treatment of alcoholism. Rather, the pattern of alcohol and opiate use by these polydrug abusers may further attest to the ability of the two types of intoxicants to substitute for each other.

Alcohol Use by Methadone-Maintenance Patients. Methadone is a syn-thetic opioid that is considered to have advantages over heroin as a drug of dependence. There are, at present, many thousand former opiate (mostly heroin) addicts enrolled in methadone-maintenance programs worldwide. Because their addition is not illicit, these methadone patients provide an opportunity to study the effects of opiate use in a readily accessible population. An opiate–alcohol reciprocity view would suggest that methadone maintenance should decrease alcohol use; however, there are reports that a high proportion of methadone-maintenance patients abuse alcohol (Green and Jaffe, 1977). Various interpre-tations of this joint use of alcohol and methadone have been offered: alcohol may affect the disposition kinetics of methadone (Tommasello and Adir, 1984) or the two drugs may affect different classes of opiate receptors (Blum et al., 1982a).

There is a tendency to stabilize methadone patients on the lowest possible dose of the narcotic, and it has been suggested that this may contribute to some patients seeking the additional stimulation provided by alcohol (Lowinson and Millman, 1979). Of course, very low doses of methadone, like very low doses of other opiates, may serve the same priming function as a low dose of alcohol with respect to further alcohol consumption.

The pharmacokinetics of a typical dose of orally administered methadone may be relevant to the use of alcohol by methadone patients. Oral methadone does not produce a "rush" immediately following administration, and (because of cross-tolerance) the individual under the effects of methadone will not ex-perience the usual euphoria following administration of another opiate. It has been suggested that alcohol may be used during methadone stabilization to simulate a heroin "high" (Belenko, 1979; Maddux and Elliott, 1975; Schut et al., 1973). Such considerations suggest that the issue of alcohol use by meth-adone-maintenance patients may not be relevant to potential complications of an opiate-substitution approach in general, or oral methadone-substitution approach in particular, to intractable alcoholism. The alcoholic has, in most cases, no extensive prior experience with opiates; there would be not invidious comparison of the effects of the orally administered synthetic opiate with the more intense effects of parenterally administered heroin.

Although reports of joint use of methadone (at substantial doses) and alcohol apparently conflict with an alcohol–opioid reciprocity view, the nature of the influence of methadone on alcohol consumption is, in fact, far from clear. Surveys indicating that methadone maintenance does not decrease, and perhaps even increases, alcohol consumption have been retrospective in design (see Green and Jaffe, 1977; Siegel, 1982*b*). Results of more recent prospective studies indicate that alcohol use *decreases* during methadone maintenance (Barr and Cohen, 1979; Jackson et al., 1982; Stimmel et al., 1982). Such findings, suggesting reciprocity between methadone and alcohol use, are further substantiated in a comparison of alcohol use by both current methadone-maintenance clients and by former clients of a comparable but terminated methadone facility (McGothlin and Anglin, 1981). It was reported that alcohol abuse was higher in the former clients of the terminated facility (in Bakersfield, California) than among the continuing clients in the ongoing facility (in Tulare, California). Thus, it has been suggested that "the myth that methadone maintenance treatment leads to a marked increase consumption of alcohol" (Jackson et al., 1982, p. 75) may be attributable to design deficiencies in studies reporting such an increase (Jackson et al., 1982; Stimmel et al., 1982).

Finally, any apparent association between alcohol abuse and methadone use may be due, in part, not to the effect of methadone, but rather to the effect of methadone withdrawal. Often methadone-maintenance patients will attempt to withdraw from the maintenance drug and consume large amounts of alcohol during periods of methadone abstinence (see Chappell et al., 1973; Schut et al., 1973; Stimmel, 1975, p. 248). For some individuals, then, the apparent joint use of methadone and alcohol may (as in the case of heroin addicts) involve sequential use, rather than concurrent use; they are using alcohol and the synthetic opioid in a reciprocal manner, further attesting to the ability of the two types of intoxicants to substitute for each other.

## Beliefs Concerning Deleterious Effects on Health

> The popular view of the effects of narcotics on addicts was eloquently expressed in a 1962 decision of the Supreme Court of the United States:
>> "To be a confirmed drug addict is to be one of the walking dead . . . The teeth have rotted out, the appetite is lost, and the stomach and intestines don't function properly . . . Such is the torment of being a drug addict; such is the plague of being one of the walking dead."
>
> The scientific basis for this opinion, however, is not easy to find (Brecher, 1972, p. 21).

If the opiate-substitution treatment were effective for alcoholics refractory to less drastic interventions, the issue is raised as to whether or not the patients would be better off with an opiate dependency than an alcohol dependency. As discussed previously, most Victorian physicians, who had ample opportunity to observe the effects of both then-freely-available substances, believed that alcoholism was more deleterious to health than opiate addiction. This conclusion, although not reflected in current popular opinion and legislation, is well supported by current research.

Physical Complications of Chronic Alcohol and Opiate Use. Clearly, both alcohol and opiates have severe physical effects if administered in near-lethal doses. The issue is, what are the chronic effects of these substances at doses administered by individuals dependent on them? The deleterious effects of alcohol are well known. As recently summarized by Keeley and Solomon (1981), "alcohol's toxicity is pervasive in comparison to most drugs" (p. 100). The parenchyma simply cannot efficiently catabolize the ethanol molecule and its oxidation product, acetaldehyde. Most of the oxidation takes place in the liver, and the major physical complications associated with prolonged, excessive alcohol consumption involve this organ. These complications are (in the usual sequence) fatty liver, alcoholic hepatitis, and portal cirrhosis (see summaries by Goodwin, 1981; Lieber, 1982). The entire sequence occurs in only a minority of alcoholics, yet alcoholism is so prevalent that liver cirrhosis is the fourth leading cause of death among men over the age of 40 in the United States (Moore, 1977). Mortality from liver cirrhosis is much higher in many other countries (e.g., about twice as high in recent years in France, Italy, and Portugal; see Alcoholism and Drug Addiction Research Foundation, 1981). The consequences of portal cirrhosis (interference with clotting mechanisms and with the portal venous return system) are such that the prognosis is very poor (see review by Galambos, 1979, Chapter 13). Alcoholism is also associated with a variety of other gastrointestinal disorders, including cancer (see Stock et al., 1980).

In addition to gastrointestinal pathology, cerebral atrophy (and resulting mental impairment) is commonly found in alcoholics (see review by Wilkinson and Carlen, 1981). Heavy drinking also increases the risk of cardiovascular diseases (Wodak and Richardson, 1982). The many other complications of alcoholism have been summarized elsewhere (see review by Popham et al., 1984). Although some of the effects, such as liver and brain alterations, may be due to the toxic effects of alcohol or its metabolites, others, such as peripheral neuropathies and Korsakoff's psychosis, may be exacerbated by the fact that alcoholics often get most of their daily caloric needs from alcohol and thereby suffer from severe deficiencies of vitamins, minerals, and amino acids. Whether due to the direct toxic effects of alcohol, or the malnutrition secondary to alcoholism, current evidence suggests that "chronic alcoholism has considerably greater and more serious medical consequences than does addiction to any other drug" (Bihari, 1974, p. 8).

Opiates, in contrast to alcohol, are not generally believed to cause cell death in the gastrointestinal system or brain (Keeley and Solomon, 1981). Moreover, a long history of opiate abuse is not associated with deficits in neuropsychological functioning (Rounsaville et al., 1982). The study of the medical effects of opiates in humans is confounded, however, by the fact that chronic opiate use is generally illegal. The addict who must obtain his or her drugs from illicit sources does indeed suffer a variety of illnesses, but these are attributable to the consequences of the illegality of nonmedical opiate use, rather than the pathological effects of the opiate molecule. Thus, nonsterile injection procedures and contaminants in illicit drug supplies are sources of severe medical problems associated with opiate addiction (see Hofmann, 1983). Similarly, malnutrition and poor health secondary to the high cost of the illegal addiction and participation in a deviant lifestyle are recognized complications of opiate dependence in contemporary Western society (Hofmann, 1983).

The classic studies of chronic opiate effects in humans, not confounded by the complications of illicit drug use, were conducted by Light and Torrance and colleagues and presented as a series of 11 papers in the *Archives of Internal Medicine* in 1929 (subsequently collected in a book, Light et al., 1930). A total of 861 male opiate addicts participated in various phases of this research program in the Philadelphia General Hospital. They were observed under controlled conditions while they were supplied with morphine and withdrawn from the drug. Light et al. (1930) concluded that opiates per se had little effect on health. Those deleterious effects noted in opiate-dependent individuals result *not* from the drug, but rather from the deviant lifestyle of the addict—a lifestyle adopted in response to the illegality of nonmedical opiate use:

> These results are of definite importance. The study shows that morphine addiction is not characterized by physical deterioration or impairment of physical fitness aside from the addiction per se. There is no evidence of change in the circulatory, hepatic, renal or endocrine functions. When it is considered that these subjects had been addicted for at least five years, some of them for as long as twenty years, these negative observations are highly significant" (Light et al., 1930, pp. 115–116).

Other studies of large populations of opiate addicts, who did not need to obtain their drugs from illicit sources, similarly indicate that the deleterious effect of opiates on health are not pronounced. For example, analyses of over 700 patient records from a morphine-maintenance clinic that operated in Shreveport, Louisiana in the early 1920s indicated that large, daily, self-injected morphine doses, over a period of many years, do not clearly cause deterioration of health and functioning (Waldorf et al., 1974). Similarly, evaluation of British addicts maintained on medically supplied opiates indicates the minimal health consequences attributable to opiate addiction per se (Trebach, 1982). Long-term methadone use is similarly not associated with clearly adverse effects (Kreek, 1978).

As summarized by Brecher (1972, Chapter 5), biographical accounts of addicts who have had adequate access to medical-grade opiates, and no need to use illicit supplies, further attest to the minimal health consequences of the drug. Such addicts are generally physicians, and perhaps the most thoroughly documented case is that of Dr. William Halsted (Penfield, 1969). Halsted was the first chief of surgery at Johns Hopkins Medical School. At the time of his appointment, and throughout his brilliant career, he was addicted to morphine. His colleagues did not suspect his dependence, despite the fact that he was taking 180 mg of morphine per day (perhaps reduced to 90 mg in later years). It is unlikely that Halsted would have made such substantial contributions if he were alcoholic. He did so while daily administering, for most of his adult life, more opiate than the average street addict in New York City. The literature describing other examples of the apparently unimpaired functioning of 20th century opiate addicts with access to the drug is summarized by Brecher (1972, Chapter 5): physicians, a congressman, a navy commander, and others, all achieving a high level of performance despite their addiction to opiates.

Even the illicit opiate user appears to suffer less in the way of physical complications than commonly believed. In one study, for example, Stevenson et al. (1956) evaluated the physical and intellectual functioning of large groups of both heroin-addicted and nonaddicted prisoners in British Columbia, Canada. In many cases, childhood records of intellectual performance were compared with psychological test results of these same people after many years of addiction. They concluded, "continued heroin addiction produces no measurable physical or mental deterioration in the user" (Stevenson et al., 1956, p. 539).

Ball and Urbaitis (1970) examined the medical records of those patients of the Lexington, Kentucky opiate-addiction treatment facility who were most frequently hospitalized for their addiction. Of the 34,043 male addicts treated during the first 32 years of the hospital's existence (1935–1966), 31 had been admitted 20 or more times. The investigators anticipated that this selected, most highly recidivistic population would be "chronically ill and debilitated"; "It seemed likely that the end result of extended opiate abuse would be physical, and perhaps mental deterioriation" (Ball and Urbaitis, 1970, p. 111). To the apparent surprise of Ball and Urbaitis, this expectation was not confirmed. With the exception of one case of malnutrition (probably attributable to the high cost of black market drugs), the remaining hard-core addicts did not seem to suffer medical complications resulting from a lifetime of opiate use.

Although the literature indicates that chronic opiate use (especially licit opiate use) is less hazardous than chronic alcohol use, it is not the case that such opiate use should be considered innocuous. Unpleasant side effects have been noted in some methadone-maintenance clients, such as constipation, reduced libido, joint pains, and impaired night vision resulting from pupillary constriction (see review by Cox et al., 1983, p. 345). In addition, recent clinical reports indicate occasional severe adverse reactions to prolonged methadone use, including pronounced fecal impaction (Rubenstein and Wolff, 1976; Spira et al., 1975) and choreic movement disorder (Wasserman and Yahr, 1980). Moreover, some investigators have suggested that methadone maintenance may be associated with deleterious effects at cytogenic, biochemical, and immunological levels (see reviews by Cushman, 1981; Falek et al., 1982; Singh et al., 1980) (although these alterations may result from the patient's premethadone illicit addiction history, and attendant exposure to a variety of infectious agents, see Cushman et al., 1977). Recall that even the Victorian practitioner, who recommended opiates to the alcoholic, did not believe that the drug was salubrious. Rather, he believed that "the use of morphine in the place of alcohol is but a choice of evils, and by far the lesser" (Black, 1889, p. 538). This conclusion is not disputed by results of more recent research concerning the relative health consequences of chronic use of these two agents.

Heroin "Overdose" and Opiate Substitution Treatment. It is possible that some reticence about recommending opiates as an alcohol substitute may be justified on the basis of well-publicized fatalities associated with heroin use. Each year, approximately 1% of American heroin addicts die, mostly from an acute reaction to the intravenously administered drug, so-called overdose death (Louria et al., 1967; Maurer and Vogel, 1973, p. 101). This might suggest that a substantial hazard is associated with regular opiate use and that opiate substitution for alcohol would be dangerous. Such a conclusion, however, is problematic; heroin "overdose" is a misnomer—it is *not* typically a pharmacological overdose (as the term is usually used). Moreover, these drug-related deaths, whatever their cause, are associated with deviant opiate use and are not a feature of medical use.

The evidence that most heroin "overdose" deaths are not precipitated by the pharmacological effects of a high opiate dose has been summarized elsewhere (Brecher, 1972, pp. 101–114; Hofmann, 1983, pp. 84–87). Briefly, (1) heroin packages found near dead addicts typically do not contain an extraordinarily high concentration of the narcotic (Baden, 1968); (2) frequently a number of addicts will inject themselves in a group, using a common drug supply, yet only one will suffer an "overdose" (Baden, 1968; Halpern and Rho, 1966); (3) postmortem examination of addicts who died immediately after heroin administration, and examination of the contents of the fluid remaining in their syringes, usually indicates that these individuals died after receiving a dose of the opiate that would not be expected to be fatal in these drug-experienced, and presumably drug-tolerant, individuals (Halpern and Rho, 1966); and (4) sometimes the victims die following self-administration of a heroin dose that was well-tolerated the previous day (Government of Canada, 1973, pp. 310–315). Thus, these so-called overdose deaths result from "an idiosyncratic reaction to an intravenous injection of unspecified material(s) and probably not a true pharmacologic overdose of narcotics" (Cherubin et al., 1972, p. 11). The basis of this "idiosyncratic reaction" is unknown: "It remains unclear why a given dose of heroin will cause this reaction at one time and not at others" (Werner, 1969, pp. 2277–2278); "the term 'overdose' has served to indicate lack of understanding of the true mechanism of death in fatalities directly related to opiate abuse" (Greene et al., 1974, p. 175).

Although various mechanisms have been proposed (Brecher, 1972; Government of Canada, 1973; Ruttenber and Luke, 1984; Siegel et al., 1982), it does appear that these enigmatic deaths are associated with *illicit* opiate abuse—fatal reactions are not a significant problem of licit opiate use as long as the patient does not have existing pathology that compromises respiratory reserve, such as emphysema (see Jaffe and Martin, 1985). Furthermore, in the 19th century era of free opiate availability, nonsuicide, adult deaths due to opiates were not substantial (Berridge and Rawson, 1979). In addition, overdoses do not appear to have been a problem during the time of operation of medically supervised morphine-maintenance clinics in the United States during the 1920s. In the study of the Shreveport clinic (Waldorf et al., 1974), it was noted that "there were *never* problems with overdose" (p. 22), despite the fact that these maintained addicts parenterally (presumably often intravenously) self-administered large morphine doses.

Effects of Withdrawal of Opiates and Alcohol. Related to the issue of the relative health consequences of chronic use of opiates or alcohol is concern about the consequences of withdrawal of these agents. Suppose the former alcoholic, converted into a teetotal opiate addict, subsequently decided to abjure all chemical dependencies. He or she would then have to be withdrawn from opiates. A prevalent conception of opiate withdrawal is that it involves life-threatening agony. In fact, the withdrawal of opiates in long-term heavy users generally causes less distress, and is certainly associated with lower mortality, than the withdrawal of alcohol in long-term alcoholics (e.g., Government of Canada, 1973; Hodding et al., 1980).

Dangers of Drugged State. Finally, there might be concern about the dangers inherent in the intoxicated state induced by narcotics. For example, the alcoholic driver is a substantial and well-publicized hazard; might not such individuals pose an even greater threat to highway safety if they continued to drive while under the effects of medically prescribed opiates rather than self-prescribed alcohol? The issue of driver performance while under the effects of opiates is probably only tangentially related to issue of opiate-substitution therapy for alcoholism—there is little likelihood that the severely debilitated end-stage alcoholics, for whom such substitution might be attempted, have ready access to an automobile. Nevertheless, it should be noted that the available data concerning the driving records of opiate-dependent individuals (driving while narcotized) are far better than the record of alcoholics (driving while intoxicated). Clients on methadone maintenance, who drive automobiles, do not have a record of driving violations, accidents, or convictions different than that of a comparison group of non-drug-using peers (Babst et al., 1973). Furthermore, large numbers of heroin addicts typically drive soon after taking heroin (some even shoot-up while driving), yet there is no evidence that these driving addicts pose a highway hazard (see summary in *Current Medicine for Attorneys,* 1973).

Summary of Health Effects. In summary, the available studies concerning the relative detrimental effects of opiates and alcohol challenge common wisdom. Although prolonged opiate use may not be innocuous, it is the alcoholic, and not the opiate addict, who is self-administering the more toxic chemical. The literature on the physical complications of opiate addiction was recently summarized by Jaffe (1985), who concluded, "good health and productive work are thus not incompatible with regular use of opioids" (p. 542). The same cannot be said of alcohol, as typically self-administered by alcoholics.

## 10. IMPLICATIONS

> Is it not the duty of the physician when he cannot cure an ill, when there is no reasonable ground for hope that it will ever be done, to do the next best thing—advise a course of treatment that will diminish to a great extent great evils otherwise irremediable? (Black, 1889, p. 540).

It would appear that there is little in the way of current data to refute the wisdom of the 19th century practice of converting intractable alcoholics into opiate addicts. In fact, results of recent research suggest that administration of substantial opiate doses (i.e., doses typically used in medical practice) does indeed greatly reduce, and perhaps eliminate, alcohol consumption. Furthermore, the effects of such sanctioned narcotic dependence are preferable to the effects of chronic alcohol abuse. These results, obtained from epidemiological studies of human addicts, and experimental studies with animals, suggest that a clinical evaluation of the opiate treatment of alcoholism would be appropriate.

Obviously, there are formidable legislative and cultural deterrents to the use of opiates in this manner (Jaffe, 1984). However, it would seem that, even in our "narcotophobic era" (Westermeyer, 1978, p. 868), ethical considerations would favor evaluation of such an innovative treatment in cases of severe alcoholics who meet certain criteria: (1) they have resisted many previous attempts at more orthodox treatments, (2) their continued drinking will likely result in increasing debilitation and death, and (3) they nevertheless continue drinking. Many alcoholics suffering from liver cirrhosis meet these criteria.

The prognosis for the patient with alcoholic cirrhosis is very poor. Although the prognosis is related to the presence or absence of a variety of diagnostic signs (e.g., jaundice and esophageal varices), some studies report a five-year survival rate of only 5–10% (see review by Galambos, 1979). It is generally reported that the survival rate for those cirrhotics who stop drinking is considerably better than for those who continue to drink heavily (Galambos, 1979). It would seem that a clinical trial of opiate maintenance in some patients suffering from alcoholic liver disease—patients who continue to drink despite the fact that the consumption of alcohol is very likely to lead to their death—would be ethically incontestable.

Although such patients would appear to constitute ethically suitable subjects in an evaluation of the efficacy of opiate substitution, their hepatic pathology does present some complications. Since the opioid drugs are (like alcohol) metabolized by the liver, there is concern about the effects of administration of these drugs to patients with compromised liver function. This concern arises because the metabolic deficiencies of the cirrhotic patient may result in elevated drug levels and excessive drug accumulation (compared with patients with normal liver function), possibly resulting in hepatic encephalopathy (see Wilkinson and Schenker, 1975). However, selection of an appropriate opioid for the cirrhotic patient can be based on consideration of the metabolic pathways of the various drugs (Jaffe and Martin, 1985). For example, experiments indicated that partially hepatectomized rats are severely deficient in the metabolism of meperidine (Way et al., 1947), together with similar findings concerning meperidine elimination in patients with cirrhosis (Klotz et al., 1974), suggest that this synthetic opioid may be contraindicated in cirrhotic patients (Anderson and Schrier, 1981, p. 203). Similar concerns have been expressed about morphine (Laidlaw et al., 1961), but recent evidence suggests that cirrhotics do not differ from controls in the disposition and elimination of morphine (Patwardhan et al., 1981). Although significant clinical problems have not been encountered in methadone-maintenance patients with liver disease (Kreek, 1978), some evidence suggests that patients with "severe" (but not "moderate" or "mild") chronic liver disease are slower to metabolize methadone than controls (Novick et al., 1980), with such evidence of increased bioavailability suggesting "cautious use" (Anderson and Schrier, 1981, p. 203). In any event, it is unlikely that reasonable use of an appropriate opioid could pose a greater hazard to the cirrhotic patient than self-administered alcohol.


## 11.   SUMMARY AND CONCLUSIONS

Is the literature worth keeping? (Maddox, 1963, p. 14).

In the last century, when both alcohol and opiates were freely available, alcohol was perceived as the more hazardous chemical. In cases of severe alcohol abuse, it was thought humane and medically appropriate to encourage the alcoholic to transform his dependence from the very dangerous alcohol to the less dangerous opiate. Early in the 20th Century, however, various political and social forces (not pharmacological considerations) caused severe restrictions to be placed on the use of opiates. Attempts to similarly restrict alcohol (e.g., the era of prohibition in the United States) were abandoned. Results of 20th Century research support the logic of the 19th Century drug-substitution practice. Data from humans and animals indicate that alcohol use decreases when opiates are used, that opiates do not pose the threat to the individual and society that alcohol does (except for the hazards of being part of the illicit opiate-use subculture), and that more conventional, currently acceptable treatments of alcoholism are, in most cases, ineffective. Moreover, very recent research indicates a possible endorphinergic basis for the substitutability of alcohol and opiates, thus providing a rationale for the empirical findings.

After heart disease, cancer, and stroke, alcoholic cirrhosis is the next leading cause of death in adult males in North America. In addition to mortality, the high degrees of morbidity associated with alcohol abuse makes it a major medical problem, perhaps "the major medical problem in our society today" (Kreek, 1978, p. 112).

A suggestion that we return to the wholesale transformation of alcoholics into opiate addicts would be preposterous, as well as advocacy of an illegal use of opiates. However, a clinical evaluation of the efficacy of an opioid substitution approach would be appropriate, especially if this evaluation were conducted with chronic, severely debilitated alcoholics, who have been refractory to all prior attempts at therapy, and who are dying as a consequence of their continued consumption of alcohol.

In conclusion, considering what we know about alcohol–opiate reciprocity, and the fact that a form of opioid maintenance is a recognized therapy for opiate addicts (i.e., methadone), it would seem reasonable to evaluate the potential value of such maintenance for people suffering from addiction to a far more pernicious substance.

## REFERENCES


Alcoholism and Drug Addiction Research Foundation of Ontario, 1981, *Statistical Supplement to the Annual Report 1979–80,* Addiction Research Foundation, Toronto.

Anderson, R. J., and Schrier, R. W., 1981, *Clinical Use of Drugs in Patients with Kidney and Liver Disease,* W. B. Saunders, Philadelphia.

Anokhina, I. P., Brusov, O. S., Nechaev, N. V., Balashov, A. M., Beligey, N. A., and Panchenko, L. F., 1983, Effect of acute and chronic ethanol exposure on the rat brain opiate receptor function, *Alcohol Alcoholism* **18**:21–26.

Anonymous, 1873, Opium-eating in Massachusetts, *London Lancet* 329–330.

Anonymous, 1876, Morphia disease, *London Lancet* 330–331.

Anonymous, 1939, *Alcoholics Anonymous: The Story of How Many Thousands of Men and Women Have Recovered from Alcoholism,* Works, New York.

Arvola, A., and Forsander, O., 1961, Comparison between water and alcohol consumption in six animal species in free choice experiments, *Nature* **191**:819–820.

Babst D. V., Newman, S., Gordon, N., and Warner, A., 1973, Driving records of methadone maintenance patients in New York State, *J. Drug Issues* **3**:285–292.

Badawy, A. A.-B., and Evans, C. M., 1983, Naloxone in ethanol intoxication, *Ann. Intern. Med.* **98**:672.

Baden, M. M., 1968, Medical aspects of drug abuse, *N.Y. Med.* **24**:464–466.

Ball, J. C., and Urbaitis, J. C., 1970, Absence of major medical complications among chronic opiate addicts, *Br. J. Addict.* **65**:109–112.

Barr, H. L., and Cohen, A., 1979, *The Problem-Drinking Drug Addict,* National Institute on Drug Abuse Services Research Report, DHEW Publication No. (ADM)79-893, U.S. Government Printing Office, Washington, D.C.

Beaman, C. M., Hunter, G. A., Dunn, L. L., and Reid, L. D., 1984, Opioids, benzodiazepines and intake of ethanol, *Alcohol* **1:**39–42.

Belenko, S., 1979, Alcohol abuse by heroin addicts: Review of research findings and issues, *Int. J. Addict.* **14:**965–975.

Bellis, D. J., 1981, *Heroin and Politicians,* Greenwood Press, Westport, Connecticut.

Berridge, V., 1977, Opium eating and life insurance, *Br. J. Addict.* **72:**371–377.

Berridge, V., and Edwards, C. T., 1981, *Opium and the People,* Allen Lane/New York, St. Martin's Press, London.

Berridge, V., and Rawson, N. S. B., 1979, Opiate use and legislative control, *Soc. Sci. Med.* **13A:**351–363.

Bihari, B., 1974, Alcoholism and methadone maintenance, *Am. J. Drug Alcohol Abuse* **1:**79–81.

Bird, K. D., Chesher, G. B., Perl, J., and Starmer, G. A., 1982, Naloxone has no effect on ethanol-induced impairment of psychomotor performance in man, *Psychopharmacology* **76:**193–197.

Black, J. R., 1889, Advantages of substituting the morphia habit for the incurably alcoholic, *Cincinnati Lancet-Clinic* **22:**537–541.

Blum, K., Wallace, J. E., Schwertner, H. A., and Eubanks, J. D., 1976, Morphine suppression of ethanol withdrawal in mice, *Experientia* **32:**79–82.

Blum, K., Briggs, A. H., Elston, S. F. A., and DeLallo, L., 1980*a,* Psychogenics of drug seeking behavior, *Subst. Alcohol Actions Misuse* **1:**255–257.

Blum, K., Briggs, A. H., Elston, S. F. A., and DeLallo, L., and Hirst, M., 1980*b,* Validation of central and peripheral models to evaluate alcohol sensitivity and opiate-membrane interactions as a function of genotype dependent responses in three different strains of mice, in: *Animal Models in Alcohol Research* (K. Eriksson, J. D. Sinclair, and K. Kiianmaa, eds.), pp. 85–91, Academic Press, New York.

Blum, K., Briggs, A. H., Elston, S. F. A., Hirst, M., Hamilton, M. G., and Verebey, K., 1980*c,* A common denominator theory of alcohol and opiate dependence: Review of similarities and differences, in: *Alcohol Tolerance and Dependence* (H. Rigter and J. C. Crabbe, Jr., eds), pp. 371–391, Elsevier/North Holland Biomedical Press, Amsterdam.

Blum, K., Briggs, A. H., and DeLallo, L., 1982*a,* On the mechanism of methadone-induced alcohol consumption in humans, *Subst. Alcohol Actions Misuse* **3:**1–4.

Blum, K., Briggs, A. H., and Verebey, K., 1982*b,* The pharmacology of addictive drugs: A review of similarities and differences, in: *Perspectives in Alcohol and Drug Abuse: Similarities and Differences* (J. Solomon and K. A. Keeley, eds.), pp. 35–57, John Wright–PSG Incorporated, Boston.

Borg, S., Kvande, H., Rydberg, V., Terenius, L., and Wahlström, A., 1982, Endorphin levels in human cerebrospinal fluid during alcohol intoxication and withdrawal, *Psychopharmacology* **78:**101–103.

Bourne, P. G., 1982, Forward to J. Westermeyer, *Poppies, Pipes, and People: Opium and Its Use in Laos,* University of California Press, Berkeley, California.

Brecher, E. M., 1972, *Licit and Illicit Drugs,* Little, Brown, Boston.

Brittan, C. S., 1878, Proceedings of the Richmond Academy of Medicine, January 15th: Opium habit, *Virginia Med. Monthly* **4:**881–884.

Brotman, R., and Freedman, A., 1968, *A Community Mental Health Approach to Drug Addiction,* U.S. Department of Health, Education, and Welfare, Social and Rehabilitation Service, Office of Juvenile Delinquency and Youth Development, Washington, D.C.

Brown, B. S., Kozel, N. J., Meyers, M. B., and DuPont, R. L., 1973, Use of alcohol by addict and nonaddict populations, *Am. J. Psychiatry* **130:**599–601.

Carroll, J. F., 1982, Personality and psychopathology: A comparison of alcohol- and drug-dependent persons, in: *Perspectives in Alcohol and Drug Abuse: Similarities and Differences* (J. Solomon and K. A. Keeley, eds.), pp. 59–87, John Wright–PSG Incorporated, Boston.

Chappell, J. N., Skolnick, V., and Senay, E. C., 1973, Techniques of withdrawal from methadone and their outcome over six months to two years, in: *Proceedings of the Fifth National Conference on Methadone Treatment,* pp. 482–489, National Association for the Prevention of Addiction to Narcotics, New York.

Charness, M. E., Gordon, A. S., and Diamond, I., 1983, Alcohol produces changes in opiate receptors which may correspond to intoxication, tolerance and withdrawal, *Neurology* **33:**175.

Cherubin, C., McCusker, J., Baden, M., Kavaler, F., and Amsel, Z., 1972, The epidemiology of death in narcotic addicts, *Am. J. Epidemiol.* **96:**11–22.

Child, G. P., 1951, The treatment of opiate addiction with intravenous alcohol, *N.Y. State J. Med.* **51:**1521–1523.

Child, G. P., Clare, F., and Farrell, J., 1950, Use of intravenous ethanol in treatment of a case of opiate addiction, and other experimental therapeutic tests with intravenous ethanol in humans, *Fed. Proc.* **9:**22–23.

Cleary, J. P., 1985, Etiology and biological treatment of alcohol addiction, *J. Neurol. Orthopaed. Med. Surg.* **6:**75–76.

Cobbe, W. R., 1895, *Doctor Judas: A Portrayal of the Opium Habit,* S.C. Griggs and Co., Chicago (reprinted, Arno Press, New York, 1981).

Cocteau, J., 1958, *Opium: The Diary of a Cure* (translated by M. Crosland and S. Road), Grove Press, New York.

Cohen, A., 1976, Alcohol and heroin: Structural comparison of reasons for use between drug addicts and alcoholics, *Ann. N.Y. Acad. Sci.* **273:**605–612.

Cohen, S., 1982, Combined alcohol-drug abuse and human behavior, in: *Perspectives in Alcohol and Drug Abuse: Similarities and Differences* (J. Solomon and K. A. Keeley, eds.), pp. 89–116 John Wright–PSG Incorporated, Boston.

Courtwright, D. T., 1982, *Dark Paradise: Opiate Addiction in America Before 1940,* Harvard University Press, Cambridge, Massachusetts.

Cox, T. C., Jacobs, M. R., LeBlanc, A. E., and Marshman, J. A., 1983, *Drugs and Drug Abuse,* Addiction Research Foundation, Toronto.

Crothers, T. D., 1902, *Morphinism and Narcomanias from Other Drugs,* W. B. Saunders, Philadelphia (reprinted, Arno Press, New York, 1981).

*Current Medicine for Attorneys,* 1973, Heroin addicts drive safely, **20:**31–32.

Cushman, P., 1981, Neuro-endocrine effects of opioids, *Adv. Alcohol Subst. Abuse* **1:**77–99.

Cushman, P., Gupta, S., and Grieco, M., 1977, Immunological studies in methadone maintained patients, *Int. J. Addict.* **12:**241–253.

Davenport, A., 1904, *China from Within: A Study of Opium Fallacies and Missionary Mistakes,* T. Fisher Unwin, London.

Davis, V. E., and Walsh, M. J., 1970, Alcohol, amines and alkaloids: A possible basis for alcohol addiction, *Science* **167:**1005–1007.

Dole, V. P., Fishman, J., Goldfrank, L., Khanna, J., and McGivern, R. F., 1982, Arousal of ethanol-intoxicated comatose patients with naloxone, *Alcohol Clin. Exp. Res.* **6:**275–279.

Eerola, R., Venho, I., Vartiainen, O., and Venho, E. V., 1955, Acute alcohol poisoning and morphine: An experimental study of the synergism of morphine and ethyl alcohol in mice, *Ann. Med. Exp. Biol. Fenniae* **33:**253–269.

Ewing, J. A., 1972, How to help the chronic alcoholic, *Am. Fam. Physician* **6:**90–97.

Falek, A., Madden, J. J. Shafer, D. A., and Donahoe, R. M., 1982, Opiates as modulators of genetic damage and immunocompetence, *Adv. Alcohol Subst. Abuse* **1:**5–20.

French, J. M., 1903, *A Text-Book on the Practice of Medicine,* William Wood, New York.

Galambos, J. T., 1979, *Cirrhosis,* Saunders, Philadelphia.

Gelfand, R., and Amit, Z., 1976, Effects of ethanol injections on morphine consumption in morphine-preferring rats, *Nature* **259**:415–416.

Genazzani, A. R., Nappi, G., Facchinetti, F., Mazzella, G. L., Farrini, D., Sinforiani, E., Petraglia, F., and Savoldi, F., 1982, Central deficiency of beta-endorphin in alcohol addicts, *J. Clin. Endocrinol. Metab.* **55**:583–586.

Gold, M. S., Dackis, C. A., and Washton, A. M., 1984, The sequential use of clonidine and naltrexone in the treatment of opiate addicts, in: *Advances in Alcohol and Substance Abuse, Vol. 3: Conceptual Issues in Alcohol and Substance Abuse* (J. H. Lowinson and B. Stimmel, eds.), pp. 19–39, Haworth Press, New York.

Goldstein, A., 1977, Future research in opioid peptides (endorphins): A preview, in: *Alcohol and Opiates: Neurochemical and Behavioral Mechanisms* (K. Blum, ed.), pp. 397–403, Academic Press, New York.

Goldstein, A., and Judston, B. A., 1971, Alcohol dependence and opiate dependence: Lack of relationships in mice, *Science* **172**:290–282.

Goodwin, D. W., 1981, *Alcoholism: The Facts,* Oxford University Press, New York.

Goodwin, D. W., Davis, D. H., and Robins, L. N., 1975, Drinking amid abundant illicit drugs, *Arch. Gen. Psychiatry* **32**:230–233.

Gottheil, E., Evans, B. D., and Verebey, K., 1982, Research relating to alcohol and opiate dependence, in: *Perspectives in Alcohol and Drug Abuse: Similarities and Differences* (J. Solomon and K. A. Keeley, eds.), pp. 179–206, John Wright—PSG Incorporated, Boston.

Government of Canada, 1973, *Final Report of the Commission of Inquiry into the Nonmedical Use of Drugs,* Information Canada, Ottawa.

Green, J., and Jaffe, J. H., 1977, Alcohol and opiate dependence: A review, *J. Stud. Alcohol* **38**:1274–1293.

Greene, M. H., Luke, J. L., and Dupont, R. L., 1974, Opiate overdose deaths in the District of Columbia, I. Heroin-related fatalities, *Med. Annu. D.C.* **43**:175–181.

Haggard, H. W., and Jellinek, E. M., 1942, *Alcohol Explored,* Doubleday, Garden City, New York.

Halushka, P. V., and Hoffmann, P. C., 1970, Alcohol addiction and tetrahydropapaveroline, *Science* **169**:1104–1105.

Helpern, M., and Rho, Y-M., 1966, Deaths from narcotism in New York City: Incidence, circumstances, and postmortem findings, *N.Y. State J. Med.* **1966**:2391–2408.

Hill, S. Y., 1978, Addiction liability of Tryon rats: Independent transmission of morphine and alcohol consumption, *Pharmacol. Biochem. Behav.* **9**:107–110.

Hill, S. Y., Cloninger, C. R., and Ayre, F. R., 1977, Independent familial transmission of alcoholism and opiate abuse, *Alcohol Clin. Exp. Res.* **1**:335–342.

Hinson, R. E., and Siegel, S., 1980, The contribution of Pavlovian conditioning to ethanol tolerance and dependence, in *Alcohol Tolerance and Dependence* (H. Rigter and J. C. Crabbe, eds.), pp. 181–189, Elsevier/North-Holland Biomedical Press, Amsterdam.

Ho, A. K. S., 1980a, Ethanol preference, withdrawal and toxicities: Modification by pharmacologic agents, in *Animal Models in Alcohol Research* (K. Eriksson, J. D. Sinclair, and K. Kiianmaa, eds.), pp. 173–178, Academic Press, New York.

Ho, A. K. S., 1980b, Interaction of alcohol and narcotics in *Alcoholism: A Perspective* (F. S. Messiha and G. S. Tyner, eds.), pp. 309–327, PJD Publications, Westbury, New York.

Ho, A. K. S., 1982, Suppression of ethanol consumption by MET-enkephalin in rats, *J. Pharm. Pharmacol.* **34**:118–119.

Ho, A. K. S., and Allen, J. P., 1981, Alcohol and the opiate receptor: Interactions with the endogenous opiates, *Adv. Alc. Sub. Abuse* **1**:53–75.

Ho, A. K. S., Chen, C.A., and Morrison, J. M., 1976, Interactions of narcotics, narcotic antagonists, and ethanol during acute, chronic, and withdrawal states, *Ann. N.Y. Acad. Sci.* **281**:297–310.

Ho, A. K. S., Chen, R. C.A., and Kreek, M. J., 1979, Morphine withdrawal in the rat: Assessment by quantitation of diarrhea and modification by ethanol, *Pharmacology* **86**:9–17.

Hodding, G. C., Jann, M., and Ackerman, I. P., 1980, Drug withdrawal syndromes: A literature review, *West. J. Med.* **133:**383–391.

Hodgson, R., Rankin, H., and Stockwell, T., 1979, Alcohol dependence and the priming effect, *Behav. Res. Ther.* **17:**379–387.

Hofmann, F. G., 1983, *A Handbook on Drug and Alcohol Abuse: The Biomedical Aspects,* 2nd ed., Oxford, New York.

Hunter, G. A., Beaman C. M., Dunn, L. L., and Reid, L. D., 1984, Selected opioids, ethanol and intake of ethanol, *Alcohol* **1:**43–46.

Isbell, H., Fraser, H. F., Wikler, A., Belleville, R. E., and Eisenman, A. J., 1955, An experimental study of the etiology of "rum fits" and delerium tremens, *Q. J. Stud. Alcohol* **16:**1–30.

Israel, F. L. (ed.), 1968, *1897 Sears Roebuck Catalogue,* Chelsea House, New York.

Jackson, C. T., Korts, D., Hanbury, R., Sturiano, V., Wolpert, L., Cohen, M., and Stimmel, B., 1982, Alcohol consumption in persons on methadone maintenance therapy, *Am. J. Drug Alcohol Abuse* **9:**69–76.

Jackson, G. W., and Richman, A., 1973, Alcohol use among narcotic addicts, *Alcohol Health Res. World* **1:**25–28.

Jaffe, J. H., 1984, Alcoholism and affective disturbance: Current drugs and current shortcomings, in: *Pharmacological Treatments for Alcoholism* (G. Edwards and J. Littleton, eds.), pp. 463–490, Croom Helm, London.

Jaffe, J. H., 1985, Drug addiction and drug abuse, in *The Pharmacological Basis of Therapeutics,* 7th ed. (A. G. Gilman, L. S. Goodman, T. W. Rall, and F. Murad, eds.), pp. 532–581, Macmillan, New York.

Jaffe, J. H., and Martin, W. R., 1985, Opioid analgesics and antagonists, in *The Pharmacological Basis of Therapeutics,* 7th Ed. (A. G. Gilman, L. S. Goodman, T. W. Rall, and F. Murad, eds.), pp. 491–531, Macmillan, New York.

Jaffe, J. H., Petersen, R., and Godgson, R., 1980, *Addictions: Issues and Answers,* Harper and Row, New York.

Jeffcoate, W. J., Hastings, A. G., Cullen, M. H., and Herbert, M., 1981, Naloxone and ethanol antagonism, *Lancet* **1:**1052.

Jeffreys, J., 1840, Observations on the improper use of opium in England, *Lancet* **1:**382–383.

Jones, M. A., and Spratto, G. R., 1977, Ethanol suppression of naloxone-induced withdrawal in morphine-dependent rats, *Life Sci.* **20:**1549–1556.

Kakihana, R., and Butte, J. C., 1980, Biochemical correlates of inherited drinking in laboratory animals, in: *Animal Models in Alcohol Research* (K. Eriksson, J. D. Sinclair, and K. Kiianmaa, eds.), pp. 21–33, Academic Press, New York.

Keeley, K. A., and Solomon, J., 1981, New perspectives on the similarities and differences of alcoholism and drug abuse, in: *Currents in Alcoholism,* Vol. VIII: *Recent Advances in Treatment and Research* (M. Galaner, ed.), pp. 99–119, Grune and Stratton, New York.

King, R., 1972, *The Drug Hang-Up: America's Fifty-Year Folly,* Charles C. Thomas, Springfield, Illinois.

Kissin, B., 1982, A historical review of drug and alcohol use, in: *Perspectives in Alcohol and Drug Abuse: Similarities and Differences* (J. Solomon and K. A. Keeley, eds.), pp. 1–14 John Wright–PSG Incorporated, Boston, Massachusetts.

Klotz, V., McHorse, T. S., Wilkinson, G. R., and Schenker, S., 1974, The effects of cirrhosis on the disposition and elimination of meperidine in man, *Clin. Pharmacol. Ther.* **16:**667–675.

Kolb, L., 1925, Types and characteristics of drug addicts, *Ment. Hygiene* **9:**300–313.

Kolb, L., 1927, Clinical contributions to drug addiction: The struggle for cure and the conscious reasons for relapse, *J. Nerv. Ment. Dis.* **66:**22–43.

Kolb, L., 1928, Drug addiction: A study of some medical cases, *Arch. Neurol. Psychiatry* **20:**171–183.

Kolb, L., 1958, Discussion, in: *Narcotic Drug Addiction Problems* (R. B. Livingston, ed.), p. 97, U.S. Department of Health, Education and Welfare, National Institutes of Health, Bethesda, Maryland.

Kreek, M. J. 1978, Medical complications in methadone patients, in: *Recent Developments in Chemotherapy of Narcotic Addiction* (B. Kissin, J. H., Lowinson, and R. B. Millman, eds.), pp. 110–132, New York Academy of Sciences, New York.

Laidlaw, J., Read, A. E., and Sherlock, S., 1961, Morphine tolerance in hepatic cirrhosis, *Gastroenterology* **40**:389–396.

Levine, A. S., Hess, S., and Morley, J. E., 1983, Alcohol and the opiate receptor, *Alcohol Clin. Exp. Res.* **7**:83–84.

Lieber, C. S., 1982, Alcohol metabolism and its interaction with drugs, liver and hormones, in: *Advances in Alcohol and Substance Abuse: Recent Advances in the Biology of Alcoholism* (C. S. Lieber and B. Stimmel, eds.), pp. 1–5, Haworth Press, New York.

Light, A. B., Torrance, E. G., Karr, W., Fry, E. G., and Wolff, W. A., 1930, *Opium Addiction*, American Medical Association, Chicago (reprinted, 1981, Arno Press, New York).

Louria, D. B., Hensle, T., and Rose, J., 1967, The major medical complications of heroin addiction, *Ann. Intern. Med.* **67**:1–22.

Lowinson, J. H., and Millman, R. B., 1979, Clinical aspects of methadone maintenance treatment, in: *Handbook of Drug Abuse* (R. I. DuPont, A. Goldstein, J. O'Donnell, and B. Brown, eds.), pp. 49–56, National Institute on Drug Abuse, Washington, D.C.

Lyon, L. J., and Antony, J., 1982, Reversal of alcoholic coma by naloxone, *Ann. Intern. Med.* **96**:464–465.

MacInnes, C., 1967, Leaves of grass (review of G. Andrews and S. Vinkenoog, *The Book of Grass*), *Encounter* **28**:67–70.

MacMartin, D. F., 1921, *Thirty Years in Hell, or, the Confessions of a Drug Fiend*, Capper Printing Co., Topeka, Kansas.

Macnish, R., 1859, *The Anatomy of Drunkenness*, W. R. McPhun, Glascow/London.

Maddox, J., 1963, Is the literature worth keeping? *Bull. Atomic Sci.* **19**:14–16.

Maddux, J. F., and Eliott, B., III, 1975, Problem drinkers among patients on methadone, *Am. J. Drug Alcohol Abuse* **2**:245–254.

Mattison, J. B., 1885, *The Treatment of Opium Addiction*, G. P. Putnam's Sons, New York.

Mattison, J. B., 1902, *The Mattison Method in Morphinism*, E. B. Treat, New York.

Maurer, D. W., and Vogel, V. H., 1973, *Narcotics and Narcotic Addiction*, Fourth Ed., Charles C. Thomas, Springfield, Illinois.

McGlothlin, W. H., and Anglin, M. D., 1981, Shutting off methadone: Costs and benefits, *Arch. Gen. Psychiatry* **34**:885–892.

McGlothlin, W. H., Anglin, M. D., and Wilson, B. D., 1977, *An Evaluation of the California Civil Addict Program*, Services Research Monograph Series, National Institute on Drug Abuse, DHEW Publication No. (ADM) 78-558, U.S. Department of Health, Education, and Welfare, Washington, D.C.

"Medicus," 1851, Teetotalism and opium taking, *London Lancet* **2**:169–170.

Mitcheson, M., Davidson, J., Hawks, D., Hitchins, L., and Malone, S., 1971, Sedative abuse by heroin addicts, *J. Psychedelic Drugs* **4**:123–131.

Moore, R. A., 1977, Dependence on alcohol, in: *Drug Abuse: Clinical and Basic Aspects* (S. N. Pradhan and S. N. Dutta, eds), pp. 211–229, Mosby, St. Louis.

Morgan, H. W., 1981, *Drugs in America: A Social History, 1800–1980*, Syracuse University Press, Syracuse, New York.

Murphy, E. F., 1922, *The Black Candle*, Thomas Allen, Toronto.

Musto, D. F., 1973, *The American Disease: Origins of Narcotic Control*, Yale University Press, New Haven, Connecticut.

Nichols, J. R., 1972, Alcoholism and opiate addiction: Theory and evidence for a genetic link between the two, in: *Biological Aspects of Alcohol Consumption* (O. Forsander and K. Eriksson, eds.), pp. 131–134, Finnish Foundation for Alcohol Studies, Helsinki.

Novick, D., Kreek, M. J., Fanizza, A., Yancovitz, S., Gelb, A., and Stenger, R., 1980, Methadone disposition in maintained patients with chronic liver disease, *Clin. Res.* **28:**622A.

O'Donnell, J. A., 1969, *Narcotic Addicts in Kentucky,* Department of Health, Education, and Welfare, National Institute of Mental Health, Chevy Chase, Maryland.

Pack, R. L., and Ferko, A. P., 1977, The effects of ethanol and pentobarbital on the naloxone precipitated escape response in morphine dependence mice, *Arch. Int. Pharmacodyn. Ther.* **228:**58–67.

Parrish, E., 1867, *A Treatise on Pharmacy,* Third Ed. Henry C. Lea, Philadelphia.

Patwardhan, R., Johnson, R., Sheehan, J., Desmond, P., Wilkinson, C. T., Hoyumpa, A., Branch, R., and Schenker, S., 1981, Morphine metabolism in cirrhosis, *Gastroenterology* **80:**1344.

Penfield, W., 1969, Halsted of John Hopkins: The man and his problem as described in the secret records of William Osler, *J. Am. Med. Assoc.* **210:**2214–2218.

Pereira, J., 1842, *The Elements of Materia Medica and Therapeutics,* Second Ed., Longman, Brown, Green, and Longmans, London.

Pescor, M. R., 1943, *A Statistical Analysis of the Clinical Records of Hospitalized Drug Addicts,* Supplement No. 143 to the Public Health Service Reports, U.S. Government Printing Office, Washington, D.C.

Peters, D., 1981, The British medical response to opiate addiction in the nineteenth century, *J. Hist. Med. Allied Sci.* **36:**455–488.

Pidduck, I., 1851, Opium taking, *London Lancet* **2:**253–254.

Popham, R. E., Schmidt, W., and Israelstam, S., 1984, Heavy alcohol consumption and physical health problems: A review of the epidemiologic evidence, in: *Research Advances in Alcohol and Drug Problems,* Vol. 8 (R. G. Smart, H. Cappell, F. Glaser, Y. Israel, H. Kalant, W. Schmidt, and E. M. Sellers, eds.), pp. 149–181, Plenum Press, New York.

Primm, B. J., and Bath, P. E., 1973, Pseudoheroinism, *Int. J. Addict.* **8:**231–242.

Rada, R. R., and Kellner, R., 1979, Drug treatment in alcoholism, in: *Psychopharmacology Update: New and Neglected Areas,* (J. M. Davis and D. Greenblatt, eds.), pp. 105–144, Grune and Stratton, New York.

Reid, L. D., and Hunter, G. A., 1984, Morphine and naloxone modulate intake of ethanol, *Alcohol* **1:**33–37.

Rosen, A., Ottenberg, D. J., and Barr, H. L., 1973, Patterns of previous abuse of alcohol in a group of hospitalized drug addicts, in: *Proceedings of the Fifth National Conference on Methadone Treatment,* pp. 306–315, National Association for the Prevention of Addiction to Narcotics, New York.

Ross, D., Hartmann, R. J., and Geller, I., 1976, Ethanol preference in the hamster: Effects of morphine sulfate and naltrexone, a long-acting morphine antagonist, *Proc. West. Pharmacol. Soc.* **19:**326–330.

Rounsaville, B., Jones, C., Novelly, R. A., and Kleber, H., 1982, Neuropsychological functioning in opiate addicts, *J. Nerv. Ment. Dis.* **170:**209–416.

Rubenstein, R. B., and Wolff, W. I., 1976, Methadone ileus syndrome, *Dis. Colon Rectum* **19:**357–359.

Rutternber, A. J., and Luke, J. L., 1984, Heroin-related deaths: New epidemiologic insights, *Science* **226:**14–20.

Satinder, K. P., 1982, Alcohol-morphine interaction: Oral intake in genetically selected Maudsley rats, *Pharmacol. Biochem. Behav.* **16:**707–711.

Schulz, R., Wüster, M., Duka, T., and Hertz, A., 1980, Acute and chronic ethanol treatment changes endorphin levels in brain and pituitary, *Psychopharmacology* **68:**221–227.

Schut, J., File, K., and Wohlmuth, T., 1973, Alcohol use by narcotic addicts in methadone maintenance treatments, *Q. J. Stud. Alcohol* **34:**1356–1359.

Seevers, M. H., 1970, Morphine and ethanol physical dependence: A critique of a hypothesis, *Science* **170**:1113–1114.

Sellers, E., Naranjo, C. A., and Peachey, J. E., 1981, Drugs to decrease alcohol consumption, *N. Engl. J. Med.* **305**:1255–1262.

Shearman, G. T., and Herz, A., 1983, Ethanol and tetrahydroisoquinoline alkaloids do not produce narcotic discriminative stimulus effects, *Psychopharmacology* **81**:224–227.

Siegel, S., 1982*a*, Drug dissociation in the nineteenth century, in: *Drug Discrimination: Applications in CNS Pharmacology* (F. C. Colpaert and J. L. Slangen, eds.), pp. 257–261, Elsevier/North Holland Biomedical Press, Amsterdam.

Siegel, S., 1982*b*, More on the "paradox" of opioid-induced alcohol consumption, *Subst. Alcohol Action Misuse* **3**:303–305.

Siegel, S., 1983, Wilkie Collins: Victorian novelist as psychopharmacologist, *J. Hist. Med. Allied Sci.* **38**:161–175.

Siegel, S., Hinson, R. E., Krank, M. D., and McCully, J., 1982, Heroin "overdose" death: The contribution of drug-associated environmental cues, *Science* **316**:436–437.

Sinclair, J. D., 1974, Morphine suppresses alcohol drinking regardless of prior alcohol access duration, *Pharmacol. Biochem. Behav.* **2**:409–412.

Sinclair, J. D., and Senter, R. J., 1968, Development of an alcohol deprivation effect in rats, *Q. J. Stud. Alcohol* **29**:863–867.

Sinclair, J. D., Adkins, J., and Walker, S., 1973*a*, Morphine-induced suppression of voluntary alcohol drinking in rats, *Nature* **246**:425–427.

Sinclair, J. D., Walker, S., and Jordan, W., 1973*b*, Behavioral and physiological changes associated with various durations of alcohol deprivation in rats, *Q. J. Stud. Alcohol* **34**:744–757.

Singh, V. K., Jakubovic, A, and Thomas, D. A., 1980, Suppressive effects of methadone on human blood lymphocytes, *Immunol. Lett.* **2**:177–180.

Smith R., 1981, Treating alcohol problems: Making ends meet, *Br. Med. J.* **2893**:1043–1045.

Smith, S. G., Werner, T. E., and Davis, W. M., 1975, Intravenous drug self-administration in rats: Substitution of ethyl alcohol for morphine, *Psychol. Rec.* **25**:17–20.

Spira, I. A., Rubenstein, R., Wolff, D., and Wolff, W. I., 1975, Fecal impaction following methadone ingestion simulating acute intestinal obstruction, *Ann. Surg.* **181**:15–19.

Stevenson, G. H., Lingley, L. R. A., Trasov, G. E., and Stansfield, H., 1956, *Drug Addiction in British Columbia: A Research Survey*, University of British Columbia, Vancouver.

Steward, J., de Wit, H., and Eikelboom, R., 1984, Role of unconditioned and conditioned drug effects in the self-administration of opiates and stimulants, *Psychol. Rev.* **91**:251–268.

Stillé, A., 1868, *Therapeutics and Materia Medica*, Third Ed., Henry C. Lea, Philadelphia.

Stimmel, B., 1975, *Heroin Dependency: Medical, Economic, and Social Aspects*, Stratton Intercontinental Medical Book Corporation, New York.

Stimmel, B., Hanbury, R., Sturiano, V., Korts, D., Jackson, G., and Cohen, M., 1982, Alcoholism as a risk factor in methadone maintenance: A randomized control trial, *Am. J. Med.* **73**:631–636.

Stock, C., Bode, J. C., and Sarles, H. (eds.), 1980, *Alcohol and the Gastrointestinal Tract*, Editions INSERM, Paris.

Straus, R., 1983, Types of alcohol dependence, in: *The Pathogenesis of Alcoholism: Psychosocial Factors* (B. Kissin and H. Begleiter, eds.), pp. 1–16, Plenum Press, New York.

Szasa, T., 1974, *Ceremonial Chemistry*, Anchor Press/Doubleday, Garden City, New York.

Tennant, F. S., 1976, Dependency traits among parents of drug abusers, *J. Drug Dep.*, **6**:83–88.

Terenius, L., 1982, Clinical aspects, in: *Endorphins: Chemistry, Physiology, Pharmacology, and Clinical Relevance* (J. B. Malick and R. M. S. Bell, eds.), pp. 229–255, Marcel Dekker, Basel.

Terry, C. E., and Pellens, M., 1928, *The Opium Problem*, Bureau of Social Hygiene, New York (reprinted 1970, Patterson Smith, Montclair, N.J.).

Tommasello, A. C., and Adir, J., 1984, Alcohol and the steady-state disposition kinetics of methadone in rats, *J. Stud. Alcohol* **45**:155–159.

Trebach, A. S., 1982, *The Heroin Solution*, Yale University Press, New Haven, Connecticut.

Van Slyke, D. C., 1945, *The Wail of a Drug Addict*, William B. Eerdmans Publishing Co., Grand Rapids, Michigan.

Verebey, K., and Blum, K., 1979, Alcohol euphoria: Possible mediation via endorphinergic mechanism, *J. Psychedelic Drugs* **11**:305–311.

Waldorf, D., 1973, *Careers in Dope*, Prentice-Hall, Englewood Cliffs, New Jersey.

Waldorf, D., and Biernack, P., 1979, Natural recovery from heroin addiction: A review of the incidence literature, *J. Drug Issues* **9**:281–289.

Waldorf, D., Orlick, M., and Reinarman, C., 1974, *Morphine Maintenance*, Drug Abuse Council, Washington, D.C.

Wasserman, S., and Yahr, M. D., 1980, Choreic movements induced by the use of methadone, *Arch. Neurol.* **37**:727–728.

Way, E. L., Swanson, R., and Gimble, A. I., 1947, Studies *in vitro* and *in vivo* on the influence of the liver on isonipecaine (Demerol) activity, *J. Pharmacol. Exp. Ther.* **91**:178–184.

Weppner, R. S., and Agar, M., 1971, Immediate precursors to heroin addiction, *J. Health Soc. Behav.* **2**: 10–18.

Werner, A., 1969, Near-fatal hyperacute reaction to intraveneously administered heroin, *J. Am. Med. Assoc.* **207**:2277–2278.

Westermeyer, J., 1978, "HMC" treatment recalled, *Am. J. Psychiatry* **135**:869.

Westermeyer, J., 1982, *Poppies, Pipes, and People: Opium and its Use in Laos*, University of California Press, Berkeley.

Westermeyer, J., 1983, Treatment of opiate addiction in Asia: Current practice and recent developments, in: *Research Advances in Alcohol and Drug Problems*, Vol. 7 (R. G. Smart, F. B. Glaser, Y. Israel, H. Kalant, R. E. Popham, and W. Schmidt, eds.), pp. 433–455, Plenum Press, New York.

Westling, A., 1958, Huumausianeiden käyttäjien suhteesta väkijuomiin (Drug addicts and alcohol) (English summary, pp. 421–422), *Duodecin* **74**:411–422.

Whitney, G., and Horowitz, G. P., 1978, Morphine preference of alcohol-avoiding and alcohol-preferring C57BL mice, *Behav. Genet.* **8**:177–182.

Wilkinson, D. A., and Carlen, P. L., 1981, Chronic organic brain syndromes associated with alcoholism: Neuropsychological and other aspects, in: *Research Advances in Alcohol and Drug Problems*, Vol. 6 (Y. Israel, F. B. Glaser, H. Kalant, R. E. Popham, W. Schmidt, and R. G. Smart, eds.), pp. 107–145, Plenum Press, New York.

Wilkinson, G. R., and Schenker, S., 1975, Drug disposition and liver disease, *Drug Metab. Rev.* **4**:139–175.

Willis, J. H., 1969, The natural history of drug dependence: Some comparative observations on United Kingdom and United States subjects, in: *Scientific Basis of Drug Dependence* (H. Steinberg, ed.), pp. 301–321, J & A Churchill, London.

Wilson, J.C., 1889, Remarks upon the treatment of the opium habit and kindred affections, *Med. Surg. Rec.* **60**:564–569.

Winter, J. C., 1975, The stimulus properties of morphine and ethanol, *Psychopharmacology* **44**:209–214.

Wodak, A., and Richardson, P. J., 1982, Alcohol and the cardiovascular system, *Br. J. Addict.* **77**:251–258.

# Index